Coastal and Estuarine Studies

Series Editors:
Malcolm J. Bowman   Christopher N.K. Mooers

Coastal and Estuarine Studies

49

Walker O. Smith, Jr. and
Jacqueline M. Grebmeier (Eds.)

Arctic Oceanography: Marginal Ice Zones and Continental Shelves

American Geophysical Union

Washington, DC

Series Editors

Malcolm J. Bowman
Marine Sciences Research Center, State University of New York
Stony Brook, N.Y. 11794, USA

Christopher N.K. Mooers
Division of Applied Marine Physics
RSMAS/University of Miami
4600 Rickenbacker Cswy.
Miami, FL 33149-1098, USA

Editors

Walker O. Smith, Jr.
Graduate Program in Ecology
University of Tennessee
Knoxville, TN 37996

Jacqueline M. Grebmeier
Graduate Program in Ecology
University of Tennessee
Knoxville, TN 37996

Library of Congress Cataloging-in-Publication Data

Arctic oceanography : marginal ice zones and continental shelves / Walker
    O. Smith and Jacqueline M. Grebmeier (eds.).
        p.   cm. — (Coastal and estuarian studies ; 49)
    Includes bibliographical references.
    ISBN 0-87590-263-4
    1. Oceanography—Arctic Ocean.   I. Smith, Walker O.
II. Grebmeier, Jacqueline M., 1955–   .   III. Series.
GC401.A755   1995
551.46'8—dc20   95-9541
                                                                                      CIP

ISSN 0733-9569

ISBN 0-87590-263-4

Printed in the United States of America.

CONTENTS

PREFACE

The Arctic Ocean is the least understood ocean on Earth, and yet its importance to the world's oceans and climate is immense. For example, it has been suggested that the Arctic is the region most likely to be affected by increased atmospheric temperatures which might occur as a result of anthropogenic releases of greenhouse gases. It also plays a critical role in global oceanic circulation, in that it modulates the formation of deep water in the North Atlantic via ice export. Despite its pivotal role in global processes, the Arctic remains poorly understood. This volume is an attempt to highlight and synthesize some of the recent advances in our knowledge of Arctic oceanography and includes topics that will interest physical, biological, chemical, and geological oceanographers as well as atmospheric scientists.

That the Arctic is so poorly known relative to other oceans is not surprising. It is largely ice-covered throughout the year, with only some of its continental shelves becoming ice-free in summer. Its ice is mostly multi-year and very thick, making penetration into the deeper portions impossible except by the most powerful ice-breakers. However, in recent years new technologies have been applied to the Arctic, and our understanding of the physical, chemical, biological and geological processes which occur within it is rapidly increasing. Satellite sensors observe the Arctic continually, allowing us to follow ice circulation, storms, and openings in the pack ice that had never been observed previously. Moorings, ships and buoys now can withstand many of the rigors of the Arctic, and observations of the water column and seabed are becoming more common. Finally, because of its importance to global processes, studies of the Arctic are attracting scientists not only from Arctic nations but from nations around the world. Arctic oceanography truly has become an international effort.

Approximately 30% of the surface area of the Arctic Ocean is occupied by its continental shelves, and hence the impact of the shelf area on the deeper regions of the basin can be expected to be greater than in other oceans. Shelves also are thought to play a major role in the production of the Arctic's shallow permanent halocline, one of its unique features. They are also heavily impacted by seasonal riverine discharges, and hence potentially can be influenced by river-borne sediments, nutrients and pollutants to a greater extent than in other ocean basins. Furthermore, the shelf region is covered and uncovered seasonally by ice, and hence the rates of biogeochemical cycling are greatest there. Shelves also are frequently the sites of marginal ice zones, which have been the focus of a substantial amount of research in the Arctic in the past decade. Marginal ice zones also occur in deep regions, such as the Greenland Sea. Regardless of their depth, marginal ice zones have been shown to be highly dynamic regions with regard to their physics and biology. For example, they can generate fronts, jets, and eddies, have unusual vortex pairs, and be the sites of intense biological activity (due to upwelling and downwelling) and material transfer. Their roles in the ice dynamics and biogeochemical cycles of polar systems are still being re-evaluated.

The purpose of this volume is to present both new results from recent studies in the Arctic, in particular the marginal ice zone and the Arctic's continental shelves, as well as to synthesize a variety of multidisciplinary studies which have been recently conducted. It is not intended by any means to be all inclusive, and readers can consult other volumes for treatment of the various sub-disciplines of Arctic oceanography which have not been treated in detail. We hope that the papers included in this volume prove to be stimulating to Arctic researchers and non-polar oceanographers alike, and that they point the way towards potentially significant research directions in the coming decades.

The editors would like to thank our colleagues (who are too numerous to mention by name) who reviewed these papers under short deadlines and whose excellent jobs insured the quality of the contributions. Partial support during the preparation of this volume was provided by the Office of Polar Programs, National Science Foundation to both of the editors.

Walker O. Smith, Jr.
Jacqueline M. Grebmeier
University of Tennessee—Knoxville

1

Satellite Remote Sensing Of The Arctic Ocean And Adjacent Seas

Josefino C. Comiso

Abstract

Many of the large-scale surface physical and biological characteristics of the Arctic Ocean and adjacent seas can best be studied with satellite remote sensing because of the vastness of, adverse weather conditions, and long periods of darkness in the region. Scientific applications specific to the polar regions are discussed in terms of the information afforded by various systems, passive or active, and by the microwave, infrared, and visible portions of the electromagnetic spectrum. In the sea-ice region, the key parameters of interest are ice concentration, ice type, surface and ice temperature, ice velocity, ice extent and area, snow cover, and melt coverage. In the open water region, the corresponding parameters are phytoplankton pigment concentration, surface temperature, wind velocity, dynamic topography, internal waves, eddies, and wave propagation characteristics. Some atmospheric parameters such as water vapor, liquid water, precipitation, and cloud cover are also of interest. The strengths and weaknesses of the various systems are considered and sources of errors and ambiguities are pointed out. Synergistic techniques, using several sensors in concert, are also discussed because they provide a means to obtain a more accurate or complete characterization of the Arctic Ocean and adjacent seas. While existing systems are already proving effectiveness, new systems planned for the next decade will even be more versatile and should usher a new era of a comprehensive and consistent coverage of Earth's systems, including the Arctic system.

Introduction

The most effective (and perhaps only) way to study large-scale surface physical and biological characteristics of the Arctic Ocean and its adjacent seas is through the use of satellite

Arctic Oceanography: Marginal Ice Zones and Continental Shelves
Coastal and Estuarine Studies, Volume 49, Pages 1–50
This paper is not subject to U. S. copyright.
Published in 1995 by the American Geophysical Union

remote sensing techniques [Onstott and Shuchman, 1990; Comiso, 1991; Thomas, 1990; Carsey, 1992]. The power of remote sensing as applied to the polar regions has been discussed and demonstrated in several publications [Zwally et al., 1983; Parkinson et al., 1987; Gloersen et al., 1992]. Mounted on satellite platforms, sensors provide the synoptic and temporal distributions that are needed in many environmental process studies, and the spatial and temporal observation consistency and detail required for climate-change studies. Sensors also provide the means for extrapolating observations from ships and surface stations and for evaluating the global (as opposed to local) impact of mesoscale processes.

A dominant aspect of the Arctic Ocean and its adjacent seas is their sea-ice cover. Although it is just a thin sheet (a few meters thick on the average) compared to the depth of the ocean, the ice cover occupies a vast area and is known to have a profound influence on the physical, chemical, and biological characteristics of the ocean. In particular, the ice cover alters surface ocean salinity through brine rejection where ice is formed and through the introduction of low-salinity water in the melt regions. It is also an effective insulating interface between the ocean and the atmosphere during winter, and a good reflector of radiation during the summer. Sea ice modifies the ocean dynamically due to friction at the ocean/ice boundary layer. Furthermore, sea ice suppresses the penetration of sunlight needed by marine life during photosynthetic processes and impedes the mixing effect of wind. Also, in the middle of the ice pack are surface features like leads and polynyas, the occurrence of which alter the heat, humidity, and salinity fluxes and the ventilation and circulation characteristics of the underlying ocean. While processes in the ice covered portion are unique and important, processes that continually occur at or adjacent to the ice edge should not be overlooked. The marginal ice zone, a very active region that includes the ice edge, is the site where interactions between the ocean, ice, and the atmosphere are most intense. Also, about half of the winter ice cover disappears during the summer, mainly in the continental shelf region, where plankton and other marine life go through massive seasonal variations. Without the aid of remote sensing, many of these and other Arctic processes can go undetected, especially in winter, when ice is most extensive and the region is in near-total darkness.

The various geophysical parameters in ice-covered regions that can be derived from satellite remote-sensing data are sea-ice concentration, ice type, ice velocity, physical surface temperature, snow cover, surface roughness, cloud cover, and surface melt conditions [Gloersen and Salomonson, 1975; Thomas, 1990; Massom, 1991; Comiso, 1991]. In the adjoining ice-free areas, the corresponding parameters are phytoplankton pigment concentration, cloud cover, sea surface temperature, wind velocity, topography, precipitation, and wave propagation characteristics [Wilheit and Chang, 1980; Wentz et al., 1986; Baker and Wilson, 1987; Apel, 1987; McClain et al., 1993]. No single sensor can provide data for all these geophysical parameters, and the accuracies in the determination of some of them may be marginal. New sensors with improved capabilities have been launched recently and even more versatile ones are expected in the near future. The accumulation of long-term data sets of parameters, including ice concentration, surface temperature, cloud cover, and phytoplankton concentration would provide an invaluable resource for Arctic- and Earth-science studies. This chapter presents a review and update of present and future capabilities.

The Arctic Environment—Physical and Radiative Characteristics

Remote sensing data are most useful if observables can be consistently and reliably interpreted in terms of geophysical parameters. This requires a good understanding of what a sensor measures and what physical property of the surface of interest is being observed. For example, it is important to recognize that over sea ice, visible sensors observe surface (i.e., skin) properties only while microwave sensors see subsurface characteristics. On the other hand, over ice-free ocean, the reverse is true. In the case of sea ice, microwave radiances may come from the snow cover, the snow-ice interface or within the ice itself depending on the wavelength of the radiation, the type of ice, and state and depth of the snow cover. In the open ocean, although the surface is optically thick for long wavelengths, its radiative and backscatter characteristics may be considerably altered by the presence of foam and big waves. To completely understand the physics behind the observations, a combination of field measurements and radiative transfer modeling is needed. The physical and electrical characteristics of the different surfaces relative to the sensor should be measured, and the propagation characteristics of radiation at different wavelengths and polarization within the media should be understood.

Efforts in this direction have been conducted in the last several years in the Arctic region (e.g., Marginal Ice Zone Experiment, Seasonal Ice Zone Experiment, and Lead Experiment). The state of the art in measurements of remote sensing parameters at high latitudes is summarized by Hallikainen and Winebrenner [1992] in the microwave region, and by Steffen et al. [1993] in the visible and infrared regions. Also, modeling studies in the microwave, infrared, and visible regions have been performed at different levels of sophistication [Fung and Chen, 1981; Tsang and Kong, 1981; Ulaby et al., 1981–86; Grenfell, 1983; Perovick, 1989], while studies of the fundamental physical and electrical properties of sea ice have been undertaken [Addison, 1969; Vant et al., 1978; Weeks and Ackley, 1982; Stogryn, 1985; Tucker et al., 1992]. In some cases, the surface and subsurface properties are so complex that correct interpretation of data may be difficult even with models. In this section, key observables in the Arctic and the environmental conditions that directly relate to remote sensing data and their observational accuracies will be discussed.

The Sea-Ice Cover

The Arctic sea-ice cover consists of two main components: the perennial ice cover and the seasonal ice cover. The perennial ice cover, which consists of predominantly multiyear ice floes, occupies a large fraction of the Arctic Ocean's deep regions during the summer. The seasonal ice cover is that which comes and goes with season and is normally located in shallow and shelf waters. It is also convenient to divide the winter ice pack into outer and inner zones. These two zones have distinct characteristics, each having a unique role in the physical and biological processes in the region. Aerial photographs showing the outer and inner zones are shown in Figures 1a and 1b, respectively. As shown, the outer zone is a very dynamic region where ice floes are distinctly separated and fragmented because of the influence of waves and atmospheric forcing, while the inner zone is an almost continuous ice sheet, except for the occasional presence of leads and polynyas.

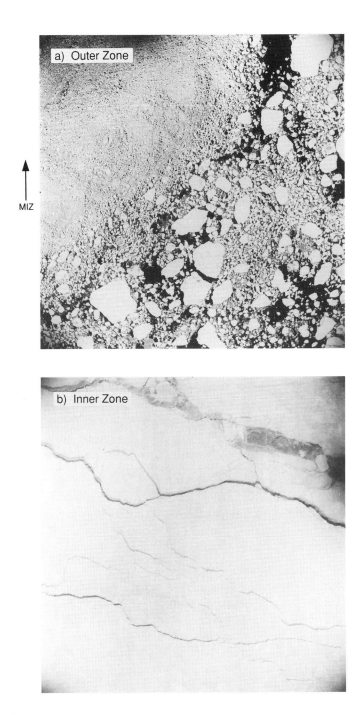

Figure 1. Photographs of Arctic sea ice in (a) the outer zone and (b) the inner zone in the Fram Strait on May 20, 1987.

Differences in emissivity and backscatter of the various types of sea ice have been observed [Eppler et al., 1992; Onstott and Shuchman, 1990] while differences in the sensitivity of vertical from horizontal polarizations have been studied by Matzler et al. [1984]. The physical characteristics that contribute most to spatial variability in radiative characteristics are thickness, salinity, snow cover, wetness, and surface roughness, which vary depending on the stage in the growth cycle (or age) of the ice. There are also seasonal variations in surface characteristics related to temperature, as will be discussed in the following section.

The ice cover can be a complex conglomerate of frazil, congelation, and snow-ice. Frazil ice is formed in relatively turbulent seawater stirred by wind. This mixing prevents the ice crystals from organizing themselves into the long columnar platelets usually found in quiescent seawater. The latter is commonly referred to as congelation ice and is usually found in frozen leads and at the bottom of thicker ice as the ice grows by thermodynamic cooling. During initial growth stages, frazil ice, which evolves to pancakes, is usually formed within the outer zone, while congelation ice, which turns to nilas, is formed in lead areas in the inner zone. These two types of new ice are different not just in structure and composition but also in roughness: nilas is horizontally homogeneous and flat except when ridged and rafted, while pancakes are noted for geometrical shapes and rough edges.

The transition from new ice to young ice (<30 cm thick with negligible snow cover) basically represents a change in thickness. Initially, the microwave emissivity and backscatter increase with thickness for thin ice not only because of the increase in emission from a thicker layer but also because of changing surface properties. For example, the presence of frost (or salt) flowers (formed by ice deposited on the surface directly from the vapor phase) enhances both the salinity and roughness of the surface and increases both the emissivity and backscatter of sea ice [Drinkwater and Crocker, 1988; Comiso et al., 1989; Onstott and Shuchman, 1990]. Once a certain thickness is reached, further increases in thickness by thermodynamic growth may no longer change the emissivity or backscatter of the ice. In many cases, thickness is increased by other factors such as rafting and ridging. Such factors in turn cause changes in surface roughness and other surface characteristics that may change emissivity and backscatter. As young ice increases in thickness and acquires snow cover, it becomes first-year ice, which is the most common form of ice in the seasonal ice region.

The salinity of sea ice is usually at its highest value during its initial stages, when the concentration of brine trapped in the ice is highest. The salinity decreases with time—a process called "brine drainage." However, during winter or cold conditions, the salinity is usually maintained at above 6% [Nakawo and Sinha, 1981]. During the summer, the drainage process is accelerated by gravitational brine migration that is facilitated due to warmer ice, and flushing through percolation of melted snow [Martin, 1979]. It is for this reason that the ice that survives the summer is significantly less saline than first-year ice. Further desalination can occur within the same ice floe during subsequent summers. The winter Arctic ice cover is thus dominated by either saline first-year ice or desalinated multiyear ice. Typical salinity, temperature, and density profiles for these two ice types are shown in Figures 2a and 2b. Large differences in the density profiles between the two ice types are obvious, especially in the upper layer.

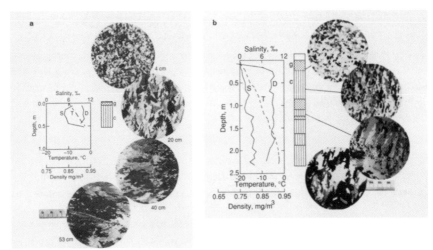

Figure 2. Salinity, density and temperature profiles of (a) first year ice and (b) multiyear ice [from Tucker et al., 1992].

Salinity influences the dielectric property, which in turn affects both the radiative and back-scatter properties of sea ice [Hallikainen and Winebrenner, 1992]. Measurements at 4 GHz show that first year ice with a surface salinity of about 5.1‰ has a dielectric constant of 3.4 ± 0.15i, while that of multiyear ice with a salinity of 1.3‰ is about 3.0 ± 0.03i [Vant, 1978]. The difference in the imaginary part makes the absorption coefficient, and therefore the opacity, of these two ice types very different. Since multiyear ice is more transparent but more inhomogeneous material, volume scattering of radiation within the ice is prevalent, causing drastic changes in both emissivity and backscatter cross-section.

Snow can change the effective emissivity and backscatter of sea ice in many ways. First, because it is such a good insulator, the ice temperature profile changes with the acquisition of the snow cover. Second, snow serves as both a scatterer and a source of radiation. Furthermore, it can change the characteristics of the snow-ice interface through formation of a saline snow-ice material. The effect of snow on the emissivity and backscatter depends on frequency and polarization [Onstott and Shuchman, 1990; Garrity, 1992; Lohanick, 1993]. During persistently cold conditions, fresh snow is transparent and the scattering effects are small especially at 19 GHz and lower frequencies because of small particle size. However, as the surface goes through melt/refreeze cycles, the snow becomes granular, and snow scattering effects become considerable. Thick snow cover can also lead to negative free-board and hence to flooding, which cause the microwave signature of sea ice to resemble that of open water or grease ice. Subsequent refreezing of the slush causes formation of the aforementioned saline snow-ice material at the snow-ice interface. Persistent flooding may also influence the ability of the ice to survive the summer by affecting the albedo and the ice structure.

Some form of melting occurs almost all year near the marginal ice zone and parts of the inner zone, where the temperature occasionally attains melting- point. The effect of surface melt on the backscatter and emissivity of sea ice has been assessed [Grenfell and Lohanick,

1985; Gogineni et al., 1992]. Surface melt can cause one of two contrasting effects. First, surface wetness of the order of 3% enhances the microwave emissivity and suppresses the backscatter of the material. Such a surface wetness condition is common over much of the Arctic region during spring and summer, as reflected by enhancements in passive microwave brightness temperatures over the entire region. Further melt effects lead to meltponding, causing surface signatures to be similar to those of water.

Sea ice can be either undeformed or deformed. Because of a constantly moving ice pack, the state and direction of which can be altered by winds, ocean current and waves, it is difficult to find sea ice that is strictly homogeneous and undeformed [Lange et al., 1989]. Undeformed ice is defined here as that with relatively flat top and bottom surfaces. Some effects that cause small-scale surface roughness have already been discussed. Deformed ice is that associated with large-scale irregularities (>1m) in surface topography such as those due to ridging and rafting of large floes. Ridging occurs when new, young, or first-year ice is caught in between colliding large and thick ice floes, while rafting is the process by which one floe overrides another. These phenomena enhance the backscatter of the surface and could also alter its emissivity.

Open Water

The seasonal ice cover and the perennial ice cover in the Arctic are about the same size at about 8×10^6 km^2 [Parkinson et al., 1987; Gloersen et al., 1992]. In the summer, much of the seasonal ice area becomes ice-free but flooded with meltwater. The volume of ice melted during the period corresponds to more than 1.2×10^4 km^3 (assuming an average thickness of 1.5 m for seasonal ice). Further freshening of the surface water is attributed to continental inputs in the form of river run-off and glacial meltwater. Inputs from major rivers alone have been estimated to be about 3.5×10^3 km^3 [Treshnikov, 1977]. In many respects, the vertical stratification, biochemical composition, optical properties, and other physical characteristics of the seasonal ice region in the Arctic are different from those of other regions including that of the Southern Ocean. This is especially the case, when compared with the latter, because its seasonal region occupies relatively shallow continental shelf areas while the seasonal region in the Southern Ocean is located in the deeper part. The open-water region adjacent to the ice edge also is the scene of many important mesoscale phenomena such as eddies, fronts, current jets, meanders, and gravity wave propagations. These are features in the ocean that are noted for singularities in lateral flow and intensified vertical motion. In autumn and winter, the ice edge also is the scene of rapid ice growth which causes formation of cold saline water and deep ocean mixing. The temporal evolution of these phenomena can be monitored either directly or indirectly by high-resolution satellite data.

Knowledge of the absorption or emission characteristics of the open water surface in the microwave, infrared, and visible regions is required to assess fluxes between the ocean and the atmosphere and to understand marine biological life cycles and distributions. Such characteristics also are crucial to the interpretation of remotely sensed data. Depending on wavelength, the signal is influenced by salinity, temperature, density, roughness, and the presence of foam, eddies, fronts and meanders. These parameters change in value with time and location because of upwelling, ice formation, meltwater production, river runoff, wind-induced mixing, and transport.

The complex dielectric constant of seawater may be calculated at any frequency at good accuracy from the equation originated by Debye [1929], which is given by

$$\varepsilon = \varepsilon_\infty + \frac{(\varepsilon_s - \varepsilon_\infty)}{1 + i\omega\tau} - \frac{i\sigma}{\omega\varepsilon_o} \tag{1}$$

where ω is the frequency, ε_∞ is the permittivity at very high frequencies, ε_s is the static permittivity, τ is the relaxation time in s, σ is the ionic conductivity in mhos m^{-1} and ε_o is 8.854 x 10^{-12} is the permittivity of free space in farads m^{-1}. The values for ε_s, σ, and τ are all dependent on salinity and temperature. The brightness temperature of calm ocean water has been shown to be sensitive to salinity and temperature [Swift, 1980; Wentz, 1983]. However, in the Arctic, the salinity varies only a few parts per thousand and the surface temperature of the water changes only by a few Kelvins. Thus, in typically calm open water regions within the ice pack the brightness temperature does not change much. In the open ocean, however, the brightness temperature could change drastically due to the presence of waves and foam that act like an extra dielectric layer on the surface [Swift, 1980]. The effect varies depending on wind speed and roughness.

At thermal infrared wavelengths, the emissivity of the water surface is similar to those of ice and snow and is consistently close to unity. Thermal infrared radiation is sensitive to the so called skin temperature and over ice, snow and land surfaces, the value is usually strongly influenced by air temperature. Over ocean water surfaces, however, the surface temperature is more stable and less sensitive to diurnal variations in surface air temperature because the temperature of the former is constantly stabilized by thermal convection with adjacent layers.

Our current knowledge about the optical characteristics of ocean water is the result of studies that started over a century ago. The theory behind ocean color is partly attributed to the work by Kalle [1938] who correctly described that the blue of clear open ocean water is due to molecular scattering and that the green color of coastal waters is due to the addition of a yellow substance (*gelstoff*) that absorbs the blue light. He also suggested that larger particles in the ocean, such as phytoplankton, could impart color to the water. It is now known that phytoplankton have specific absorption characteristics that normally change the color to that of greenish hue but that there are some types that change the color to red, yellow, blue-green, or mahogany. Two general classes of ocean water have been identified: class 1 for those whose reflectance is determined by photosynthetic pigments and class 2 for those whose reflectance are caused by suspended particulates and dissolved organic matter. As a rule, oceanic waters belong to class 1. The physical characteristics of the water that affects the optical properties include the downwelling irradiance as measured at various depths (i.e., photosynthetically available radiation (PAR) in quanta/m^2/sec); the total incoming radiation rate (300–4000 nm spectral range); and the scattering coefficient at wavelengths of interest. The corresponding biological and chemical characteristics are the particulate organic carbon concentration, the pigment chlorophyll a + pheophytin a concentration, and the primary production at different depths. Modeling studies [Morel, 1988] indicated non-linear relationships between these parameters. This has been attributed to the change of the living-to-detrital organic carbon ratio and therefore to the respective contributions of absorption and of scattering.

To gain insight into the spatial variability of the bioptical characteristics of class 1 water, it is useful to know the growth rate, lifespan, species, and distribution patterns of phytoplanktons. Phytoplankton growth is influenced by several factors, including light, temperature, supply of nutrients, and grazing of pelagic animals [Heimdal, 1983]. The lifespan has been estimated to be about 1–10 days [Lewis, 1992] while the growth rate has been estimated at 0.7 doublings per day [Smith et al., 1987]. The spring/summer season in the Arctic is marked by a drastic increase in light availability and a milder increase in surface temperature. Large areas of high concentrations of phytoplankton have been observed in the Arctic during this period, especially in areas previously covered by sea ice. The high values have been postulated as partly due to meltwater production [Müller-Karger and Alexander, 1987; Niebauer and Alexander, 1985]. Because of its lower density than seawater, meltwater stabilizes the water column and allows phytoplankton and forms of other marine life to grow in a highly stable, high-irradiance environment. Stability is further enhanced by solar warming of the low albedo surface waters. In addition, algal cells released from the ice contribute to extant phytoplankton, formerly under the ice, to seed the nutrient-rich and now well-lighted water column. Ice algae commonly achieve large standing stocks; for example, Garrison et al. [1987] found up to 50 mg/m^3 of chlorophyll in some sea-ice samples. Hence, it is clear that ice algae potentially provide a large inoculum for water-column phytoplankton, particularly if the volume through which the algae are distributed is small. Bioptical characteristics of polar waters different from those of tropical waters have indeed been observed by Mitchell et al. [1991]. Regional variability also is apparent in their study because the waters in the Barents and Kara Sea are found to be different from those in the Bellingshausen Sea in the Southern Hemisphere.

Remote Sensing Techniques and Applications

The sensor best suited for a particular application in the Arctic depends on the parameter of interest, the spatial resolution requirements, and the temporal and spatial coverage needed. Several remote sensing systems, active and passive, have been developed using frequencies from the visible to the microwave, and using varied polarizations [Elachi, 1987]. A display of the electromagnetic spectrum together with atmospheric transmission at the various wavelengths is shown in Figure 3. Channel locations (in wavelength and frequency) for some of the most popular sensors are also given. A more comprehensive list and description of past, present, and future sensors is provided by Massom [1991]. The set of channels for a particular sensor is chosen to optimize accurate extraction of desired geophysical parameters. For example, to obtain surface information, channels with a minimum of atmospheric attenuation are chosen. In the visible-to-near-infrared wavelengths, the attenuation is caused largely by water vapor and carbon dioxide. The atmosphere is almost opaque in the far-infrared due to the presence of absorption bands associated with various atmospheric constituents, but there are some windows in the thermal-infrared region that have been utilized for surface temperature measurements. In the microwave region, attenuation is caused mainly by the strong absorption bands brought about by oxygen and water vapor. Overall,

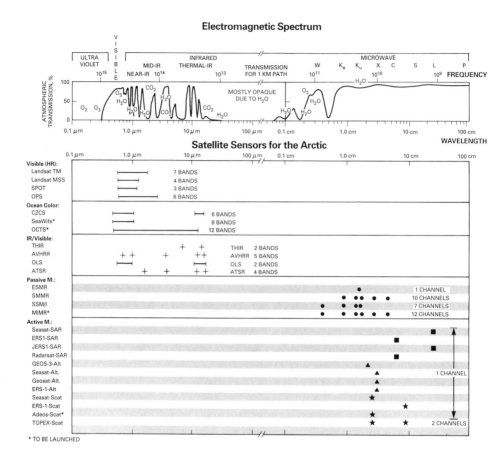

Figure 3. The electromagnetic spectrum and atmospheric transmission for 1-km path length from ultraviolet through microwave frequencies. The wavelengths corresponding to each channel of satellite sensors used in the Arctic are indicated.

the sensors that have been used successfully in the polar regions are the passive and active microwave sensors, the ocean-color (visible) sensor, the high-resolution visible sensors, and the medium-resolution visible and infrared sensors. The characteristics of these sensors and the geophysical quantities derived from the data over sea ice and ice-free ocean will be discussed in the following sections.

Passive Microwave Systems

The first commercial space-based imaging passive microwave system was the 19-GHz Electrically Scanning Microwave Radiometer (ESMR) launched onboard the Nimbus–5 satellite on December 28, 1972. This sensor provided more than four years of good data and was followed by the Scanning Multichannel Microwave Radiometer (SMMR) on board Seasat which was launched in July 1978 and the Nimbus–7 satellite which was launched in

October 1978. The Seasat SMMR was operational for only three months while the Nimbus–7 SMMR remained in operation until 1987. SMMR was a dual-polarized system operating at 6.6, 10.7, 18.0, 21.0, and 37.0 GHz. Since July 28, 1987, a number of the Special Sensor Microwave Imagers (SSM/I) have been launched onboard a satellite of the Defense Meteorological Satellite Program (DMSP), continuing the time series of the microwave data set established by ESMR and SMMR. With no channels at 6.6 and 10.7 GHz, the SSM/I sensor has less capability than SMMR at the lower frequencies but has more capability at the higher frequencies with the addition of a dual-polarized 85-GHz channel. Also, the SSM/I sensor collects data continuously, whereas the Nimbus–7 SMMR sensor operated only every other day. Furthermore, SSM/I has a larger swath width than SMMR (1394 km compared to 780 km) giving it better daily spatial coverage. The large swath width of the SSM/I provides data covering the entire arctic region each day, except the area above 88° N, which is not covered by the system (the satellite inclination angle being 98.7°). SSM/I data mapped in polar grid is now routinely available in CD-ROM format as described by Barry et al. [1993]. A new generation of passive microwave sensors is being planned, including the Multifrequency Imaging Microwave Radiometer (MIMR), which is a dual-polarized system at six frequencies (6.8, 10.6, 18.7, 23.8, 36.5, and 90 GHz). This system has significantly better spatial resolution at each channel (4–60 km, depending on frequency) than either SMMR or SSM/I and should allow for more extensive applications. A version of this sensor is expected to be onboard European and Japanese satellites as well as the U.S. Earth Observing System (EOS) which are scheduled for launch within the coming decade.

The basic parameter measured by passive microwave sensors is the brightness temperature which is the radiative flux expressed in temperature emitted from the surface per unit solid angle per wavelength. The parameter that provides information about the surface is the emissivity, which is the ratio of its brightness temperature and the radiative flux from a blackbody of the same physical temperature. Eppler et al. [1992] compiled averages of microwave emissivities of open water and different types of sea-ice surfaces observed from limited Arctic regions at various frequencies and polarizations. The surface emissivity increases from open water through new ice to young ice showing effects of ice thickness. The older and thicker ice types have a larger range of emissivities because of their snow cover and salinity fluctuations. The emissivity of first-year ice is less varied since it is modified by scattering in the snow cover only, while that of multiyear ice fluctuates more since it is further modified by scattering within the ice material. Surface wetness also modifies the emissivity of all ice types.

The sea ice region

Two-day averages of brightness temperatures (T_B) from SSM/I mapped in a standard polar stereographic format for SSM/I data are shown in Figure 4. The color-coded images show in a broad perspective the entire Arctic region using 19-, 37-, and 85-GHz T_B's at vertical (V) and horizontal (H) polarizations averaged over a period of two days. A data element (pixel) of each image matrix corresponds to a 25- by 25-km area. Such an area is large compared to the size of some ice cover components and may represent a surface consisting of hummocks, frozen meltponded ice, ridged ice, first-year ice, newly frozen ice, and open water. The fraction of each component varies from one pixel to another. The images show relatively low values in the open ocean, slightly enhanced values within the marginal ice zone where a predominance of new ice is expected, and high values in the seasonal ice region—consistent with the aforementioned emissivities of the corresponding surfaces. Relatively low brightness temperatures in the Central Arctic region, where there is a predominance of multiyear ice, is also apparent while in some areas in the open ocean (e.g., upper left corner in the Pacific Ocean), the values are considerably enhanced because of strong atmospheric or weather effects.

To gain insight into the utility of multichannel data, 3-D scatterplots of co-registered brightness temperatures in the Arctic region are shown for three different frequencies (85V, 37V, and 18V) in Figure 5a, and for two different frequencies and another polarization (19V, 37V, and 37H) in Figure 5b. The scatter plots provide information on how the different ice surfaces are represented in the multichannel microwave data. Clusters of data points labeled A, B, C, and D are all in consolidated ice areas (e.g., near 100% ice concentration). They are believed to be different types of surfaces (including first-year and older ice types) or mixtures of these surfaces. The cluster of points along the label O to W corresponds to ice-free ocean data. Data points between the consolidated ice clusters (AD) and the open water cluster (OW) are either mixtures of ice and water or new ice and are labeled E. A principal component analysis of multichannel microwave data revealed that the multispectral data are basically two-dimensional, since most of the channels are highly correlated with each other [Rothrock and Thomas, 1988]. Thus, using conventional algorithms, three equations can be set up (two for the two independent channels and the mixing equation) to allow at most, three different types of surfaces to be identified. This creates a limitation in sea-ice applications, since there are usually more than three types of radiometrically different surfaces, including open water, within the ice pack.

To illustrate the seasonality of sea-ice emissivity, 2-D scatterplots of SMMR monthly brightness temperature data in the Arctic during different seasons (from March, June, September, and December) are shown in Figure 6. The plots utilized vertically polarized 18- and 37-GHz data which provide information similar to the 19- and 37-GHz SSM/I data. In autumn and winter, the 18-GHz data show considerable contrast between open water and ice. Although the 37-GHz data show significant contrast between open water and first-year ice, overlapping values for open water and central Arctic sea ice (which is mainly multiyear ice) are obvious. Also, the 37-GHz data show a significant contrast between first-year ice and Central Arctic multiyear ice while the 18-GHz data show considerably smaller variations in these areas. This phenomenon is due to greater sensitivity to internal scattering in multiyear ice at 37 GHz than at 18 GHz because of the shorter wavelength of the former. In late spring and early summer (June), the brightness temperatures in the central Arctic are much more uniform in both channels due to melt effects that virtually make the multiyear ice indistinguishable from first year ice. In midsummer, the data points for consolidated ice are much more dispersed, due to surface effects including meltponding and wetness [Comiso, 1990]. In September, a large spread of data points at 37 GHz is apparent due to refrozen surfaces that makes internal scattering a factor again.

Ice concentration has been derived from satellite multichannel passive microwave data using various techniques [Svendsen et al., 1983; Cavalieri et al., 1984; Swift et al., 1985;

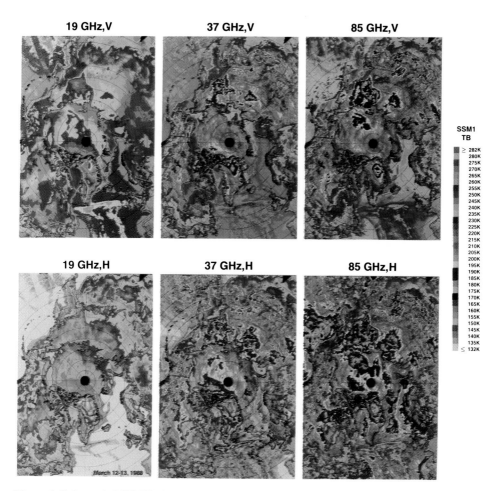

Figure 4. Color coded SSM/I brightness temperature maps at 19-, 37-, and 85-GHz channels at vertical (V) and horizontal (H) polarization on March 12–13, 1988.

Comiso, 1986; Gloersen et al., 1992]. All the algorithms utilize the radiative transfer equation for the surface and the atmosphere in combination with a mixing formalism that assumes that only sea ice and open water are present within the footprint of the sensor. They differ in the parametrization as well as in the set of channels utilized. A summary of the techniques and some comparisons of results were provided by Steffen et al. [1992]. A less conventional approach makes use of Kalman filtering and time series of data in an ice model [Thorndike, 1988; Thomas and Rothrock, 1988]. The technique is promising but may have as one of its shortcomings the lack of adequate and evenly spaced buoys for temperature and velocity data needed in the parameterization of the model.

Monthly ice concentration maps for an entire annual cycle of SSM/I data using the bootstrap technique as described in Comiso [1986] are shown in Figure 7 for the Arctic. The set

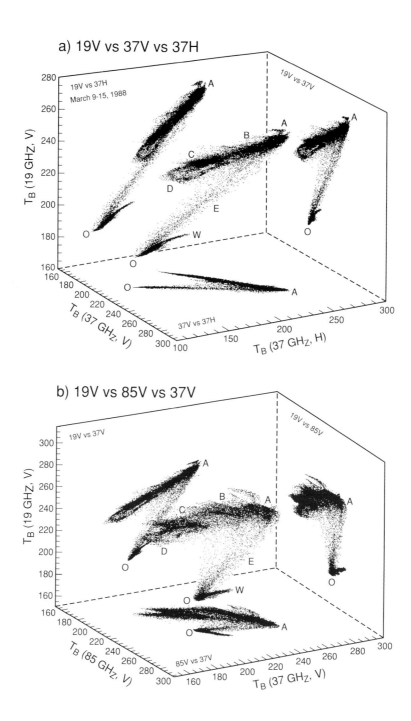

Figure 5. 3-D scatterplots of SSM/I brightness temperatures using: (a) 19 GHz(V) vs. 37 GHz(V) vs. 37 GHz(H) and (b) 19 GHz(V) vs. 85 GHz(V) vs. 37 GHz(V).

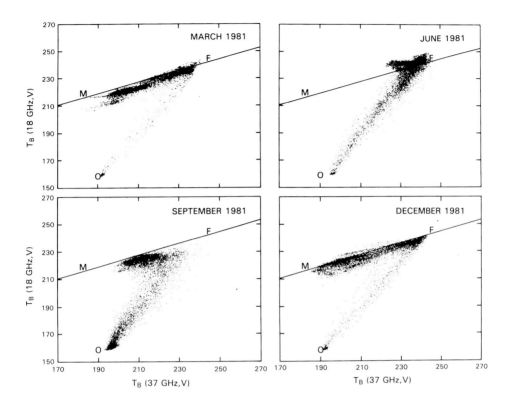

Figure 6. Scatterplots of 18 GHz(V) vs. 37 GHz(V) using SMMR data for March, June, September, and December 1981 [from Comiso, 1990].

of images basically shows the large-scale seasonal growth and decay characteristics of the sea-ice cover. It also illustrates the strength of the passive microwave sensor at providing global coverage—the temporal resolution of which can be as good as less than a day. Comparison of ice concentrations and ice edges derived from this technique have shown good general agreement with in situ and other satellite data. However, in areas of lead and polynya formations and ice edges, larger errors are incurred in the calculation of ice concentration because the emissivity of new ice varies continuously with thickness up to several centimeters [Grenfell and Comiso, 1986; Comiso et al., 1989; Wensnahan et al., 1993]. Thus, in the ice concentration maps, new ice cover with high concentration is represented as a lower concentration of open water and thick ice. Other ambiguities are associated with surface melt during early spring which enhances ice emissivities and causes abnormally high values of ice concentration [Grenfell and Lohanick, 1985]. Also, flooding or melt-ponding lowers the emissivity of the surface [Grenfell and Lohanick, 1985; Lohanick 1993] and can cause ice concentrations to be underestimated [Comiso and Kwok, 1993]. Spatial variations in ice temperatures are approximately taken into account in the bootstrap technique with slight errors in the Central Arctic as described in Comiso [1986], and signifi-

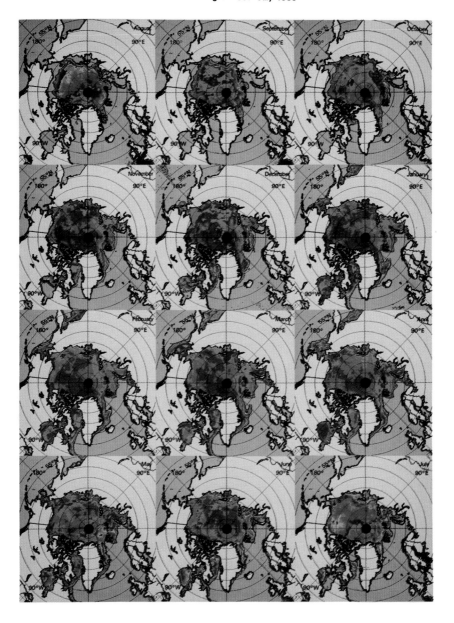

Figure 7. Color coded monthly ice concentration maps using SMM/I data from January through December 1988 (from Comiso, 1991] (reprinted with permission from Elsevier.)

cantly larger errors in the seasonal ice region. However, in the latter, the temperature of the snow/ice interface, the main source of radiation for first-year ice, does not change much because its snow cover is such a good insulator. Previous in situ measurements in seasonal ice regions gave average snow/ice interface temperatures of 7°C with a standard deviation of about 4°C [Comiso et al., 1989]. When used in the formulation for the determination of ice concentration the uncertainties due to variations in physical temperature are less than 3%. Furthermore, errors due to greater sensitivity of the horizontal than vertical polarization to various surface effects [Matzler et al., 1984] are avoided by the use of vertical polarization channels only in the seasonal ice region.

A supervised cluster analysis technically implemented by Comiso [1986] showed that there are a few large areas in the Central Arctic region with persistent signatures during the winter period. The basic results have been reproduced using an automated cluster technique on six SMMR channels [Comiso, 1990]. While interpretation of the clusters is still preliminary, there is good coherence of the geographical location of data points within each cluster with identifiable features of the ice pack. For example, the outer zone (including the ice edge and the marginal ice zone) and the seasonal ice region are well defined by corresponding clusters. In the perennial ice zone, the geographical locations of data points from identified clusters are consistent with advection characteristics of ice of different ages as predicted by the analysis of buoy data [Colony and Thorndike, 1985]. In situ and aircraft measurements confirm the basic interpretations [Comiso et al., 1989; Comiso et al., 1991] but additional research is necessary to ensure correct interpretation of the significance of each cluster of data points. The cluster approach, however, has already been useful for identifying regions and time periods where local (as opposed to global) algorithms should be used to obtain accurate results. Further refinement of the analysis may help identify zones of flooding, areas of thick snow cover, areas of second year ice and areas of highly deformed ice cover.

Open-water region

In the open ocean, passive microwave data have been used for quantifying large-scale distribution of sea-surface temperature, wind speed, water vapor, and precipitation [Wilheit and Chang, 1980; Wentz et al., 1986; Spencer et al., 1989]. The advantage is good synoptic and global coverage even in darkness and cloudy conditions. The drawbacks are coarse resolution, spatial variations in surface emissivity, and antenna patterns that make it difficult to obtain good results at the land/ocean and ice/ocean boundaries. Adverse weather conditions also affect retrieval of some of the parameters.

Sea-surface temperatures (SST) have been determined from SMMR data using regression techniques with ship measurements [e.g., Wilheit and Chang, 1980]. The 6.6-GHz channels are used primarily to obtain SST but other channels must be used to correct for variations in surface emissivity on account of environmental effects. For example, to correct for effects of surface roughness and foam the 10.7-GHz channels must be utilized, while the 18-GHz and 21-GHz channels are used to correct for atmospheric opacity and water vapor. Also, additional corrections were applied to account for apparent bias in the absolute calibration and shortcomings of the radiative transfer model used. Gridded maps of SST from

SMMR data have been produced with a resolution of 150 km and an estimated rms-error of about 0.9 K [Milman and Wilheit, 1985].

Passive microwave data also have been used to estimate integrated atmospheric water vapor [Wentz, et al., 1986; Petty and Katsaros, 1990], cloud liquid water content [Alishouse et al., 1990; Curry et al., 1990; Petty and Katsaros, 1990], and precipitation [Prabhakara et al., 1982; Spencer et al., 1989; Adler et al., 1993; Liu et al., 1992; Chang et al., 1993]. The first two parameters are calculated in conjunction with the calculation of wind speed to minimize errors due to effects of the latter. The integrated water vapor emission is measured by channels close to the water vapor band (i.e., 22.235 GHz) and is estimated by some from the difference between the 22- and 19-GHz channels at vertical polarization. These two channels are utilized to minimize the effect of cloud and precipitation on the retrieval. Cloud liquid has been inferred from 18-, 21-, and 37-GHz data using either multiple regression techniques or an approximate solution to the radiative transfer equation. Others choose to include the use of the 85-GHz channels because of their strong sensitivity to integrated liquid water. For rain rate estimates, the combined use of the 37- and 85-GHz channels provides the information for locating raining areas but other channels are needed for optimum accuracy (e.g., to minimize effects of water vapor). The microwave data tend to underestimate the rate because of inhomogeneity of rain within the field of view of the sensor and the nonlinearity in the brightness temperature-to-rain rate relations. Statistical estimates indicate that a factor of 1.5 is needed to account for this bias [Chang et al., 1993].

Estimates have been made of near-surface scalar winds using all 10 SMMR channels in combination with in situ data (from buoys and ocean stations). In this method, the 6.6-GHz channels were used to account for surface temperature effect in the wind estimates. However, Wentz et al. [1986] showed that without compromising accuracy, a set of six channels using 19, 22, and 37 GHz available in the SSM/I sensor is just as effective as the set of 10 in the SMMR sensor if climatological temperatures are used. SSM/I data have been used successfully in producing scalar wind data at an accuracy of about 2 m/sec in the range of 3–25 m/sec [Goodberlet et al., 1989]. However, the retrieval of wind is possible only where the effect of precipitation and liquid water do not mask out the ocean surface signal.

Although only the scalar values are derived, the wind speed data set has been utilized effectively for the study of some polar processes. For example, it has been used to monitor the temporal development of polar lows, which are regions of large temperature gradients at the boundary between sea ice and ocean [Claud et al., 1993]. The SSM/I data have also been combined with the European Climate Model Wind Field (ECMWF) wind speeds through the use of variational analysis techniques by Atlas et al. [1991] to generate improved geostrophic wind fields. The latter products have been used successfully to explain short-term variability of ice edge positions [J. Comiso and R. Atlas, private communication, 1992].

Active Microwave Systems

Among the active microwave systems, the most useful for Arctic process studies are the Synthetic Aperture Radar (SAR), the radar altimeter, and the scatterometer. Of the three, SAR is the only system with imaging capabilities and is especially attractive because of fine resolution (about 18–30 m) that affords the detection of many surface features. Such features are strongly represented in the images but some ambiguities in the interpretation are apparent as will be discussed in the following sections. Radar altimeter and scatterometer data are primarily for open ocean topography and wind velocity studies, respectively, but they may also be useful for studying sea ice surface characteristics as well.

SAR is a highly improved version of an older system called Side-Looking Airborne Radar (SLAR), which utilizes a pulse compression technique to obtain good cross-track resolution. However, unlike the SLAR, SAR also has a good along-track resolution. Using a pulse coherent radar, the Doppler-shifted radar echoes are recorded and played back through a coherent image processor to synthesize an along-track antenna length, which is much longer than the antenna's physical length. The along-track resolution is then vastly improved because it is determined by the wavelength multiplied by the slant range divided by the synthesized along-track antenna length.

The first civilian spaceborne SAR was an L-band system aboard the Seasat satellite, which was launched in July 1978, but operated for only three months. This system was horizontally polarized (HH) and operated at a fixed wavelength of 23 cm and at a fixed look angle of 20° from nadir. The swath width was 100 km and the resolution was approximately 25 m. The immediate successor to Seasat SAR on a satellite platform is the ERS–1 SAR launched by the European Space Agency in July 1991. This system is a vertically polarized C-band system with a swath width of 100 km located 250 km to the right of the satellite track. At a ground resolution of 30 m, the flow of data is so large that onboard recordings are not utilized and instead, about 15 stations worldwide are used to collect data when the satellite is in sight. This C-band SAR system was followed by the JERS–1 SAR, a horizontally polarized L-band system, launched in February 1992 by NASDA of Japan. The latter has a ground resolution of 18 x 18 m and a swath width of 75 km located 35° off nadir (at 570 km altitude). In 1995, RADARSAT, a horizontally polarized C-band SAR, scheduled to be launched by NASA for the Canadian Space Agency. Designed for improved global coverage, it will operate at available swath between 50–500 km at a resolution of 30–100 m located at 20° to 50° to the right of nadir.

Radar altimeters are designed to measure surface elevation and were initially used for ocean topography studies. However, they also have found applications in sea ice and continental regions (e.g., ice sheets). The first spaceborne radar altimeter was the GEOS–3, which operated from 1975–78. This was followed by the Seasat altimeter, which had a much shorter lifetime (July–October 1978), and then the Geosat, which operated from 1985–89. Most recently, the ERS–1 and TOPEX/POSEIDON altimeters were launched in July 1991 and August 1992, respectively, with significantly improved capabilities. Basically, the range from satellite to the surface is inferred from the time it takes for short radar pulses to reach the surface and return to the sensor. At a pulse rate of approximately 1000 per second, 50 consecutive pulses are summed to form composite pulse waveforms which are used as the basis for evaluating the range. The vertical accuracy of the ERS–1 altimeter is designed to be about 50 cm in the measurement of significant wave height. The TOPEX/POSEIDON altimeter, which is a dual frequency system (13.6 and 5.3 GHz) and a built-in multichannel passive microwave system to correct for atmospheric effects, is expected to have a vertical accuracy approaching 2 cm. However, the spatial coverage for the latter excludes a large fraction of the Arctic region since it collects data only up to 66° N.

The radar scatterometer is designed to measure the normalized radar backscattering cross-section as accurately as possible. A scatterometer has been onboard Skylab (1974), Seasat (1978) and most recently ERS–1. Scatterometers will also be onboard Radarsat and the Advanced Earth Observation Satellite (ADEOS), which is expected to be launched in 1996. The primary use of this system is to measure wind speed and direction. The principle is based on the dependence of the normalized radar cross-section on ocean roughness, which in itself is dependent on surface wind speed. The scatterometer operates by measuring the change in radar reflectivity of the ocean surface due to the small ripples caused by pertur-bation of the wind on the surface. The energy of these ripples increases with wind velocity. With the ERS-1 version, there are three sideways looking antennae with incident angles at 18°– 47°, 25°–59°, and 25°–59° and aspect angles of 0, +45°, and -45°. The antenna beams successively sweep over a single point on the surface as the platform moves along its orbit with a 500 km swath width. The basic parameter measured by the scatterometer is radar scattering cross-section per unit area, simply called $\sigma^°$, and expressed in units of decibels (dB). With several $\sigma^°$ measurements of the same area from different measurement direc-tions, the surface wind vector can be determined in terms of speed and direction.

Sea-ice regions

Among the most important surface characteristics associated with backscatters from active microwave systems are roughness, salinity, thickness, wetness, and the presence of volume scatterers. As discussed earlier, these characteristics are closely linked with properties of different types of sea ice. The extremely good resolution makes the SAR potentially able to detect and characterize most ice pack features (e.g., leads, ridges, and pancakes), without the application of a mixing algorithm. However, it has not been easy to design unsupervised techniques that could unambiguously identify each of these features. One problem is that the current SAR sensors are just one-channel systems that usually operate at very long wavelengths. Thus, apparently distinct features may have similar backscatter and end up as belonging to the same surface type when conventional techniques are used. The application of sophisticated pattern-recognition techniques may be useful in such cases to accurately interpret the data.

Ice concentration can be derived from SAR by estimating the fraction of open water within refrozen leads, between divergent floes, and in the marginal ice zone. In a given image, lead regions can be identified by inspection. Thin ice is usually difficult to discriminate from open water in lead areas but the former oftentimes show spatial features associated with rafting and small-scale ridging. A lower limit in the value of ice concentration percentage can be estimated from the areal extent of undeformed lead areas. This can be done either by identifying these regions interactively in a computer or by training the computer system to make the identification. If the fraction of such lead areas is very small (e.g., <5%), as is mainly the case in the Central Arctic in winter, then the error in the determination cannot be greater than this fraction and the ambiguities associated with new ice/open water dis-crimination may not be a big factor. Errors can be further reduced by incorporating meteo-rological wind data in the analysis [Kwok et al., 1992; Comiso and Kwok, 1993]. Wind speeds greater than 4 m/sec cause enhanced backscatter for open water and allow better discrimination of the latter when some of the ambiguities occur.

The various ice types that may be identified with SAR are new ice, first-year ice, undeformed multiyear ice, and ridged ice. Previous techniques for analyzing SAR at various frequencies and polarizations have had mixed success [Campbell et al., 1977; Burns et al., 1987; Onstott et al., 1987; Comiso et al., 1991]. An approach used for analyzing ERS–1 SAR data has been developed by Kwok et al. [1992]. This approach utilizes a lookup table for the backscatter of the different surfaces during different seasons based on surface measurements. Good calibration of the sensor is required for a good match of satellite data with the backscatters of the different surfaces as given by the lookup table. Ranges of the backscatter of each ice type are determined by cluster analysis that takes into account regional changes in backscatter and imperfections in the calibration. A sample geophysical product of ERS–1 data collected and processed by the Alaska SAR Facility (ASF) is shown in Figure 8. The original image coded in gray level backscatter values (with white corresponding to high backscatters) is shown in Figure 8a while a color coded image of the ice classes is shown in Figure 8b (white for multiyear ice, blue for first year ice, and red for new/thin ice). The results in Figure 8b show a general consistency with features in the original data shown in Figure 8a. However, the classification may not be comprehensive enough, since there are usually more ice types than what the system expects.

Ambiguities in the interpretation of SAR images are common in first year ice and lead areas where the top layer of the ice surface is lossy. Since the incidence angle is about 23°, the strength of the backscatter measured by the SAR instrument is mainly influenced by the roughness of the surface. If the surface is smooth, the backscatter is weak because much of the original signal is reflected in the other direction. If the surface is rough, the backscatter is strong because of the presence of surfaces perpendicular to the direction of the incident signal. The relative surface roughness can be quantified in terms of the average height of surface irregularities, h, which is equal to $\lambda/(8 \cos \theta)$ for smooth surfaces and $\lambda/(4.4 \cos \theta)$, for rough surfaces [Peak and Oliver, 1971], where λ is the wavelength and θ is the angle of incidence. Since water surfaces within the ice pack are usually smooth and are excellent specular reflectors, the SAR backscatter from these surfaces is usually low. However, new, young, and first year ice surfaces are also normally smooth surfaces and are represented by very weak backscatter as well. Thus, discrimination among calm water, new ice, young ice, and first year ice surfaces would be difficult. It should also be pointed out that the presence of frost flowers on the surface of new and young ice causes some effects (e.g., surface roughness) that makes the backscatter strong while open water areas can have ripples because of strong winds that cause the backscatter to be strong as well. All these factors complicate the interpretation of SAR data.

The backscatter of multiyear ice in winter is usually predictable because of its generally rough low-loss surface compared with that of first-year and new ice. Thus, in general the fraction of multiyear ice can be derived at a high accuracy. It should be noted, however, that there can be large spatial variations in structure and surface roughness of this ice type and flooding can occur in some areas because of snow loading. The range of backscatter values can in fact be larger than what the classifier can tolerate as previously reported for X-band SAR [Comiso et al., 1991]. In such cases, automated classification schemes that makes use

(a)

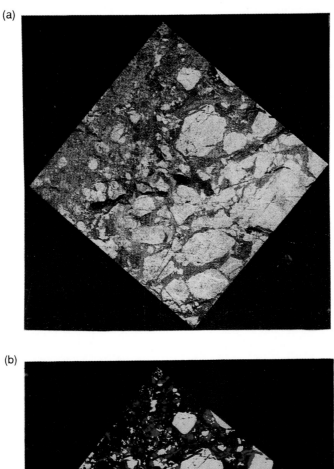

(b)

Figure 8. ERS-1 SAR product from the Alaska SAR Facility: (a) low resolution image and (b) color coded ice types [from Kwok et al., 1992].

of look-up tables would have some problems. The use of time series information [Fily and Rothrock, 1990] or pattern-recognition techniques [Holt et al., 1989] may be required to minimize such ambiguities.

Because of their unique topography, ridges have been postulated as having significantly higher SAR backscatter than other surfaces. Using image enhancement techniques, statistics of ridges from SAR have been studied [Vesecky et al., 1988]. However, analysis of ridges using X-band SAR data show significant variations from one ridge to another. Observations during the Coordinated Eastern Arctic Research Experiment (CEAREX) indicate that the backscatter changes depending on the ridge orientation [Lepparanta and Thompson, 1989]. Results by Holt et al. [1990] also show that lower frequency SAR (e.g., P-band at 0.440 GHz) is superior in the delineation of ridges than the higher frequency versions (e.g., C-band at 5.35 GHz). This is clearly illustrated in the multifrequency polarimetric SAR data shown in Figure 9, which were taken from an aircraft. Images from the various channels are shown to have different contrast for different surfaces, with the lower frequency data showing more consistent backscatters for ridges. Also, the lower frequency channels have lower backscatter values over multiyear ice because of less sensitivity to the air bubbles (wavelengths very large compared to scatterers), while more details are apparent in the first year ice region. Multichannel SAR systems could thus provide more accurate classification of the ice cover as suggested by some studies [Drinkwater, 1990]. While no satellite version of the polarimetric SAR is planned for the near future, the set of images could provide guidance on which of the different SAR systems currently available (i.e., L-band SAR in JERS–1 versus the C-band SAR in ERS–1) would be most useful for a particular study.

Some other applications of SAR data provide less ambiguity. By analyzing time-sequenced SAR images of the same area, rotational and transverse displacements of each ice floe can be calculated [Hall and Rothrock, 1981; Leberl et al., 1983]. As an example, two Seasat SAR images over the Beaufort Sea, collected three days apart, are shown in the first two images of Figure 10. A large change in ice conditions is apparent from one period to another. The ice vectors derived from the two images are shown in the third image (bottom right). Similar vector fields have been successfully derived using ERS–1 SAR data [Holt et al., 1992]. With several years of SAR data already available, such vector fields can be used to better understand large-scale ice dynamics in the Arctic.

The satellite radar altimeters have been designed primarily for open-ocean applications but since the return pulse has a significantly different shape for sea ice than for open ocean, the data have been studied for potential application over sea-ice regions [Hawkins and Lybannon, 1989; Fetterer et al., 1992]. The altimeter data are believed to be most useful for detecting the ice edge, small fractions of open water in the inner zone, and swell penetration. A parameter called ice index, which is a value derived from the automatic gain control signal and the average altitude/specular gate signal [Laxon, 1989], was found to be useful for surface classification since it has a value ranging from 0.6–0.7 over water and greater than 1 over ice. Generally, the ice indices have been found to compare favorably with results from SMMR data. However, there are mismatches in the location of ice edges and in high concentration areas. The altimeter data may provide accurate information about the location of the ice edge because of sensitivity to the presence of pancakes or brash ice usually located

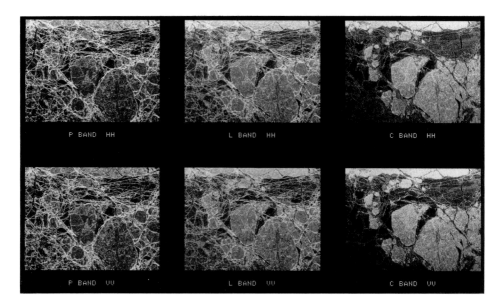

Figure 9. JPL-SAR multifrequency/polarimetric (P-band at 0.440 GHz, 68.18 cm; L-band at 1.25 GHz, 23.5 cm; and C-band at 5.35 GHz, 5.7 cm) images (courtesy of J. Crawford of the NASA/Jet Propulsion Laboratory).

in the ice edge region. Also, because of high sensitivity to the presence of even a small fraction of open water in consolidated ice region, it may be able to detect the presence of leads more accurately than other sensors. Nevertheless, the application of the altimeter for process and climate studies is limited because of its narrow beam size (it takes several days of orbital altimeter data to cover the entire polar region). Since the sea-ice cover is constantly changing, a several day composite of altimeter data would be difficult to interpret.

Open-water regions

Active microwave systems have been employed to characterize processes in the open ocean and to quantify wave propagation, wind velocities, and surface heights. As an imaging system over the ocean, SAR can provide information about mesoscale ocean current circulation patterns including frontal boundaries and eddies, slicks, internal waves, and rapid mesoscale wind - field variations [Apel, 1987; Raney, 1993]. The SAR backscatter responds primarily to surface roughness in the form of small ripples generated by wind and currents. The small ripples are modulated by longer gravity waves and the tilting of the small ripples reflects the wave pattern of the larger waves on the SAR images. Studies have shown that even at moderate wind speeds of 4–10 m/s, frontal current boundaries, meanders, and eddies are detectable [e.g., Johannessen et al., 1993]. However, at about 12 m/s, the spatial distribution of short gravity waves are predominantly reflecting wind stress and eventual swell fields, while waves of about 0.07 m wavelengths disappear below 3 m/s.

Waves and mesoscale features in the marginal ice zones have also been studied by Liu and Peng [1993]. They showed that ERS–1 SAR data provide wave spectra that are accurate

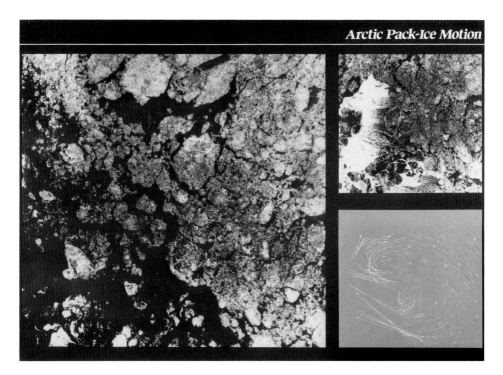

Figure 10. SAR time sequential images of ice vector determination (from NASA, Oceanography from Space brochure).

enough for use in areas where there are no other data available. An example of a SAR image in the marginal ice zone with a corresponding wave spectra product for ice and ocean is shown in Figure 11. Comparison of their results with those provided by in situ buoy data and the hindcast results of numerical wave models showed very good agreement. It is thus expected that SAR data can routinely provide dominant wavelengths and directions of wave systems.

Synoptic measurement of wind velocities over the oceans at a relatively good time resolution can be obtained from scatterometers. Although it was active for only 3 months, the Seasat scatterometer provided much more comprehensive data than those previously available from ship-based measurements. The Active Microwave Instrument (AMI) onboard the ERS–1 satellite was designed to be accurate to within 2 m/sec and 20°, within 50 km diameter cells and an interval of 25 km. This sensor has already provided a few years of global wind data that are potentially invaluable for ocean circulation studies including those in the Arctic. The conversion of the σ° value into wind data is performed with a mathematical model, that defines the relationship between σ°, wind speed, wind direction, incidence angle of the scatterometer pulse and the signal polarization. Preliminary analysis shows that retrievals from daily average data have an average rms difference from buoy data of 3.0 m/s while the rms difference for monthly data is about 1.2 m/s [Halpern et al., 1993].

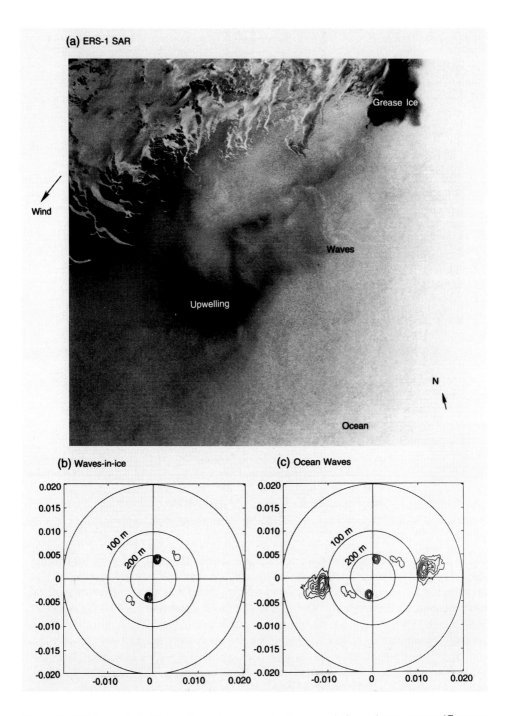

Figure 11. SAR image in the Bering Sea and wave spectra for waves in ice and ocean waves (Courtesy of A. Liu, NASA/GSFC). (SAR image, ESA Copyright, 1992.)

The radar altimeter has provided global maps of sea surface topography for several periods [e.g., Douglas and Gaborski, 1979]. The sea surface data can be combined with gravitational topography data to generate dynamic topography [Marsh et al., 1990; Koblinsky, 1993]. The latter could then provide ocean-current velocities that could enhance our understanding of surface ocean circulations including that in the Arctic. Altimeter data have been used for process studies in mid- to high-latitude regions. For example, Bhaskaran et al. [1993] used Geosat altimeter data to study the variability of the Gulf of Alaska over a 3.75-year period. They used empirical orthogonal function (EOF) analysis and showed that distinct zonal bands can be identified whose location and orientation agrees with historical depiction of the gyre in the region.

Ocean Color Systems

The first imaging ocean-color sensor was the Coastal Zone Color Scanner (CZCS) onboard the Nimbus–7 satellite which was in operation from November 1978 through June 1986. The sensor had five visible channels centered at the following wavelengths: 0.443, 0.520, 0.550, 0.670, and 0.750 μm; and had a resolution of about 825 m over a swath width of 1566 km (1970 pixels). Unlike those for other visible sensors meant primarily for studies over land or the atmosphere, these channels were chosen specifically for ocean color studies. Also, since the ocean reflectance is low, the set was chosen such that the signal-to-noise ratios of those channels sensing reflected solar radiances from the surface were high. The sensor had considerable flexibility built into it to improve performance, such as an adjustable gain to provide a dynamic range suitable for measurements from the ocean. Also, the scan mirror could be tilted from nadir by angles up to $\pm 20°$ to avoid sun glint.

The CZCS has not been replaced since mechanical problems caused its demise in 1986. However, a new satellite-based ocean color sensor called SeaWiFS is scheduled for launch by Orbital Systems for NASA in June 1995 [SeaWiFS Working Group, 1987]. In its present configuration, SeaWiFS will have eight channels (0.412, 0.443, 0.490, 0.510, 0.555, 0.670, 0.765, and 0.865 μm) with a resolution of about 1 km. Only a limited amount of data at this resolution will be collected automatically, but these data can be downloaded directly through the HRPT direct broadcast. Moreover, global data at a resolution of 4 km (to be derived from the 1 km data by subsampling) will be archived, calibrated, and processed to pigment concentration values. A similar sensor called the Ocean Color and Temperature Scanner is also scheduled for launch onboard the ADEOS satellite in 1996 by the National Space Development Agency (NASDA) of Japan.

The ocean-color data provide estimates of the near-surface concentrations of phytoplankton pigments (chlorophyll a and its associated degradation product, phaeophytin a). To extract pigment concentrations from ocean color data, the total radiance, $L_t(\lambda)$, is decomposed into its components. For the CZCS, the standard approach [Gordon et al., 1988] is as follows:

$$L_t(\lambda) = L_r(\lambda) + L_a(\lambda) + t(\lambda)L_w(\lambda) \qquad (2)$$

where L_r is the contribution due to scattering from air molecules (Rayleigh scattering), L_a is that due to scattering from particles (i.e., aerosol) suspended in air, L_w is the water leaving radiance diffusely transmitted to the top of the atmosphere while t is the diffuse transmittance of the atmosphere. The contribution from atmospheric scattering, which is responsible for about 80–90% of the satellite-received radiance, is quantified using a radiative transfer algorithm developed by Gordon et al. [1983]. The multiple Rayleigh scattering correction developed by Gordon et al. [1988] was later implemented and applied to the CZCS data to address the problems caused by large solar zenith angles, such as those encountered in high latitudes. The near-surface pigment concentrations were then computed from corrected water-leaving radiances by a second, bio-optical algorithm developed by Gordon et al. [1983]. The bioptical algorithm utilizes an empirical equation that makes use of the ratios of water leaving radiances observed at 443 nm (low pigment) or 520 nm (high pigment) and those at 550 nm and in situ pigment concentration data. At high concentrations, water leaving radiances at 443 nm are so small that spatial details cannot be derived from the 8-bit data recorded by the satellite. To improve the situation and minimize errors, a switch was applied in which the ratio of 520-nm to 550-nm radiances was used for concentrations >1.5 mg/m^2. This scheme and the following equations were adapted to obtain pigment concentration, C, in the global processing of CZCS data [McClain et al., 1993]:

$$C = 1.13[L_w(443)/L_w(550)]^{-1.71} \text{ for C} \leq 1.5 \text{mg/m}^3, \text{ and} \qquad (3)$$

$$C = 3.33[L_w(520)/L_w(550)]^{-2.44} \text{ for C} > 1.5 \text{mg/m}^3. \qquad (4)$$

In the interpretation of pigment values, however, it should be noted [McClain et al., 1993] that the surface reflectance is not always dominated by chlorophyll, the specific absorption coefficient of phytoplankton can vary by an order of magnitude, and that the concentration estimate is an optically weighted average over the first optical depth which depends on the vertical distribution of pigments.

Certain problems have been pointed out in the use and interpretation of CZCS data at high latitudes [Clark and Maynard, 1986; Maynard, 1986; Gordon et al., 1988; Comiso et al., 1993]; most notably, these include low solar illumination caused by large solar zenith angles, and persistent cloud cover. Application of the multiple scattering Rayleigh radiance model of Gordon et al. [1988], as indicated earlier, makes possible a more consistent pigment retrieval at any latitude. However, new studies show that Gordon's technique may not be the best to use in the polar regions; local algorithms [Mitchell et al., 1991] were found to be more consistent with in situ data [Sullivan et al., 1993].

It is also known that immediately after the CZCS scanned over highly reflective targets (e.g., clouds, land or ice), the instrument experienced saturation-induced errors [Mueller, 1988]. Depending upon spatial extent and brightness of the surface, the data are unusable for distances up to 100 pixels from the saturation point [McClain and Yeh, 1994]. The effect of this saturation, known as electronic "overshoot," differed between the individual bands and was not consistent within or between scenes. Because the CZCS scanned from left to right (west to east), it was always observed to the right of the bright target [Mueller, 1988]. Therefore, data from leads in the ice or along the ice edge, clouds, and land must be cautiously interpreted by examining radiance levels adjacent to highly reflective targets in all visible spectral bands. It should be pointed out, however, that the electronic design of the SeaWiFS sensor is such that this "overshoot" problem is generally reduced.

The CZCS data have been used in several studies in the northern and southern hemispheres [Maynard and Clark, 1987; Müller-Karger et al., 1990; Mitchell et al., 1991; Comiso et al., 1993; Sullivan et al., 1993]. The average of all pigment concentrations in the entire Arctic region collected from November 1978 through June 1986 is shown in Figure 12. The data, earlier presented in a different image format by Mitchell et al. [1991], clearly illustrates that satellite data are needed to have a good understanding of global biological processes. The image shows a strong dominance of the mid-to-high latitude region by greatly enhanced pigment concentration values. The meridional zonation, as shown, has been found consistent with available in situ data. However, care should be exercised in the detailed interpretation of the data. While blooms are expected in the marginal ice zone where meltwater stabilizes the water column, some of the values may have been enhanced due to electronic overshoot, as discussed earlier, erroneous cloud masking, or possible contamination of the signal by sea ice. Also, the low pigment region in the Barents Sea may be due to a bias in the temporal sampling or the inability of the sensor to detect massive subsurface blooms. It should be noted in this regard that much of the CZCS signal is derived from one optical depth, which is typically less than 20 m. If the mixed layer in the region is very deep, column integrated biomass may be poorly represented in the ocean color data.

High-Resolution Visible Systems

The launch of the first Earth Resources Technology Satellite (later renamed Landsat-1) in July 1972 started detailed coverage of Earth's surface in the visible channels. The primary sensor onboard this satellite (and later versions) has been the MultiSpectral Scanner (MSS) which has a resolution of 80 m and a swath width of 185 km. The wavelengths of the spectral bands are centered at 0.55, 0.65, 0.75. and 0.95 μm. Since the launch of Landsat-4, the Thematic Mapper (TM), which has a resolution of 30 m and a swath width of 185 km, has become part of the system for enhanced capabilities. The latest version of the series is Landsat-5, which is similar to Landsat-4. The TM has seven bands centered at 0.48, 0.56, 0.66, 0.83, 1.65, 11.45, and 2.22 μm. The first of a new generation of Landsat sensors incorporating the Enhanced Thematic Mapper (ETM), which has eight spectral bands from 0.45–12.5 μm at 15 m resolution and a panchromatic band in the range from 0.5–0.86 μm was launched in October 1993 onboard Landsat-6. Unfortunately, the launch was not successful and such capability would have to wait until the launch of Landsat-7.

In addition to Landsat, there are other satellite systems with similar capabilities. Among them are two high-resolution visible scanners (10 m in panchromatic mode and 20 m in multispectral mode) both with a swath width of 60 km onboard the Systeme Pour l'Observation de la Terre (SPOT) satellite, launched on February 26, 1986. The SPOT images have a higher resolution, pointing capability, and better spatial coverage in the polar regions than the Landsat TM but the latter offers appropriate wavelengths for snow/cloud discrimination and night (or winter) viewing. Another instrument operating in the same spectral range is the Multispectral Electronic Self Scanning Radiometer (MESSR) on the Japanese Marine Observation Satellite-1 (MOS-1) which has four channels from the visible to the near-infrared. This system is similar to the Landsat MSS but has higher spatial resolution (50 m) and a narrower swath (100 km). Another system is the Optical Line Scan System (OLS) on

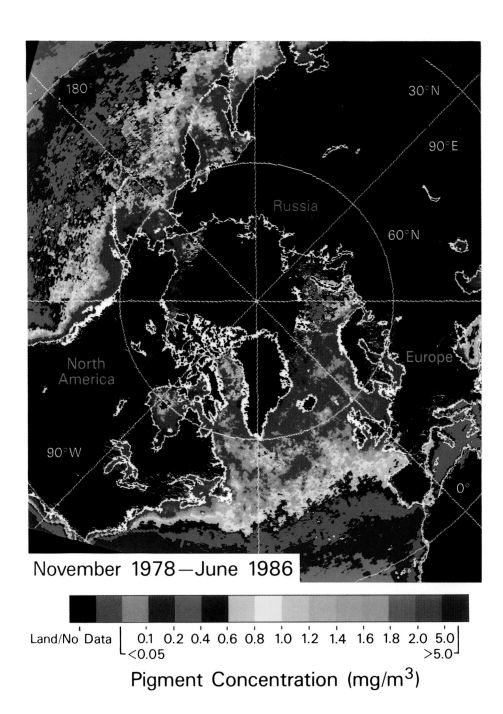

Figure 12. Color coded average pigment concentration of CZCS data [Mitchell et al., 1991] in the Arctic from November 1978 through June 1986.

board the JERS–1 satellite, which has several channels at a resolution of 12 m and has a 3-D imaging capability.

High resolution optical and photographic images of the Arctic are noted for showing remarkably detailed features of ice floes, ridges, leads, and new ice. Persistent cloud cover and long periods of darkness (especially in winter), however, limits their capability. Nevertheless, in cloud-free and illuminated areas, they can provide excellent information about ice conditions and processes in the marginal ice zone and adjacent open ocean. They also provide good complementary information to the SAR and passive microwave coverage for more complete characterization of the ice cover. This is especially the case because different ice types have different albedos [Allison et al., 1993] and in areas where the microwave sensors have problems with ice classification, the visible channels may be able to provide the critical information.

Landsat images have been used for Arctic applications since its inception. Time series of images have been used by Nye [1975] to generate ice vectors and study ice dynamics and deformation characteristics of sea ice. Useful statistics of ridging, rafting, and lead sizes can be generated if cloud cover, darkness and different sun elevations do not cause problems. The data have also been used to generate maps of ice classes in the north water polynya by Ito [1982] for the period from 1973–81. In this study, Landsat data were compared qualitatively with photographs from aircraft to provide verification of the interpretation of the former. A time series of vectors of short-term ice movements in the region were also calculated but analysis of such ice vectors is made difficult by the lack of consistency in the way they were generated. Someday they were processed from data two days apart while in other days they are processed from data three to five days apart because consecutive data cannot be obtained due to persistent cloud cover conditions.

These data have also been used for comparative analysis with ice concentrations derived from passive microwave data [Steffen and Maslanik, 1988; Steffen et al., 1992]. While general agreement is observed in consolidated ice regions, large discrepancies are evident in predominantly new ice areas mainly because of the large range in the emissivity of new ice and partly because new ice and open water are sometimes impossible to discriminate in the Landsat data. Also, mixtures of small floes and open water may look identical to new ice in the Landsat data.

The high resolution visible system may also provide useful information from the open ocean as well. Although these systems are not designed for the ocean, they can provide some information not available in the ocean color systems because of better resolution. Also in the marginal ice zone, dominant features such as eddies, fronts, and icebergs can be readily identified and studied. Statistical analysis of these features, during daylight and cloud free conditions, can be used to evaluate frequency of occurrences, sizes, and population.

For several months of the year, darkness and twilight dominate in the Arctic, and for much of the year, clouds and fog persist. Furthermore, Landsat coverage has a repeat cycle of 18 days at the equator. This allows repeat coverage of a given location in the Arctic during 3 consecutive days but for about 15 days thereafter, there is no coverage at all in the same area. Thus, the repeat cycle and broad coverage needed to analyze ice-cover changes during various time periods, especially during spring breakup, cannot be carried out as effectively as with microwave sensors. The utility of Landsat and other high-resolution systems for large-scale process and climate studies in the Arctic may thus be limited.

Infrared Systems

The sensors most suitable for measuring ice or sea surface temperatures are those with thermal infrared channels. Among these systems are the Temperature Humidity Infrared Radiometer (THIR), the Advanced Very High Resolution Radiometer (AVHRR), and the Along Track Scanning Radiometer (ATSR). A version of the THIR was flown on every Nimbus satellite from Nimbus–4 through Nimbus–7. It is a scanning radiometer with two channels: a 10.5-mm–12.5-mm (11.5mm) thermal window channel to image temperatures of ocean surfaces, land, and cloud tops, and a 6.5-μm–7.0-μm (6.7μm) channel to provide moisture and cirrus cloud content of the upper troposphere and stratosphere. The ground resolution is 6.7 km for the 11.5-μm channel and 20 km for the 6.7-μm channel.

The AVHRR is a multipurpose sensor with spectral bands from the visible through the near infrared to the thermal infrared channels (0.72, 0.91, 3.74, 10.55, and 12.0 μm). Since it was first introduced in 1978 as part of NOAA's polar orbiting operational satellite system, it has undergone only minor modifications. The operational system consists of two satellites in complementary near-polar orbits, with one crossing the Equator at local solar times of approximately 0730 and 1930, and the other at 0230 and 1430. With a swath width of 3000 km and a resolution of 1 km, the sensor provides very good spatial coverage. The temporal resolution is also good since a surface area at 60° and higher latitudes is covered by two or more consecutive orbits twice daily.

The ATSR onboard the ERS–1 satellite uses spectral channels that are very similar to those on AVHRR–2 with many improvements in accuracy. Its four channels, at wavelengths of 1.6, 3.7, 10.8, and 12 μm, are fully coregistered by a common field stop. The sensor has a resolution of 1 x 1 km at nadir and 1.5 x 2 km at forward view, and a swath width of 500 km. A unique feature is the viewing of the same area through a near vertical atmospheric path and through an inclined path of different length some way along the satellite track. This technique allows for better determination of the atmospheric correction than previous methods.

Monthly surface temperature maps in the Arctic region have been derived from the THIR data using the thermal channel only, as described by Comiso [1994]. A set of these maps for 1980–85 during a winter month (January) is shown in Figure 13. The images show details of interannual variability of cold regions in the Arctic not previously observed. Monthly averages were generated using a special filter that minimizes contamination of the data by cloud-top temperatures and that deletes data in stormy areas. The results were found to be qualitatively in agreement with climatological data but reveal many detailed features not found in the latter. Agreement of satellite data with station measurements on the ice sheets is good to within about 3 °C with rms error of about 2°. It is also highly correlated with drift stations over sea ice in the Arctic, but a slight bias in these areas is observed.

January NP Temperatures from THIR (11.5μm)

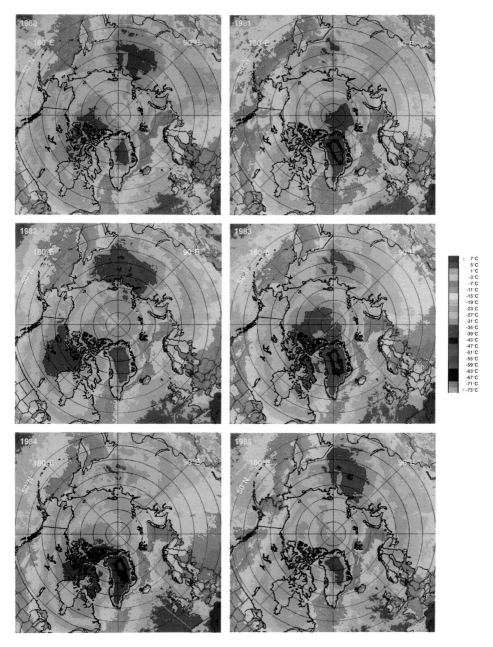

Figure 13. Arctic monthly surface temperatures from 1980–89 during a winter month—January [from Comiso, 1994].

More accurate surface temperature maps can be obtained from multichannel- IR-data as is afforded by AVHRR and ATSR. With the AVHRR, cloud masking can be done with limited success with the aid of the 3.74-μm channel and two split-thermal channels. With the ATSR, the addition of the 1.6-μm channel further improves cloud masking capabilities even during darkness. Various techniques have been developed for estimating temperatures over sea ice. One technique developed by Key and Haefliger [1992] models the relationship of AVHRR radiances with surface temperature of sea ice through a radiative transfer formulation. In this technique, seasonally adjusted coefficients to account for changing atmospheric conditions are utilized. Another technique used by Massom and Comiso [1994] relates surface in situ observations with AVHRR radiances using a regression model. The temperatures generated from these techniques are believed to have accuracies ranging from 0.5 °C–2.5 °C. However, they have been tried only in limited areas and it is not certain how well these techniques perform when utilized over the entire Arctic region. In the open ocean, the algorithms are more mature and rms errors are sometimes better than 0.5 °C [Bernstein, 1982; Viehoff and Fischer, 1988]. However, the effect of the intervening atmosphere (mostly due to water vapor) is to lower the actual value by a few tenths of a Kelvin in very cold and dry atmospheres to nearly 10 K in very warm and moist atmospheres [McClain, 1980]. Thus, the atmospheric effects in the Arctic region may not be as important as in other regions because of usually dry and cold atmosphere in the former.

While cloud masking is necessary to be able to characterize the surface effectively, the spatial distribution of cloud cover is by itself essential in many studies (e.g., radiation budget and surface irradiance studies). Global cloud distribution has been assessed using the Nimbus–7 THIR and Total Ozone Mapping Spectrometers (TOMS) data by Hwang et al. [1988] and the AVHRR data by Rossow and Schiffer [1991]. In the polar regions, however, derived cloud distributions are not as accurate as in other areas because of difficulties in the discrimination of some types of clouds from snow covered areas [Schweiger and Key, 1992].

Synergistic Use of Combined Systems

A summary of parameters that can be derived from satellite data in the Arctic as well as a qualitative assessment of capabilities of each type of sensor is presented in Table 1. While the capabilities and potentials of these sensors are obviously strong, as shown in the table, limitations exist, as described in previous sections, in the ability of some to provide geophysical parameters that are accurate enough to be useful. Even redesigning instruments to include additional channels may not help resolve the problem because the channels within a certain spectral band may be too closely correlated so as to make additions to the number of channels immaterial [Comiso, 1986; Rothrock and Thomas, 1988]. An alternative approach is to use data from two or more sensors in concert to study properties of the problem surfaces. The information content of a combined data set will certainly be enhanced, especially if the different sensors detect different physical characteristics of the surface. New insights into fusion techniques (including the use of neural network and expert systems) for processing and analysis of multispectral and multisensor data are provided by Collins [1992] and Dawson et al. [1993]. Also, the use of powerful statistical tools such the multivariate adaptive regression splines (MARS) developed by Friedman [1991] can pro-

TABLE 1. Qualitative assessment of satellite sensor capabilities for Arctic applications using microwave, infrared, and visible channels. The score ranges from 0–10, with 10 being excellent.

	Passive Mw	SAR	Alt	Scatt	IR	Vis	Ocean Color
SEA ICE:							
ice conc.	9	9	5	-	7	8	7
ice type	6	6	5	-	6	6	6
ice edge	8	8	7	-	6	6	6
ice area	9	8	8	-	-	-	-
ice motion	3	8	2	2	6	7	6
ice temp	7	4	1	1	4	-	-
melt onset	8	8	6	-	6	5	5
melt ponding	3	3	3	-	3	5	5
surf temp	2	2	1	1	9	-	-
snow on ice	5	1	-	-	3	3	3
cloudiness	-	-	-	-	7	6	6
precipitation	6	-	-	5	5	2	2
OPEN OCEAN:							
pigment conc.	-	-	-	-	-	-	9
SST	7	-	-	-	9	-	-
topography	-	9	-	-	-	-	-
water vapor	7	-	-	-	-	-	-
waves	-	7	3	5	-	4	-
wind	7	-	-	8	-	-	-
stress	-	-	-	8	-	-	-
cloudiness	2	-	-	-	7	6	6
precipitation	6	-	-	-	6	-	-
GENERAL:							
resolution	4	10	7	7	7	10	7
in darkness	10	10	10	10	10	-	-
through clouds	10	10	10	10	-	-	-
spatial coverage	10	7	2	2	8	5	8
temporal cov.	10	6	2	2	8	5	6

vide a means to sort out the most important parameter in a multi-parameter data, as demonstrated by De Veaux et al. [1993].

Many satellites have already been equipped with several sensors, an example of which is the Nimbus–7, that was launched with passive microwave, infrared, ocean color, ozone, aerosol, and atmospheric sounding sensors aboard. In this case, the research goal behind each sensor was achieved, but the utility of the data extended beyond expectations. It also provided an opportunity for the joint analysis of data from different sensors [Comiso, 1986; Comiso et al., 1993]. NASA's Earth Observing System and similar systems planned by

ESA and NASDA for the next decade will provide excellent mixtures of complementary systems. While economics might be the main driving force behind such systems, they afford unique research opportunities including: intercomparison of images of the same region by different sensors, using data from different sensors to derive a geophysical parameter, and using derived parameters from different sensors to study cause and effects behind a phenomenon. An example of Arctic data from two different sensors but from the same satellite is shown in Figure 14. One image shows monthly brightness temperatures (18 GHz, V) from the SMMR while the other shows corresponding surface temperatures derived from the THIR observation for September 1981. The two images are consistent in locating the ice/ocean boundary, with the SMMR data showing a more defined ice edge than the THIR data because the temperature of the ice at this time of the year is close to that of the adjacent water. Surface temperature data, however, provide valuable information about the surface needed to adequately interpret the passive microwave data. This is especially the case when the sea ice emissivity changes quite substantially on account of onset of melt, meltponding, freeze/thaw periods, and onset of refreezing.

Images from two different sensors and from different satellites are shown in Figure 15. The top image in Figure 15 shows an ice concentration map derived from SSM/I while the bottom image shows the 0.37-μm-channel image from the AVHRR for June 13, 1988, in the Arctic. The two images illustrate the advantage of having images from two systems, with the passive microwave data showing the relatively coarse but complete coverage of the entire region, while the AVHRR data show a detailed characterization of the ice cover where it is cloud free but no information about ice at all in the cloud-covered areas. The availability of co-registered cloud-covered and cloud-free data in the AVHRR image can be useful for modeling cloud effects, if any, on passive microwave signatures. Also, cloud-free data in AVHRR can be used to test retrievals of geophysical parameters from the passive microwave data. For example, around the Banks Island (to the right) the AVHRR image shows near 100% ice cover while the corresponding SSM/I values are about 75%. This can be an indication of flooding or melt effects which are known to affect the radiometer signal. The additional information can thus be used to adjust reference parameters in the passive microwave algorithm to obtain better ice concentration values.

Comparative analysis of passive microwave and SAR images have been done previously [Campbell et al., 1977; Carsey, 1985; Burns et al., 1987; Livingstone, 1989; Comiso et al., 1991]. The value of combining the two data sets is illustrated in Figure 16. Color-coded SSM/I brightness temperature data at 19 GHz (vertical polarization) for June 11 and 20, 1992, are shown in Figure 16a and 16b, respectively. On June 11, the brightness temperatures of sea ice in the Central Arctic are relatively low as expected for a predominantly multiyear ice region. On June 20, the same region has much higher brightness temperature. This illustrates the effect of onset of melt, as discussed previously, on the multiyear ice passive microwave signature. Figure 16c shows SAR images for the corresponding days including that for June 14 to better illustrate the temporal evolution of the backscatter. The geographical locations of the SAR images are indicated in the SSM/I images (see data points labeled E, F, and G). In the SAR images, a reversal in the backscatter of the surface during the same period is shown indicating sensitivity to the same change in the physical property of the surface. Because of much higher resolution, however, the SAR images provide detailed infor-

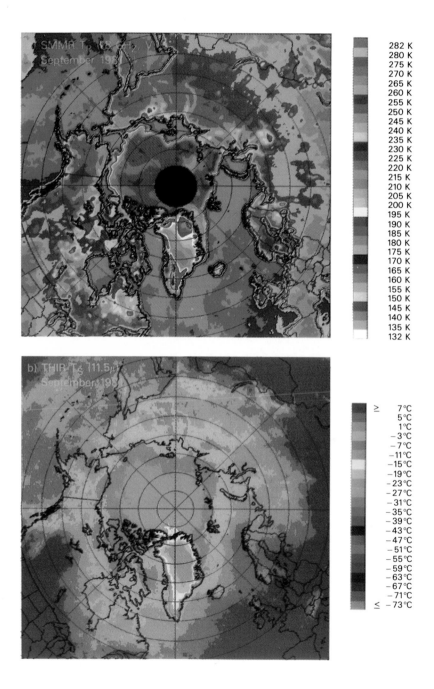

Figure 14. (a) SMMR brightness temperatures in the Arctic region in summer (September 1981), and (b) Physical temperatures derived from THIR.

CANADA BASIN
JUNE 13, 1988

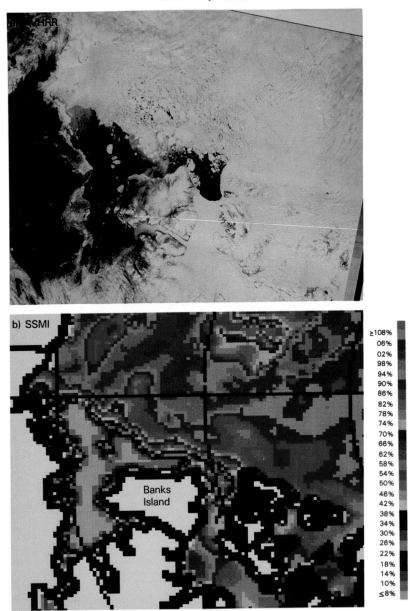

Figure 15. AVHRR visible channel image and SSM/I ice concentration map in the Canadian Basin.

data points in the algorithm, ice concentrations derived from SSM/I are substantially lower than those obtained from the SAR data [Comiso and Kwok, 1993]. The reference brightness temperatures used in the SSM/I algorithm can again be adjusted such that the ice concentration values compare better with the SAR values. Such process would make the passive microwave data set, which has a long history and comprehensive global coverage, even more useful.

Comparison of Landsat TM and SAR images is intriguing because both systems have very high resolution (~30 m). One such set of images over the sea ice region is shown in Figure 17. The two near-coincident images of Landsat and SAR show various features in the ice including that of a refrozen lead. The area labeled "1" in the images is dark gray in the Landsat image suggesting that the region is covered by thin ice. However, the same area has high backscatter values in the SAR image, likely because of salt flowers, and could be misclassified as multiyear ice. Also, the area labeled "2" appears to be a portion of a thick floe in the Landsat image but shows a significantly different backscatter than that of the rest of the floe in the SAR image. Furthermore, in the area labeled "3", what appears to be a well-defined lead (containing open water or grease ice) in the Landsat image is not apparent in the SAR image. The set of images thus indicates that while SAR has good resolution, extreme care should be exercised in the interpretation of the data. Also, care in the interpretation of Landsat images is needed since the discrimination of some surfaces (e.g., between open water and new ice) may not be possible with the data.

Needless to say, the use of more than two sensors would lead to even better characterization of physical surface conditions. For example, knowledge of surface temperature in the Landsat versus SAR study would help resolve some of the ambiguities (e.g., thick versus thin ice or barren versus snow-covered ice). The interdependence of the different geophysical parameters, however, should not be overlooked when using data from different satellite sensors. Wind velocity and temperature drives the ice margin, and having these parameters overlaid on each other would also enable analysis of cause and effect of processes in the region, as discussed earlier. Furthermore, altimeter data can be used to infer surface current movements in the open ocean area, while ocean color data can provide ensuing biological activity. Even data of different parameters derived from the same sensor can be very useful. Monthly winter and summer images of water vapor and wind speed derived from SSM/I data, using the technique of Wentz et al. [1986] are shown in Figure 18. The images show very large changes in water vapor and wind speed from summer to winter. They also show, as expected, that water vapor and wind speed are not correlated in either season even though the same source of data is used. Using such data in conjunction with IR radiance data would enable more accurate determination of surface temperature from the latter. Also, when used in algorithms for sea ice, better weather filters can be developed and atmospheric effects over sea ice can be either assessed or taken into account.

Using several types of sensors to study a certain region brings about the issue of scaling. Good resolution is highly desirable but care should be exercised in the interpretation of the data. For example, a SAR image consists of pixels about 30 m in size. Using the data at the finest resolution may have problems for some applications because of occasional occurrences of speckles, the physical significance of which is usually questionable. Also, even at this resolution, the system does not always resolve a particular type of surface and some

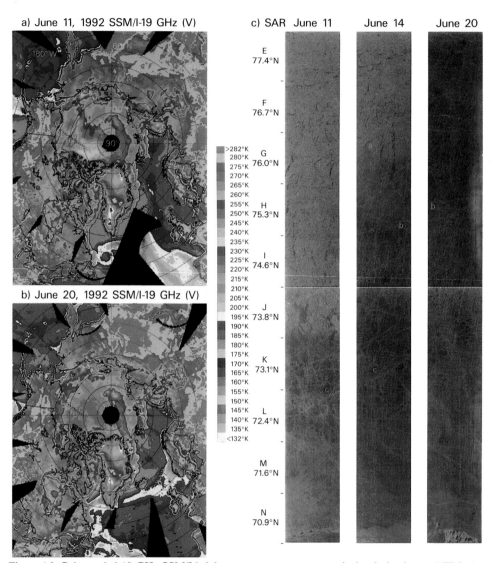

Figure 16. Color-coded 19-GHz SSM/I brightness temperatures at vertical polarization and ERS–1 SAR images during onset of spring. SSM/I data for (a) June 11, 1992, and (b) June 20, 1992, are compared with corresponding SAR images shown in (c). The geographical location of the SAR data (labeled E, F, G, etc.) are indicated in the SSM/I images. The June 14, 1992, SAR data show intermediate conditions.

mation about the transformation of the backscatter each floe and other surfaces during the period. The time series of SAR images indicate consistency in the relative location of ice floes (e.g., see same areas labeled a, b, c, in different days). The set of images establishes that the effect cannot be due to divergence or a change in ice concentration. Again, this information provides a means to better interpret the SSM/I images. Using winter reference

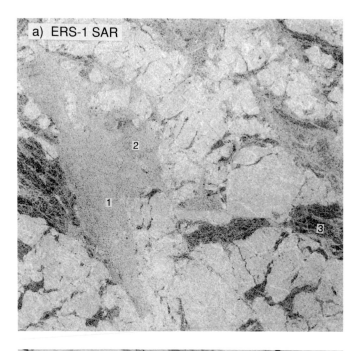

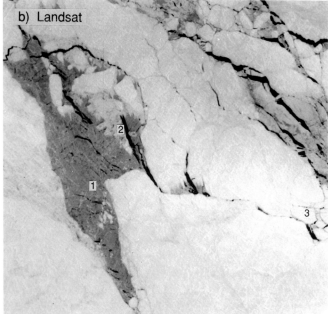

Figure 17. Concurrent Landsat and SAR images in the Beaufort Sea at 71.77 N, 145.28 W, on April 16, 1992. The images cover an approximately 15 by 15 km area (Courtesy of Koni Steffen of the University of Colorado and ASF). (SAR image, ESA Copyright, 1992.)

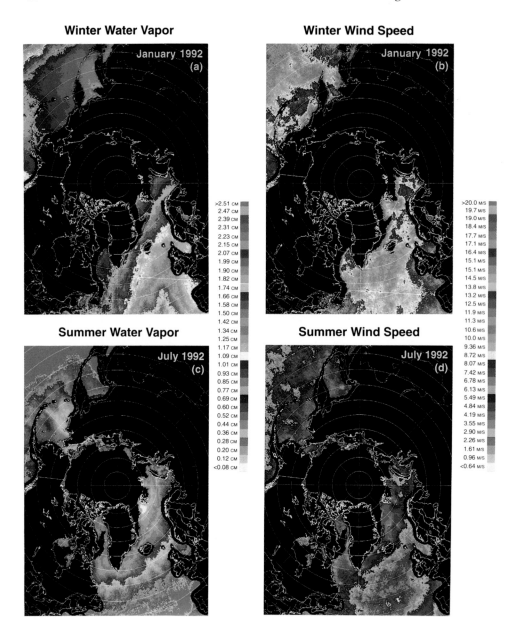

Figure 18. Water vapor and wind speed maps for summer and winter in the Arctic open ocean region (Courtesy of Frank Wentz).

pixels may actually represent a mixture of different surfaces. It should be noted as well that multiyear ice floes are normally very large compared to the 25-m pixel size and it is only by averaging the data points to generate a coarser resolution image that the basic character- istics of the ice floe corresponding to the data are revealed. With the passive microwave data, the situation is just the opposite. Each data element measured by the passive sensor is so large that sometimes it is not possible to interpret the data accurately because there are too many types of surface within each pixel. It is thus important to know what information is available at each scale, how such information may vary from one scale to another, and how to treat the data such that the same information is inferred from the same surface regardless of scale.

Conclusions

Satellite remote sensing provides a unique and powerful means of studying processes in the Arctic region. The coverage is comprehensive, the temporal sampling is very good, and the measurements are consistently made. Furthermore, there is now good international partic- ipation in the launching of new satellites and collaborations between countries are expected to improve and expand. New and innovative sensors have been launched in the last few years. Better ones with improved capabilities are also expected in the next decade.

The time history of some data sets is becoming long enough to allow trend studies of parameters in the region to be made. Some historical data have been reprocessed or recali- brated to make them more useful while data archives have been established to make data more readily available to the public. An intriguing question associated with climate change has been "How stable is the Arctic sea ice cover?" A modeling study by Budyko [1966] has indicated that a 4°C rise in summer temperatures would be sufficient to completely melt an Arctic sea ice cover, 4 m thick on the average, within 4 years. Using passive microwave sat- ellite data, it has been possible to calculate actual ice extent consistently since 1973. Such a study has been made by Gloersen and Campbell [1988]. Their analysis suggests a slight decrease in the total extent in the Arctic during the SMMR period. Also, an evidence that the ice cover is thinning in the Arctic has been reported [Wadhams, 1990]. More detailed work is needed to verify these reports but a good requirement would be to continue with the time series satellite studies until more concrete answers becomes evident.

In the utilization of satellite data, there are, however, caveats that should be noted. First, these data do not (and may never) replace surface measurements. Satellite sensors do not have the capability of observing many of the parameters now routinely measured from ship or surface stations. Also, the satellite data tend to degrade with time and the geophysical quantities derived from them should always be checked or compared with in situ measure- ments. Second, derived products are sometimes generated using satellite data and a model. There could be hidden errors associated with shortcomings of the model, and appropriate error analysis should be done. Third, the validation of geophysical parameters that are used for large-scale and global applications must be done in several regions and for all seasons. The integrity of the satellite data should be established before they can be used for serious scientific studies.

The synergistic use of multisensor data should also be a strong part of remote sensing activ- ities. Data from as many sensors as possible should be used and if possible, interactively, to improve understanding of processes being studied. Complementary information from different sensors are oftentimes needed to resolve some of the ambiguities in the interpre- tation of the data. If data of the same region from different sensors are shown to be consistent with each other and provides new insights about processes in the region, the value of satellite remote sensing would be greatly enhanced.

Acknowledgments: I wish to thank Rico Allegrino of Hughes STX and Eueng-nan Yeh of GSC for help in the processing of some of the satellite data presented. Also, I am grateful to Chuck McClain, Neal Sullivan, Rob Massom, Claire Parkinson, and two anonymous reviewers for valuable comments and suggestions, and to Luann Bindschadler for editorial help. This work was supported by the Cryosphere Processes Program at NASA Headquarters.

References

Addison, J.R., Electrical properties of saline ice, *J. Appl. Physics, 40(8)*, 3105–3114, 1969.

Adler, R.F., A.J. Negri, P.R. Keehn, I.M. Hakkarinen, Estimation of monthly rainfall over Japan and surrounding waters from a combination of low-orbit microwave and geosynchronous IR data, *J. Applied Met., 32(2)*, 335–356, 1993.

Alishouse, J.D., J.B. Snyder, E.R. Westwater, C.T. Swift, C.S. Ruf, S.A. Snyder, J. Vongsathorn, and R.R. Ferraro, Determination of cloud liquid water content using the SSM/I, *IEEE Trans. Geosc. Remote Sens., 28(5)*, 817–821, 1990.

Allison, I., R.E. Brandt, and S.G. Warren, East Antarctic sea ice: albedo, thickness, distribution, and snow cover, *J. Geophys. Res., 98*, 12,417–12,429, 1993.

Apel, J.R., Principles of Ocean Physics, *International Geophysics Series, Vol. 38*, Academic Press, 631 pp., Orlando, FL, 1987.

Atlas, R., S.C. Bloom, R.N. Hoofman, J.V. Ardizzone, and G. Brin, Space based surface wind vectors to aid understanding of air-sea interactions, *EOS, 72(18)*, 201, 204, 205, 208, 1991.

Baker, D.J., and W.S. Wilson, Spaceborne observations in support of earth science, *Oceanus, 29(4)*, 76–85, 1987.

Barry, R.G., J. Maslanik, K. Steffen, R.L. Weaver, V. Troisi, D.J. Cavalieri, and S. Martin, Advances in sea-ice research based on remotely-sensed passive microwave data, *Oceanography, 6(1)*, 4–12, 1993.

Bernstein, R.L. Sea surface temperature estimation using the NOAA–6 satellite Advanced Very High Resolution Radiometer, *J. Geophys. Res., 87(C12)*, 9455–9465, 1982.

Bhaskaran, S., G.S.E. Lagerloef, G.H. Born, W.J. Emery, and R.R. Leben, Variability in the Gulf of Alaska from Geosat altimetry data, *J. Geophys. Res., 98(C9)*, 16311–16330, 1993.

Budyko, M.I., Polar ice and climate, In *Proceedings of the Symposium on the Arctic Heat Budget and Atmospheric Circulation*, ed. by J.O. Fletcher (ed), RM 5233-NSF, Rand Corporation, Santa Monica, CA, 3–21, 1966.

Burns, B.A., D.J. Cavalieri, M.R. Keller, W.J. Campbell, T.C. Grenfell, G.A. Maykut, and P. Gloersen, Multisensor comparison of ice concentration estimates in the marginal ice zone, *J. Geophys. Res., 92*, 6843–6856, 1987.

Campbell, W.J., P. Gloersen, and H.J. Zwally, R.O. Ramseier, C. Elachi, Simultaneous passive and active microwave observations of nearshore Beaufort sea ice, *Proceedings of the 9th Annual Offshore Technology Conference*, May 2–5, 1977, Houston, 287–294, 1977.

Carsey, F., Summer Arctic sea ice character from satellite microwave data, *J. Geophys. Res., 90*, 5015–5034, 1985.

Carsey, F.(ed.), *Microwave Remote Sensing of Sea Ice, AGU GM–68*, Washington, DC, 462 pp., 1992.

Cavalieri, D.J., P. Gloersen, and W.J. Campbell, Determination of sea ice parameters with the Nimbus–7 SMMR, *J. Geophys. Res., 89*, 5355–5369, 1984.

Chang, A.T.C., L.S. Chiu, and T.T. Wilheit, Oceanic monthly rainfall derived from SSM/I, *EOS, Transactions, AGU, 74(44)*, 505, 513, 1993.

Clark, D.K., and N.G. Maynard, Coastal zone color scanner imagery of phyto-plankton pigment distribution in Icelandic waters, *Proc. SPIE Int. Soc. Opt. Eng., 637*, 350–357, 1986.

Claud, C., N.M. Mognard, K.B. Katsaros, A. Chedin, N.A. Scott, Satellite observations of a polar low orbit over the Norwegian Sea by Special Sensor Microwave Radiometer, *J. Geophys. Res.*, 98(C8), 14,487–14,506, 1993.

Collins, M.J., Information fusion in sea ice remote sensing, Chapter 25, in *Microwave Remote Sensing of Sea Ice*, ed. by F. Carsey, AGU GM–68, Washington DC, 431–441, 1992.

Colony R., and A. Thorndike, Sea ice motion as a drunkard's walk, *J. Geophys. Res.*, 90, 965–974, 1985.

Comiso, J.C., Characteristics of Arctic winter sea ice from satellite multispectral microwave observations, *J. Geophys. Res., 91*, 975–994, 1986.

Comiso, J.C., Multiyear ice classification and summer ice cover using Arctic passive microwave data, *J. Geophys. Res., 95*, 13411–13422, and 13593–13597, 1990.

Comiso, J.C., Satellite remote sensing of the Polar Oceans, *J. Mar. Sys., 2*, 395–434, 1991.

Comiso, J.C., Surface Temperatures in the polar regions using Nimbus–7 THIR, *J. Geophys. Res., 99(C3)*, 5181–5200, 1994.

Comiso, J.C., and R. Kwok, Summer Arctic ice concentration and characteristics from ERS–1 SAR and SSM/I data, *Proceedings of the First ERS–1 Symposium: Space at the Service of our Environment, Cannes, France, November 4–6, 1992, ESA SP-359*, 367–372, 1993.

Comiso, J.C., T.C. Grenfell, D. Bell, M. Lange, and S. Ackley, Passive microwave in situ observations of Weddell Sea Ice, *J. Geophys. Res., 88*, 7686–7704, 1989.

Comiso, J.C., C. McClain, C. Sullivan, J. Ryan, and C. L. Leonard, CZCS pigment concentrations in the Southern Ocean and their relationships to some geophysical parameters, *J. Geophys. Res., 98(C2)*, 2419–2451, 1993.

Comiso, J.C., P. Wadhams, W.B. Krabill, R.N. Swift, J.P. Crawford, and W.B. Tucker III, Top/Bottom multisensor remote sensing of Arctic sea ice, *J. Geophy. Res., 96(C2)*, 2693–2711, 1991.

Curry, J.A., C.D. Ardeel, and L. Tian, Liquid water content and precipitation characteristics of stratiform clouds as inferred from satellite microwave measurements, *J. Geophys. Res., 95(D10)*, 16,659–16,671, 1990.

Dawson, M.S., A.K. Fung, and M.T. Manry, Surface parameter retrieval using fast learning neural networks, *Remote Sensing Rev., 7*, 1–18, 1993.

De Veaux, R., A.L. Gordon, J.C. Comiso, and N.E. Chase, Modeling of topographical effects on Antarctic sea ice using multivariate adaptive regression splines, *J. Geophys. Res., 98(C11)*, 20207–20319, 1993.

Debye, P., Polar molecules, Dover Reprint, originally published by Chemical Catalog Company, Reinhold, New York, 172 pp., 1929.

Douglas, B.C., and P.D. Gaborski, Observation of sea surface topography with GEOS-3 altimeter data, *J. Geophys. Res., 84*, 3893–3896, 1979.

Drinkwater, M.R., Multi-frequency imaging radar polarimetry of sea ice, *Proceedings of ITC '90*, Cambridge, 18–20, 1990.

Drinkwater, M.R. and G.B. Crocker, Modelling changes in the dielectric and scattering properties of young snow-covered sea ice at GHz frequencies, *J. Glaciol., 34(118)*, 274–282, 1988.

Elachi, C., Introduction to the physics and techniques of remote sensing, John Wiley and Sons, New York, 413 pp., 1987.

Eppler, D., M.R. Anderson, D.J. Cavalieri, J.C. Comiso, L.D. Farmer, C. Garrity, P. Gloersen, T. Grenfell, M. Hallikainen, A.W. Lohanick, C. Matzler, R.A. Melloh, I. Rubinstein, C.T. Swift,

Passive microwave signatures of sea ice, Chapter 4 in *Microwave Remote Sensing of Sea Ice*, ed. by Frank Carsey, AGU, Washington, DC, 47–71, 1992.

Fetterer, F.M., M.R. Drinkwater, K.C. Jezek, S.W.C. Laxon, R.G. Onstott, L.M.H. Ulanders, Sea Ice Altimetry, Chapter 17, in *Microwave Remote Sensing of Sea Ice*, ed. by Frank Carsey, AGU GM–68, Washington, DC, 111–135, 1992.

Fily M., and D.A. Rothrock, Opening and closing of sea ice leads: Digital measurements from synthetic aperture radar, *J. Geophys. Res.*, *95*, 789–796, 1990.

Friedman, J.H., Multivariate adaptive regression splines, *The Ann. of Stat.*, *19(1)*, 1–141, 1991.

Fung, A.K., and M.F. Chen, Emission from an inhomogeneous layer with irregular interfaces, *Radio Sci*, *16*, 289–298, 1981.

Garrison, D.L., K.R. Buck, and G.A. Fryxell, Algal assemblages in Antarctic pack ice and in ice edge plankton, *J. Phycol.*, *23*, 564–572, 1987.

Garrity, C., Characterization of snow on floating ice and case studies of brightness temperature changes during the onset of melt, Chapter 16 in *Microwave Remote Sensing of Sea Ice*, ed. by Frank Carsey, AGU GM–68, Washington, DC, 313–326, 1992.

Gloersen, P., and V. V. Salomonson, Satellites-new global observing techniques for ice and snow, *J. Glaciol.*, *15*, 373–389, 1975.

Gloersen, P., and W.J. Campbell, Variations in the Arctic, Antarctic, and Global sea ice covers during 1978–1987 as observed with the Nimbus–7 Scanning Multichannel Microwave Radiometer, *J. Geophys. Res.*, *93*, 10666–10674, 1988.

Gloersen, P., W. Campbell, D. Cavalieri, J. Comiso, C. Parkinson, H.J. Zwally, Arctic and Antarctic Sea Ice, 1978–1987: Satellite Passive Microwave Observations and Analysis, *NASA Spec. Publ. 511*, 1992.

Gogineni, S.P., R.K. Moore, T.C. Grenfell, D.G. Barber, S. Digby, and M. Drinkwater, The effects of freeze-up and melt processes on microwave signatures, Chapter 17 in *Microwave Remote Sensing of Sea Ice*, ed. by Frank Carsey, AGU GM–68, Washington, DC, 313–326, 1992.

Goodberlet, M.A., C.T. Swift, and J.C. Wilkerson, Remote sensing of ocean surface winds with the Special Sensor Microwave Imager, *J. Geophys. Res.*, *94(C10)*, 14,547–14,555, 1989.

Gordon, H.R., D.K. Clark, J.W. Brown, O.B. Brown, R.H. Evans, and W. W. Broenkow, Phytoplankton pigment concentrations in the Middle Atlantic bight: comparison of ship determinations and CZCS estimates, *Appl. Opt.*, *22*, 20–36, 1983.

Gordon, H.R., J. W. Brown, and R.H. Evans, Exact Rayleigh scattering calculations for use with Nimbus–7 Coastal Zone Color Scanner, *Appl. Opt.*, *27*, 862–871, 1988.

Grenfell, T., A theoretical model of the optical properties of sea ice in the visible and near-infrared, *J. Geophys. Res.*, *88*, 9,723–9,735, 1983.

Grenfell, T., and A.W. Lohanick, Temporal variations of the microwave signatures of sea ice during the late spring and early summer near Mould Bay, NWT, *J. Geophys. Res.*, *90*, 5063–5074, 1985.

Grenfell, T., and J.C. Comiso, Multifrequency passive microwave observations of first year sea ice grown in a tank, *IEEE Trans. on Geoscience and Remote Sensing*, *GE–24*, 826–831, 1986.

Hall, R.T., and D.A. Rothrock, Sea ice displacement from Seasat Synthetic Aperture Radar, *J. Geophys. Res.*, *86*, 11078, 1981.

Hallikainen, M., and D.P. Winebrenner, "The physical basis for sea ice remote sensing," Chapter 3, *Microwave Remote Sensing of Sea Ice*, ed. by Frank Carsey, AGU, Washington, DC, 29–46, 1992.

Halpern, D., M.H. Freilich, and R.S. Dunbar, Evaluation of two January–June 1992 ERS–1 AMI wind vector data sets, *Proceedings of the First ERS–1 Symposium: Space at the Service of our Environment, Cannes, France, November 4–6, 1992, ESA SP-359*, 135–140, 1993.

Hawkins, J.D., and M. Lybannon, Geosat altimeter sea ice mapping, *IEEE, J. Ocean Eng.*, *14(2)*, 139–148, 1989.

Heimdal, B.R., Phytoplankton and nutrients in the waters north-west of Spitzbergen in the autumn of

1989, *J. Plank. Res.*, *5*, 901–918, 1983.

Holt, B., D.A. Rothrock, and R. Kwok, Determination of sea ice motion from satellite images, Chapter 18, in *Microwave Remote Sensing of Sea Ice*, ed. by Frank Carsey, AGU GM–68, Washington, DC, 343–354, 1992.

Holt, B., J. Crawford, and F. Carsey, Characteristics of sea ice during the Arctic winter using multifrequency aircraft radar imagery, *Sea Ice Properties and Processes: Proceedings of the W.F. Weeks Symposium, CRREL Monogr. 90-1*, ed. by S. Ackley and W.F. Weeks, Cold Reg. Res. and Eng. Lab., Hanover, NH, 224 pp., 1990.

Holt, B., R. Kwok, and E. Rignot, Ice classification algorithm development and verification for the Alaska SAR Facility using aircraft imagery, *Proceedings of IGARRS '89*, *2*, 751–754, 1989.

Hwang, P.H., L.L. Stowe, H.Y.M. Yeh, H.L.Kyle, and the Nimbus–7 Team, The Nimbus–7 global cloud climatology, *Bul. of the Am. Meteor. Soc.*, *69(7)*, 743–752, 1988.

Ito, H., Sea ice atlas of Northern Baffin Bay, *Zurcher Geographische Schriften, 7(ZGS 7)*, Zurich, 142 pp, 1982.

Johannessen, J.A., R.A. Shuchman, K. Davidson, O. Frette, G. Digranes, and O.M. Johannessen, Coastal ocean studies with ERS–1 SAR during NORSEX '91, *Proceedings of the First ERS–1 Symposium: Space at the Service of our Environment, Cannes, France, November 4–6, 1992, ESA SP-359*, 113–117, 1993.

Kalle, K., Zum Problem der Meereswasserfarbe, *Ann. de Hydrogr. Und Mar. Meteor.*, *66(5)*, 1–13, 1938.

Key, J., and M. Haefliger, Arctic ice surface temperature retrieval from AVHRR thermal channels, *J. Geophys. Res.*, *97(D5)*, 585, 1992.

Koblinsky, C.J., Ocean surface topography and circulation, in *Atlas of Satellite Observations Related to Global Change*, ed. by R.J. Gurney, J.L. Foster, C.L. Parkinson, Cambridge Univ. Press, New York, 251–264, 1993.

Kwok, R., G. Cunninghan, and B. Holt, An approach to identification of sea ice types from spaceborne SAR data, Chapter 19, in *Microwave Remote Sensing of Sea Ice*, ed. by Frank Carsey, AGU GM–68, Washington, DC, 355–360, 1992.

Lange, M.A., S.F. Ackley, P. Wadhams, G.S. Diekman, and H. Eicken, Development of sea ice in the Weddell Sea, *Ann. Glaciol.*, *12*, 92–96, 1989.

Laxon, S.W.C., Satellite radar altimetry over sea ice, Ph.D. Thesis, Mullard Space Science Laboratory, University College, London, 246 pp., 1989.

Leberl, F., J. Raggam, C. Elachi, and W.J. Campbell, Sea ice motion measurements from SEASAT SAR Images, *J. Geophys. Res.*, *88*, 1915-1928, 1983.

Lepparanta, M., and T. Thompson, BEPERS–88 Sea ice remote sensing with Synthetic Aperture Radar in the Baltic Sea, *EOS 70(28)*, 698–699, 708–709, 1989.

Lewis, M.R., Satellite ocean color observations of global biogeochemical cycles, in *Primary Productivity and Biogeochemical Cycles in the Sea*, ed. by P.G. Falkowski and A.D. Woodhead, Plenum Press, New York, 139–153, 1992.

Liu, A.K., and C.Y. Peng, Waves and mesoscale features in the marginal ice zone, *Proceedings of the First ERS–1 Symposium: Space at the Service of our Environment, Cannes, France, November 4–6, 1992, ESA SP-359*, 343–347, 1993.

Liu, W.T., W. Tang, and F.J. Wentz, Precipitable water and surface humidity over global oceans from special sensor microwave imager and European Center for Medium Range Weather Forecasts, *J. Geophys. Res.*, *97(2)*, 2251–2264, 1992.

Livingstone, C.E., Combined active/passive microwave classification of sea ice, *Proc. of IGARRS' 89*, *1*, 376–380, 1989.

Lohanick, A., Microwave brightness temperatures of laboratory-grown undeformed first-year ice with an evolving snow cover, *J. Geophys. Res.*, *98(C3)*, 4667–4674, 1993.

Marsh, J.G., C.J. Koblinsky, F. Lerch, S.M. Klosko, J.W. Robbins, R.G. Williamson, and G.P. Patel, Dynamic sea surface topography, gravity, and improved orbit accuracies from the direct evaluation of seasat altimeter data, *J. Geophys. Res.*, *95*, 13,129–13,150, 1990.

Martin, S., A field study of brine drainage and oil entrainment in first year sea ice, *J. Glaciol.*, 22, 473–502, 1979.

Massom, R., Satellite remote sensing of polar regions: Applications, limitations, and data availability, Belhaven Press, London, 307 pp., 1991.

Massom, R., and J.C. Comiso, Sea ice classification and surface temperature determination using Advanced Very High Resolution Radiometer satellite data, *J. Geophys. Res.,99(C3)*, 5201–5218, 1994.

Matzler, C., R.O. Ramseier, and E. Svendsen, Polarization effects in sea ice signatures, *IEEE J. Oceanic Eng.*, *OE-9(5)*, 333–338, 1984.

Maynard, N.G., Coastal Zone Color Scanner imagery in the marginal ice zone, *Mar. Technol. Soc. J.*, *20*, 14–27, 1986.

Maynard, N.G. and D. Clark, Satellite color observations of spring blooming in Bering Sea shelf waters during the ice edge retreat in 1980, *J. Geophys. Res.*, *92*, 7127–7139, 1987.

McClain, C.R., and E. Yeh, CZCS sensor ringing mask comparison, in Case Studies for SeaWIFS calibration and validation, 13, *NASA TM 104566*, 21–29, 1994.

McClain, C.R., G. Feldman, and W. Esaias, Oceanic biological productivity, in *Atlas of Satellite Observations Related to Global Change*, ed. by R.J. Gurney, J.L. Foster, C.L. Parkinson, Cambridge Univ. Press, New York, 251–264, 1993.

McClain, E.P., Passive radiometry of the ocean from space-an overview, *Boundary Layer Meteorology*, *18*, 7–24, 1980.

Milman, A.S., and T.T. Wilheit, Sea surface temperatures from the scanning multichannel microwave radiometer on Nimbus–7, *J. Geophys. Res.*, *90(C7)*, 11631–11641, 1985.

Mitchell, B.G., E.A. Brody, E.N. Yeh, C. McClain, J.C. Comiso and N.G. Maynard, Meridional zonation of the Barents Sea ecosystem inferred from satellite remote sensing and in situ bio-optical observations, Pro Mare Symposium, *Polar Research*, *10(1)*, 147–162, 1991.

Morel, A., Optical modeling of the upper ocean in relation to its biogenous matter content (Case 1 waters), *J. Geophys. Res.*, 93C, 10749–10768, 1988.

Mueller, J.L., Nimbus–7 CZCS: Electronic overshoot due to cloud reflectance, *Appl. Opt.*, *27*, 438, 1988.

Müller-Karger, F.E., and V. Alexander, Nitrogen dynamics in a marginal sea ice zone, *Cont. Shelf Res.*, 7, 805–823, 1987.

Müller-Karger, F.E., C.R. McClain, R.N. Sambrotto, and G.C. Ray, A comparison of ship and Coastal Zone Color Scanner mapped distribution of phytoplankton in the southeastern Bering Sea, *J. Geophys. Res.*, *95(C7)*, 11483–11499, 1990.

Nakawo, M., and N.K. Sinha, Growth rate and salinity profile of first year ice in the high Arctic, *J. Glaciol.*, *27*, 315–330, 1981.

Niebauer, H.J., and V. Alexander, Oceanographic frontal structure and biological production at an ice edge, *Cont. Shelf Res.*, *4*, 367–388, 1985.

Nye, K.F., The use of ERTS photographs to measure the moment and determination of sea ice., *J. Glaciol.*, *(15)73*, 429-436, 1975.

Onstott, R.G., and R.A. Shuchman, Remote Sensing of the Polar Ocean, in *Polar Oceanography*, ed. by W.O. Smith, Jr., Academic Press, Inc., San Diego, CA, 123–169, 1990.

Onstott, R.G., T.C. Grenfell, C. Matzler, C.A. Luther, and E.A. Svendsen, Evolution of microwave sea ice signatures during early and mid summer in the marginal ice zone, *J. Geophys. Res.*, *92*, 6825–6837, 1987.

Parkinson, C.L., J.C. Comiso, H.J. Zwally, D.J. Cavalieri, P. Gloersen, and W.J. Campbell, Arctic sea ice, 1973–1976: Satellite passive microwave observations, *NASA SP489*, 296 pp., 1987.

Peak, W.H., and T.L. Oliver, The response of terrestrial surfaces at microwave frequencies, The Ohio State University Electroscience Laboratory, *Technical Report AFAL-TR-70-301*, 1971.

Perovick, D., A two-stream multilayer, spectral radiative transfer model for sea ice, *CRREL Report 89-15*, Hanover, NH, 17 pp., 1989.

Petty, G.W., and K.B. Katsaros, New geophysical algorithms for the Special Sensor Microwave Imager, Paper presented at the 5th International Conference on Satellite Meteorology and Oceanography, London, September 3–7, 1990.

Prabhakara, C., I. Wang, H.D. Chang, and A.T.C. Chang, Remote sensing of precipitable water over the oceans from Nimbus–7 measurements, *J. Appl. Meteorol., 21*, 59–68, 1982.

Raney, R.K., An overview of spaceborne SAR ocean observation, in *Modern Radio Science 1993*, ed. by Hiroshi Matsumoto, Oxford University Press, New York, 141–158, 1993.

Rossow, W.B., and R.A. Schiffer, ISCCP cloud data products, *Bull. Amer. Meteor. Soc., 72*, 2–20, 1991.

Rothrock, D.A., and D.R. Thomas, Principal component analysis of satellite passive microwave data over sea ice, *J. Geophys. Res., 93*, 2321–2332, 1988.

Schweiger, A.J., and J.R. Key, Arctic cloudiness: comparison of ISCCP-C2 and Nimbus–7 satellite-derived cloud products with a surface-based cloud climatology, *J. of Climate, 5*, 1514–1527, 1992.

SeaWIFS Working Group Report, NASA Goddard Space Flight Center, Greenbelt, MD, 200 pp., 1987.

Smith, W.O., Jr., M.E.M. Baumann, D.L. Wilson, and L. Aletsee, Phytoplankton biomass and productivity in the marginal ice zone of the Fram Strait during summer 1984, *J. Geophys. Res., 92(C7)*, 6777–6786, 1987.

Spencer, R.W., H.M. Goodman, and R.E. Hood, Precipitation retrieval over land and ocean with the SSM/I: Identification and characteristics of the scattering signal, *J. Atmos. Oceanic Technol., 6*, 254–273, 1989.

Steffen, K. and J.A. Maslanick, Comparison of Nimbus–7 Scanning Multichannel Microwave Radiometer radiance and derived sea ice concentrations with Landsat imagery for North water areas of Baffin Bay, *J. Geophys. Res., 93*, 10769–10781, 1988.

Steffen, K., D. J. Cavalieri, J. C. Comiso, K. St. Germain, P. Gloersen, J. Key, and I. Rubinstein, The estimation of geophysical parameters using passive microwave algorithms, Chapter 10 in *Microwave Remote Sensing of Sea Ice*, ed. by Frank Carsey, AGU, Washington, DC, 201–231, 1992.

Steffen, K., R. Bindschadler, G. Casassa, J. Comiso, D. Eppler, F. Fetterer, J. Hawkins, J. Key, D. Rothrock, R. Thomas, R. Weaver, and R. Welch, Snow and ice applications of AVHRR in Polar Regions, *Ann. Glaciol., 17*, 1–16, 1993.

Stogryn, A., A study of some microwave properties of sea ice and snow, *Rep. 7788*, Aeroject Electrosyst. Co., Azusa, CA, 1985.

Sullivan, C.W., K.R. Arrigo, C.R. McClain, J.C. Comiso, and J. Firestone, Distributions of phytoplankton blooms in the Southern Ocean, *Science, 262*, 1832–1837, 1993.

Svendsen, E., K. Kloster, B. Farrelly, O.M. Johannessen, J.A. Johannessen, W.J. Campbell, P. Gloersen, D.J. Cavalieri, and C. Matzler, Norwegian Remote Sensing Experiment: Evaluation of the Nimbus–7 Scanning Multichannel Microwave Radiometer for sea ice research, *J. Geophys. Res., 88*, 2781–2791, 1983.

Swift, C.T., Passive microwave remote sensing of the ocean—a review, *Boundary Layer Met., 18*, 25–54, 1980.

Swift, C.T., L.S. Fedor, and R.O. Ramseier, An algorithm to measure sea ice concentration with microwave radiometers, *J. Geophys. Res. 90*, 1087–1099, 1985.

Thomas, D.R., and D.A. Rothrock, Blending sequential Scanning Multichannel Microwave

Radiometer and buoy data into a sea ice model, *J. Geophys. Res.*, 93(C3), 2321–2332, 1988.

Thomas, R.H., Polar research from satellites, Joint Oceanographic Inst., Washington, DC, 91 pp., 1990.

Thorndike, A., A naive zero-dimensional sea ice model, *J. Geophys. Res.*, *93(C5)*, 5093–5099, 1988.

Treshnikov, A.F., Water masses of the Arctic basin, in *Polar Oceans*, ed. by M. Dunbar, Arctic Inst. North Am., Calgary, Alberta, Canada, 17–31, 1977.

Tsang, J., and J.A. Kong, Scattering of electromagnetic waves from random media with strong permittivity fluctuations, *Radio Science*, *16*, 303–320, 1981.

Tucker, W.B. III, D.K. Perovich, and A.J. Gow, Physical properties of sea ice relevant to remote sensing, in *Microwave Remote Sensing of Sea Ice*, ed. by Frank Carsey, AGU, Washington, DC, 9–28, 1992.

Ulaby, F.T., R.K. Moore, and A.K. Fung, Microwave Remote Sensing Active and Passive, *1-3*, Addison-Wesley Publishing Company, Reading, MA, 2161 pp., 1981, 1982, and 1986.

Vant, M.R., R.O. Ramseier and V. Makios, The complex-dielectric constant of sea ice at frequencies in the range 0.1–40GHz, *J. Appl. Phys.*, *49(3)*, 1234–1280, 1978.

Vesecky, J.F., R. Samadani, M.P. Smith, J.M. Daida, and R.N. Bracewell, Observation of sea-ice dynamics using synthetic aperture radar images: automated analysis, *IEEE Trans. on Geoscience and Rem. Sensing, GE-26(1)*, 38–48, 1988.

Viehoff, T., and J. Fischer, Satellite sea surface temperatures at the North Atlantic polar front related to high-resolution towed conductivity-temperature-depth data, *J. Geophys. Res.*, 93, 15551–15560, 1988.

Wadhams, P., Evidence for thinning of the ice cover north of Greenland, *Nature*, *345*, 795–797, 1990.

Weeks, W.F., and S.F. Ackley, The growth, structure, and properties of sea ice, *CRREL Monogr, 82-1*, U.S. Cold Reg. Res. and Eng. Lab., Hanover, NH, 136 pp., 1982.

Wensnahan, M., G.A. Maykut, and T.C. Grenfell, Passive microwave remote sensing of thin sea ice using principal component analysis, *J. Geophys. Res.*, *98(C7)*, 12453–12468, 1993.

Wentz, F.J., Model of ocean microwave temperatures, *J. Geophys. Res.*, *88*, 1892–1908, 1983.

Wentz, F.J., L.A. Mattox, and S. Peterherych, New algorithms for microwave measurements of ocean winds: Applications to Seasat and the Special Sensor Microwave Imager, *J. Geophys. Res.*, *91*, 2289–2307, 1986.

Wilheit, T.T., and A.T.C. Chang, An algorithm for retrieval of ocean surface and atmospheric parameters from the observations of the scanning multichannel microwave radiometer, *Radio Sci.*, *15*, 525–544, 1980.

Zwally, H.J., J.C. Comiso, C.L. Parkinson, W.J. Campbell, F.D. Carsey, and P. Gloersen, Antarctic sea ice 1973–1976 from satellite passive microwave observations, *NASA Spec. Publ. 459*, 224 pp., 1983.

2

Atmosphere-Ocean Interactions in the Marginal Ice Zones of the Nordic Seas

Peter S. Guest, Kenneth L. Davidson, James E. Overland, and Paul A. Frederickson

Abstract

Atmospheric-ocean interactions occur on a variety of time and space scales in marginal ice zones (MIZs). The most important momentum and heat flux events in the MIZs of the Nordic Seas occur during the passage of synoptic-scale cyclones, many of which follow the general axis of the warmest waters in the Norwegian Sea and cross the MIZ near Svalbard or in the Barents Sea. If deep layers of mid-latitude marine air penetrate over the ice, intense cyclogenesis can occur *in situ*. The low-level baroclinicity associated with MIZs alone is not enough to cause strong cyclogenesis, but it can generate moderate mesoscale cyclones. If other forcing mechanisms are present, these mesoscale cyclones may become intense polar lows which affect the MIZ, but usually polar low development occurs a few hundred kilometers or more seaward of the ice edge. Based on our measurements, we find that during the spring, the median height of the temperature inversion base varies from 155 m over the pack ice to 1240 m over the open ocean for off-ice winds, and 160 m to 1490 m for on-ice winds. Corresponding median 10 meter air temperatures are -13 °C to 0 °C and -13 °C to -7 °C and median wind speeds are 4.5 ms^{-1} to 9 ms^{-1} and 3 ms^{-1} to 8 ms^{-1}. The median, upward, bulk-derived, combined sensible and latent, surface heat fluxes are 178 Wm^{-2} just off ice edge to 136 Wm^{-2} at locations between 100 km and 150 km seaward of the ice edge during off-ice winds and 101 Wm^{-2} to 41 Wm^{-2} during on-ice winds. During the summer, the median height of the temperature inversion base varies from 160 m over the pack ice to 505 m over the open ocean for off-ice winds,

Arctic Oceanography: Marginal Ice Zones and Continental Shelves
Coastal and Estuarine Studies, Volume 49, Pages 51–95
This paper is not subject to U. S. copyright.
Published in 1995 by the American Geophysical Union

and 155 m to 680 m for on-ice winds. Summer median air temperatures are -2 °C to 2 °C and -1 °C to 4 °C and median wind speeds are 3 ms^{-1} to 7 ms^{-1} and 4.5 ms^{-1} to 9 ms^{-1}. The systematic wind speed gradients caused by boundary layer depth and surface heat flux gradients have effects on wind stress that are as significant as surface roughness gradients. As a result, there is typically a wind stress minimum over the inner pack ice and two maxima, one over the rough MIZ ice and one in the open ocean, where the wind speed is highest.

Introduction

Marginal ice zones (MIZs) are regions with intense air-sea interactions generated by the contrasts between a well-insulated ice-covered surface and an ocean directly exposed to the air. The goal of this paper is to examine how various atmospheric and surface features interact to produce surface momentum, heat and radiation exchanges in MIZ regions. These exchanges represent the means by which the atmosphere influences ice, oceanographic, chemical and biological processes in MIZs.

It is not possible to examine all aspects of air-sea interactions in MIZs; the most important features will be examined. We include past studies in the discussion, but special emphasis is given to previously-unpublished results from several field programs during the 1980's in the MIZs of the Nordic Seas, which are the Greenland Sea, the Norwegian Sea, and the Barents Sea (Figure 1). We define two MIZs in the Nordic Seas, the Fram Strait/Greenland Sea MIZ (FSMIZ) and the Barents Sea MIZ (BSMIZ). The northern part of FSMIZ is associated with an ice edge that has modest seasonal changes in location while the BSMIZ ice edge advances and retreats across much of the Barents Sea during the annual cycle (Figure 1).

The atmospheric forcing of the Nordic Sea MIZs depends on processes such as the location of the major storm tracks and seasonal temperature trends that are driven by hemispherical radiation and temperature gradients. But mesoscale processes driven by local MIZ gradients also have a strong effect on air-sea interactions. Microscale features, such as the depth of atmospheric boundary layer or the surface roughness, also affect the exchange of momentum and heat with the ocean. Thus, a discussion of phenomena on a variety of time and space scales is required to get a "big picture" of the important air-sea interactions in MIZs.

The first section will analyze the climatology of synoptic-scale cyclones and present a case study of a cyclogenesis over sea ice near the MIZ. Next, relationships between the MIZ location and mesoscale features such as small cyclones, including polar lows, and katabatic winds will be analyzed, followed by a presentation of atmospheric boundary layer (ABL) characteristics in MIZs, based on statistics from field program measurements and theoretical results. Finally, some quantifications of microscale heat and momentum flux parameters are presented.

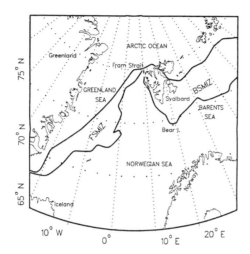

Figure 1. Map of major geographical features of the Nordic Seas. Median ice edge locations for the months of April (lower thick line) and November (upper thick line) represent maximum and minimum ranges during a normal year [Stringer *et al.*, 1984]. The Fram Strait/East Greenland Sea MIZ (FSMIZ) and the Barents Sea MIZ (BSMIZ) are the zones adjacent to the ice edges and also change location seasonally.

Synoptic-Scale Processes

Cyclone Tracks

Synoptic-scale phenomena are resolvable by the standard atmospheric observation network of ships, buoys and land stations. In the Nordic Seas region, these include cyclones, anticyclones and atmospheric waves with horizontal scales greater than about 800 km. Mesoscale phenomena, with 2 km to 800 km scales, are considered in the next section. Synoptic-scale phenomena play a dominant role in controlling the air flow and surface forcing locally within marginal ice zones. The high winds and heat fluxes associated with a single intense cyclone can have a long-lasting effect on the ice cover and physical and biological processes within the underlying water column. For example, the upper ocean turbulence generated by a storm can deepen the mixed layer. This may decrease biological productivity by preventing plankton from remaining near the surface, where energy from the sun is greatest. Nature cannot "unmix" the upper ocean after a strong mixing event. Even if a new mixed layer forms, a remnant of the storm-generated mixed layer may persist below the new mixed layer. In some situations, cyclones may increase productivity because large gradients in surface wind stress can generate vertical velocities and bring nutrients near the surface where they can be utilized by phytoplankton.

Although Arctic MIZs are north of the prevailing westerlies and associated frequent cyclonic activity, several significant synoptic-scale cyclones affect the MIZ regions in the Nordic Seas each season. Many of these are occluded storms originating from the Icelandic Low region. According to Serreze and Barry [1988], hereafter SB, approximately three to five (two to four) cyclones each winter (summer) cross the MIZ from the North Atlantic into the central Arctic basin. SB's mean storm tracks for both summer and winter are oriented southwest to northeast across the Norwegian Sea about 300 km southeast of the FSMIZ ice edge before curving north over Svalbard into the Arctic Ocean. Other studies, e.g. Keegan [1958], Reed and Kunkel [1960], U.S. Navy [1963], Wilson [1967], Sater *et al.* [1971], Gorshkov [1983] and Sechrist *et al.* [1989] described a similar storm track in the Norwegian Sea, but with most storms moving into the Barents Sea rather than over Svalbard. Generally, during all seasons, this major storm track was found over, or slightly north of, the central axis of a broad "tongue" of warm water which protrudes into the Norwegian Sea from the southwest. We define this as the Norwegian Sea Cyclone Track (NSCT). The storms travelling along the NSCT create a mean sea level pressure trough which extends from the Icelandic Low (just south of Iceland proper) across the Norwegian Sea toward the Barents Sea [Walsh and Chapman, 1990].

Synoptic-scale cyclones centered along the NSCT are large enough to affect FSMIZ regions hundreds of kilometers to the north and east. For example, Rutherford [1993], using synthetic aperture radar (SAR) imagery from the ERS-1, processed by Nansen Remote Sensing Center, Bergen Norway, found that during the Seasonal Ice Zone Experiment from January 7 to January 16, 1992, the FSMIZ ice morphology characteristics were determined by the presence of strong cyclones centered roughly along the NSCT. During three cyclone events, strong northerly and easterly winds along the FSMIZ created a compact and relatively featureless ice edge. Between cyclone events, when atmospheric forcing was weak, the ice edge became diffuse and complicated patterns in the ice field were created as a result of oceanic circulations.

In addition to the NSCT, several minor cyclone tracks affect the MIZs of the Nordic Seas. Greenland, with elevations above 3000 m, is a major obstacle to storm movement from the west. However, during the winter, cyclones or troughs in the Baffin Bay region sometimes cross over the Greenland ice cap and become re-organized cyclones over the FSMIZ before joining the NSCT (for example see SB's Figure 6). Occasionally, this will also occur in the summer (SB). Gorshkov [1983] identified a large number of other mean cyclone tracks which can affect the MIZs of the Nordic Seas. Gorshkov's tracks depended on the month of the year and a classification of the large scale synoptic situation.

Individual cyclone tracks can be quite different from the climatological averages. For example, none of the cyclones which occurred during the summer Marginal Ice Zone Experiment (MIZEX84), the spring Marginal Ice Zone Experiment (MIZEX87), or the fall and winter Coordinated Eastern Arctic Region Experiment (CEAREX) followed the NSCT closely (Figure 2). Cyclones moved into the Nordic Sea MIZs from many directions. More cyclones than expected, from the above studies, formed

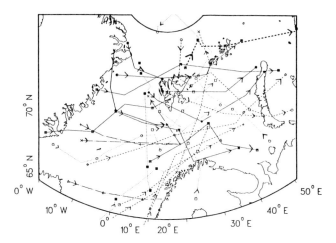

Figure 2. Tracks of all detected synoptic-scale cyclones in the Nordic Seas region which occurred during the authors' field programs. Periods represented are the CEAREX drift phase from September 17 to October 31 (dashed) and November 1 to January 7 (solid), 1988-1989, respectively [Hamilton, 1991]. Also represented are the MIZEX87 period, March 20 to April 10, 1987 (dash-dot) [Schultz, 1987] and the MIZEX84 period, June 8 to July 17, 1984 (dotted) [Lindsay, unpublished manuscript]. The thick dashed line indicates the track of the October 12, 1988 storm described in the text. Symbols are to distinguish tracks; they do not represent a regular time step.

along the East Greenland Coast. Also, unlike the SB storm tracks, but similar to other studies, most of the observed cyclones traveled toward the Barents Sea rather than Svalbard. However, the most intense and longest-lived cyclones occurred in the general region of the NSCT or the southern Barents Sea. A notable exception, shown as a thick dashed line in Figure 2, is described later.

To summarize, most cyclones that affect the MIZs of the Nordic Seas come from southerly marine regions. A few cyclones cross the ice edge from the ice pack after crossing, or forming over, the eastern slopes of Greenland. Occasionally, cyclones move over the MIZs from the central ice pack, or form *in situ* (see below). Individual cyclone tracks (Figure 2) depend upon the particular overall synoptic situation and are much more variable than climatological tracks.

Synoptic-scale Cyclogenesis Near the MIZ

Although we know that cyclones can exert a strong influence on sea ice characteristics in MIZs, the reverse effect, i.e. the effect of sea ice on synoptic storms is less well-understood. Carleton (1985) found that, during the months of January and April, representing mid-winter and mid-spring, cyclones in the northern hemisphere were more common and had more southerly tracks during periods of greater ice extent. There are reports of synoptic-scale cyclogenesis or storm re-intensification in MIZ regions [Sechrist *et al.*, 1989, p. 3-32 to 3-33 and p. 4-1 to 4-14; Fett, 1992, p. 1a-2 to

1a-36] and we have observed a few cases of cyclogenesis in the FSMIZ (Figure 2). But, in general, these storms are relatively weak and the Nordic Sea MIZs do not seem to be particularly associated with strong synoptic-scale cyclogenesis compared to regions further south in the NSCT region. In his review article of meteorology in seasonal ice zones, Barry (1986) concludes that changes in ice edge location modulate the occurrence of cyclonic activity, but the resulting climatological effects are not yet known.

The opportunity to document cyclogenesis over sea ice near the MIZs of the Nordic Seas occurred during the CEAREX project. The authors and associates measured meteorological parameters from the R/V Polarbjoern in the Svalbard region from September 4, 1988 to May 18, 1989. On October 12, 1988, an intense cyclogenesis event almost directly over the Polarbjoern 240 km NE of Svalbard caused the highest wind event (at least 25 ms^{-1}) recorded during CEAREX. Unfortunately, our valiant attempts at launching rawinsondes were unsuccessful during the storm; therefore there is no in situ upper-level air mass information, but there is continuous surface information.

Late on October 11, 1988 (all times are UT), strong southerly winds in advance of a large-scale upper-level (500 mbar) trough brought a deep mass of warm moist air into the Arctic basin northeast of Svalbard, creating a baroclinic zone between a warm marine air mass northeast and east of Svalbard and a cold Arctic air mass north of, and over, Svalbard. Upper-level winds were unusually strong for high-latitude regions, indicating that deep baroclinicity, more typical of mid-latitudes was present. The movement of the trough and associated upper-level positive vorticity advection (PVA) within the baroclinic zone provided essential ingredients for baroclinic instability and cyclogenesis. An array of six temperature-pressure buoys surrounding the ship at a 60 km spacing were used to locate a forming low pressure center about 120 km SW of the Polarbjoern at 2100 October 11. The initial formation location and track of this cyclone is indicated by the thick dashed line in Figure 2.

For the next 11 hours, the pressure dropped 8 mb, the air temperature increased from -26 °C to -8 °C, and the easterly wind increased from calm to 10 ms^{-1} at the Polarbjoern (Figure 3). After the passage of a distinct warm front marked by a wind direction shift at 0800 October 12, winds remained moderate (10 ms^{-1}) from the SE and the temperature was constant while the pressure continued to drop rapidly as the low deepened and approached from the SW.

During this initial period, an upper-level "baroclinic leaf" formed over the cyclogenesis region (Figure 4). This is a large mass of upper-level clouds with a leaf-like structure which is commonly observed in mid-latitudes. A baroclinic leaf is a valuable indicator for predicting cyclogenesis, especially over data-poor marine regions [Weldon, 1979]. Its occurrence in high-latitude regions has not previously been documented, but this example, which was identified by R. Fett [pers. comm.], shows that the detection of a baroclinic leaf from satellite imagery could potentially be a valuable tool for the prediction of cyclogenesis in the Arctic. The "baroclinic leaf" obscured the low-level circulation features associated with the forming cyclone.

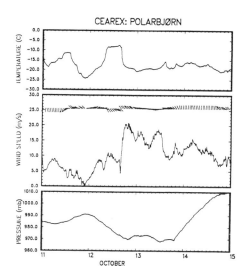

Figure 3. Time series of air temperature, wind vector, and pressure at 14 m elevation from the *R/V Polarbjoern* during an intense cyclogenesis event in 1988. The wind direction barbs are oriented in the meteorological sense, so that an upward pointing barb represents a wind direction from the north. The maximum winds, at 1600 October 12, are probably too low by 10% - 30% due to the sheltered location for northwest winds of our only operational anemometer at this time.

At 1530 October 12, an intense cold front passed over the ship (Figure 3). The sudden 180-degree shift in the wind direction suggests that the center of the cyclone passed close to the *Polarbjoern* at this time. Our anemometer indicated a 22 ms^{-1} sustained wind from the NW after the cold front passage. Actual winds were probably considerably higher, but our anemometer at this time was sheltered by the ship's bridge and bow mast. The temperature dropped 6 °C during the 20 minutes following the cold front passage and another 4 °C in the next two hours. The pressure continued to drop until reaching a minimum of 972 mb at 1930 October 12. The 4-hour delay in the bottoming-out of the pressure, compared to the frontal passage, suggests that the cyclone was continuing to strengthen; any increase in pressure associated with increasing distance from the cyclone center was counteracted by a temporal decrease in pressure of the central regions of the cyclone.

After passing over the ship, the cyclone slowly moved to the NE and remained very strong for 24 hours (Figure 5), as indicated by the low pressure and high winds at the ship (Figure 3). After this, the cyclone began to weaken as cold air filled in the center. By October 15, the cyclone was not detectable on the satellite images, nor the surface analyses.

The circulation associated with this storm during its most intense period extended north to the pole, east to Novaya Zemlya, south to Norway and west to Greenland, a diameter of about 3000 km, (Figure 5). A storm of this size and magnitude has a large effect on biological and physical processes within the MIZs because of the

Figure 4. NOAA-10 channel 4 (infrared) satellite image for 1012 UTC October 12, 1988, showing the baroclinic leaf associated with cyclogenesis. Svalbard is located in the center, the baroclinic leaf is just to the right or east. Note the cloud streets to the west of Svalbard, indicating off-ice winds in the Fram Strait.

intense turbulence and vertical motions which are generated in the ocean. The relative scarcity of strong synoptic-scale cyclogenesis over pack ice or MIZ regions, and the characteristics of the rare case described above, provide observational evidence that a relatively deep baroclinic layer is required to induce synoptic-scale

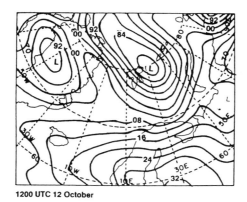

1200 UTC 12 October

Figure 5. Sea-level pressure map for 1200 UTC October 12, 1988, [Lackmann *et al.*, 1989].

cyclogenesis over MIZs. The local turbulent surface fluxes alone are not sufficient to create a deep baroclinic zone. In order for synoptic-scale cyclogenesis to occur over an MIZ, a deep baroclinic zone, i.e. the main polar front, must be present.

Mesoscale Processes

Overview

The mesoscale (2 km - 800 km horizontal scales) encompasses a wide variety of atmospheric phenomena which cannot be resolved by the standard *in situ* surface and upper-air observation network. These phenomena must be detected from cloud features on satellite images or with *in situ* measurements from field programs. In general, mesoscale phenomena cannot be simulated or predicted using operational numerical weather prediction models. Important mesoscale phenomena affecting MIZ regions are vortices, fronts and katabatic flows. In addition to being affected by mesoscale atmospheric features, MIZs are themselves mesoscale features which have an effect on the atmosphere, particularly the ABL. We will discuss variations in the ABL across MIZs in the next section.

Mesoscale Vortices

Visible and infrared satellite images of cloud features in the Nordic Sea MIZs and adjacent ocean areas reveal a plethora of mesoscale vortices, ranging in size from 20 km diameter island-wake eddies [e.g. Sechrist *et al.*, 1989, p. 5-9 to 5-10.] to 800 km diameter baroclinic comma-clouds. Using satellite imagery alone, Wos [1992] identified 941 individual cyclones in the Nordic Seas (Greenland Sea, Norwegian Sea and Barents Sea) region with diameters 100 km to 1000 km during the time period from September 1988 to May 1989, inclusive, an average of almost 3.5 per day. The median lifetime of these disturbances was 18 to 24 hours. In just one month, virtually all MIZ and open ocean areas in the Nordic Seas region were affected by mesoscale cyclones at some time (Figure 6, bottom). Many of these mesoscale cyclones were first detected near the ice edge (Figure 6, top), suggesting that the baroclinicity associated with the ice edge makes it a preferred location for mesoscale cyclogenesis.

Although mesoscale cyclogenesis is quite common in or near the MIZs of the Nordic Seas region during all seasons, most of the resulting circulations are weak and short-lived. (Surface wind speed was not a criterion used in Wos' [1992] analyses.) For example, during the MIZEX84 project from June 8 to July 17, 1984, about half of the significant wind events, i.e. sustained surface wind speed greater than 5 ms^{-1}, to affect the FSMIZ were associated with mesoscale cyclones that formed near the MIZ [R. Lindsay, University of Washington, unpublished manuscript]. But none of the mesoscale cyclones produced surface winds greater than about 12 ms^{-1}. The five events with wind speeds greater than 12 ms^{-1} were associated with synoptic-scale cyclones that had formed to the south of the MIZ.

As a result of mesoscale surface and upper-air measurements from two drifting ice camps and a research vessel during the spring phase of CEAREX, the authors and Erik Rasmussen, University of Copenhagen, were able to document a case of mesoscale cyclogenesis near the FSMIZ (Figure 7) that was clearly due to a low-level baroclinic instability. In this case, a baroclinic wave developed along a strong low-level baroclinic zone which formed over the pack ice near the FSMIZ as a northward moving marine air mass with surface temperatures near 0 °C came in contact with a southward moving shallow (200-400 meter thick), cold (-30 °C) Arctic air mass. The cyclone shown in Figure 7 developed at the northern crest of the wave in the pack ice 250 km from the ice edge. An intense surface cold front (12 °C drop in less than 30 minutes) passed the ice camps as the winds veered from the south to the northeast, reaching a maximum speed of 10 ms^{-1} (Figure 8). Because the disturbance was shallow, the pressure drop associated with the frontal passage at the ice camps was modest (Figure 8).

The baroclinic wave and cyclone formed underneath a large upper-level (500 mbar) ridge; no upper-level short waves or jet streaks were detected on satellite images (not shown). Because the cyclone formed over the pack ice some distance from the ice edge, the development was attributable entirely to the dynamic properties of the flow; the direct thermal and frictional effects of the underlying surface were small [Rasmussen, 1992a]. This case shows how low-level baroclinicity, such as exists over MIZ regions and sometimes nearby pack ice, is able to create moderate mesoscale disturbances, even in the absence of other instability mechanisms.

Polar Lows

One type of mesoscale cyclone that has received considerable recent attention [e.g. Twitchell *et al.*, 1989] is the polar low. The terms "polar low" or "arctic low" have been

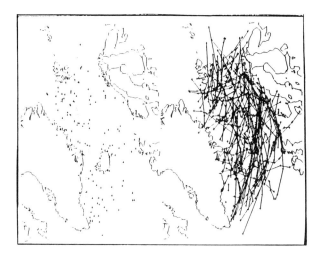

Figure 6. Mesoscale cyclone initial detection points (left) and cyclone tracks (right) for March 1988, according to Wos [1992]. Approximate ice edge is shown on left panel (dots).

used to describe a wide variety of mesoscale vortices, but most experts now agree that the terms should apply only to mesoscale, maritime cyclones that form poleward of the main baroclinic zone and produce at least gale-force (15 ms^{-1}) surface winds [Rasmussen, 1992b]. Polar lows occur when cold air is advected over relatively warm water. The "arctic-front" type of polar low [Businger and Reed, 1989] is associated with shallow ABL fronts that form in MIZ regions and propagate downwind. The initiation of deep vertical convection that is associated with arctic-front type polar lows requires some time for the surface heating to break through the inversion capping the ABL. For this reason, air parcels travel some distance over open ocean

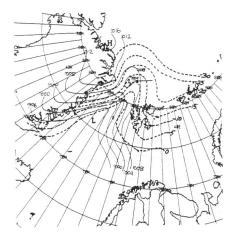

Figure 7. Near-surface temperature (°C) and pressure (mb) analyses of the Fram Strait and surrounding regions showing a mesoscale cyclone that has just formed over the pack ice [Rasmussen, 1992a]. The ice edge was located between the 0 °C and -5 °C isotherms.

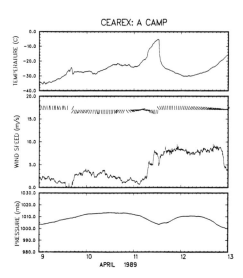

Figure 8. Same as Figure 3, but for the CEAREX "A" camp from April 9 to April 13, 1989. Note the intense frontal passage at 1230 April 11 associated with an over-ice mesoscale cyclogenesis event.

before entering the polar low circulation or genesis areas. Arctic-front type polar lows share many characteristics with tropical cyclones [e.g. Emanuel and Rotunno, 1989; Rasmussen, 1989; Businger, 1991], such as their spiral shape and warm cores. Similar to tropical cyclones, conditional instability of the second kind (CISK) [e.g. Rasmussen, 1979; Craig and Cho, 1989] and/or air-sea interaction instability (ASII) [Emanuel, 1986] are probably crucial mechanisms for creating intense arctic-front type polar lows. CISK involves a positive feedback between small scale processes, such as cumulus convection, and large-scale (storm size) moisture convergence. Cumulus convection releases latent heat, which decreases the central pressure of a storm which then increases the moisture convergence into the storm, which further increases cumulus convection and release of latent heat. The ASII feedback loop occurs when strong upward fluxes of sensible and latent heat from the surface increase the average temperature, and thus decrease the central pressure of a storm. The resulting stronger pressure gradients will increase the surface winds, which increases the surface heat fluxes and closes the feedback loop. The ASII instability mechanism requires a surface sensible heat and/or moisture source, while CISK is also aided by a moisture source; this may explain why polar lows (and tropical cyclones) rapidly dissipate over pack ice or land.

The case study in the previous sub-section showed that low-level baroclinicity in or near MIZs is a sufficient condition for moderate mesoscale cyclogenesis. The low-level baroclinicity formed in MIZ regions is important in the initial phases of arctic-front type polar low formation. But it appears that a CISK or ASII is required for a weak or moderate cyclone generated by low-level baroclinicity to become an arctic-front type polar low. MIZs are not favored regions for the CISK or ASII mechanisms compared to warm moist open ocean areas, and arctic fronts are advected downwind from the MIZs before polar low cyclogenesis occurs. For these reasons, arctic-front type polar lows rarely affect Nordic Sea MIZs.

An exception is the MIZ region near Bear Island in the winter, which is often downwind of low-level fronts formed just to the north. During northerly winds, Svalbard acts as barrier which allows the creation of a relatively warm air mass over the open water to the west, while the air flowing around the east side of Svalbard remains cold over the ice-covered Barents Sea. A front forms where the air masses re-join in the region between Svalbard and Bear Island. Vortices along the front may develop into polar lows [Fett, 1989a; Shapiro and Fedor, 1989], which can strongly affect the BSMIZ near Bear Island.

Another type of polar low, identified by Businger and Reed [1989], is the "short-wave/jet-streak" type that is associated with comma-clouds and upper-level baroclinic waves in the troposphere. This type is a more pure baroclinic instability than the arctic-front type, but most occurrences of short-wave/jet-streak polar lows are at lower latitudes, just poleward of a pre-existing polar front (jet stream) region. Occasionally, the main polar front and associated deep baroclinic zone will move to the north over a MIZ, similar to the synoptic-scale case described previously. If an upper-level shortwave within the baroclinic zone traverses a MIZ, a combined "short wave/jet streak" and "arctic-front" polar low may form [Businger and Reed, 1989].

To summarize, polar lows result from baroclinic instability and CISK or ASII mechanisms, or a combination thereof. Because polar lows form in cold air outbreaks, they are usually advected in directions away from the pack ice. Therefore, the polar lows most likely to affect MIZ regions must form *in situ* or over the pack ice. The little observational evidence available suggests that the low-level baroclinicity produced by a MIZ is not capable of producing a cyclone with gale force winds, unless upper-level PVA, i.e. deep baroclinic waves, CISK or ASII also contribute to the dynamic instability. Because CISK and ASII favor the warm, moist open ocean surfaces found equatorward of the MIZ, we theorize that a polar low in a MIZ of the Nordic Seas region (with the exception of the MIZ near Bear Island) must be a primarily baroclinic disturbance with low-level baroclinicity (e.g. arctic fronts) and deep baroclinicity (e.g. upper-level short waves or jet streaks) acting in concert. Confirmation of this theory will require further observations.

Katabatic flows

Katabatic flows are caused by radiational cooling of a high-elevation surface, which cools the near-surface air by contact; the air then flows downhill as a shallow gravity current. The winds from katabatic flows are strongest on the fringes of glacial ice caps where the terrain creates converging flow. There have been several studies of the intense katabatic flows produced by the continental ice cap of Antarctica. The effect of the katabatic flows can extend for at least 1000 km beyond the sloping regions [Bromwich *et al.*, 1992a and have a significant impact on ice and oceanographic processes in the MIZ regions which surround Antarctica [e.g. Bromwich *et al.*, 1992b. Also, katabatic flows can spawn mesoscale cyclones which affect Antarctic MIZ regions [Bromwich, 1991].

The world's other continental ice cap, on Greenland, also produces katabatic flows, but these have received little scientific attention. These flows can extend at least as far as 100 km from the mouth of the fjords on Greenland's west coast, as indicated by clouds lines observed on satellite imagery [Sechrist *et al.*, 1989]. We suspect that the associated winds can have a significant physical impact on the mass of sea ice and associated MIZ region which extend down the west coast of Greenland, but there have been no studies to verify this.

By using a backtrack procedure, Wos [1992] traced the origin of over half of the mesoscale cyclones which he detected in the Nordic sea region to converging glacial mouth regions along the east coast of Greenland, where katabatic flows are especially common. However, Wos had no *in situ* verification data to prove the link between katabatic winds and mesoscale cyclogenesis in this region. Unlike Antarctica, mesoscale *in situ* measurement arrays for studying katabatic flows or resulting cyclones have never existed in eastern Greenland or the adjacent pack ice and MIZ regions. Our measurements in the FSMIZ [Lindsay, 1985; Guest and Davidson, 1988, 1989; Frederickson *et al.*, 1990] have not detected any apparent katabatic flows, although all of these measurements were further than 100 km from the Greenland coast. Until a mesoscale measurement program is undertaken in the vicinity of suspected Greenland katabatic flows and nearby MIZ regions, the impact of katabatic flows and related phenomena on the meteorology and oceanography of the Greenland Sea will remain unknown.

Atmospheric Boundary Layer

Definitions

The ABL is the part of the atmosphere in direct turbulent contact with the surface. Turbulent fluxes closely link the surface interfacial temperature, T_s, to the ABL air temperature, T_a [e.g. Overland and Guest, 1991]. A low-level temperature inversion is almost always present in central Arctic and MIZ regions. The suppression of turbulence and mixing above the inversion base height, Z_i, isolates the ABL from the rest of the atmosphere, thus allowing local heat sources to dominate T_a. In the mid-latitude cyclone belt, most local non-diurnal T_a changes are caused by the movement of large-scale, relatively deep air masses, while in the central Arctic, temperature changes are generally associated with changes in local radiation conditions or mixing with warmer air from above the inversion [Vowinckel and Orvig, 1970]. Although the horizontal advection of heat into the Eastern Arctic Ocean associated with cyclones

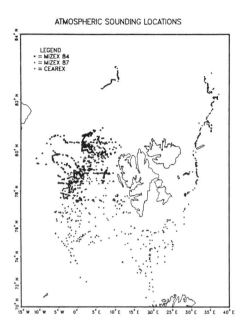

Figure 9. Location of rawinsonde soundings contained in the MIZCLM data set.

causes brief winter warm spells [Overland and Guest, 1991] and is important to the overall Arctic heat budget [Nakamura and Oort, 1988], local radiation conditions dominate T_a characteristics over most pack ice locations. In contrast, advection and turbulent fluxes usually dominate near-surface temperatures within MIZs, due to the large surface temperature gradients. But unlike the mid-latitude cyclone belt, where moving air mass boundaries cause advective heating or cooling, advection in MIZs depends on the average ABL wind direction relative to a quasi-stationary ice edge. In this section, we use rawinsonde sounding data collected by the authors and colleagues during several field experiments to examine statistical relationships between variables such as distance from the ice edge, X, inversion base height, Z_i, near-surface wind speed, U_s, cloud coverage and average ABL air temperature, T_a. The ice edge is defined as the 50% ice concentration isopleth. Depending on availability, we determined the ice edge location using infrared and visible satellite imagery, aircraft SAR imagery, ship logs, and maps prepared by the Naval Polar Oceanography Center and the Norwegian Meteorological Institute. Due to uncertainties in the exact location (or definition) of the ice edge, the data were grouped into 50 km X location bins for the results presented here, with negative X representing over-ice locations. Over 1300 soundings were obtained in the MIZ regions of the Nordic Sea and in the Eastern Arctic Ocean during the MIZEX and CEAREX projects (Figure 9). The MIZ statistics presented here are based on the approximately 950 soundings which were obtained within 200 km of the ice edge. Approximately 10 of these soundings were obtained after the ship had moved to avoid high winds and seas. Also, approximately eight more potential soundings were not attempted or were unsuccessful due to bad conditions. Therefore, there is a slight bias in the data against stormy periods. But this bias has a negligible effect on median values of the various parameters discused below because the number of affected data points is so small. The bias may have had a more significant effect on mean surface heat flux estimates (discussed in a later section) because individual storms can have very large fluxes which are important to the overall average. But because no standard upper-air observing stations are located in MIZ regions and no other long-term field programs have been undertaken, this data represents the only "climatology" available for a MIZ region and we will refer to the data set as MIZCLM. MIZCLM is not a true climatology because it is based on only one or two years, which may not be typical years, but nonetheless, enough soundings are available to detect statistically significant spatial and seasonal trends in various physical parameters within the MIZ ABL. The MIZ soundings obtained during June and July, 1984 as part of the MIZEX84 project are classified as "summer" while those obtained in March or April during MIZEX87 or CEAREX are classified as "spring". The inversion base height, Z_i, is classified as the lowest layer where the temperature increases with height, assuming that there is at least a 2.5 °C increase within the entire inversion layer. The rawinsondes were not able to resolve the vertical temperature structure below about 100 meters. If the lowest points on the rawinsonde increased with height by at least 2.5 °C, the inversion was classified as "surface-based" and Z_i was set to 0 m, even though it may actually be anywhere below 100 meters.

Wind direction regimes

Because of the importance of advection in controlling ABL characteristics within MIZs, it is useful to classify the MIZ ABL into four regimes according to the average relative wind direction within the ABL, WD_a. These regimes are off-ice (ice-to-ocean airflow), on-ice, parallel-right (ice to the right looking downwind) and parallel-left. We define the relative wind direction, WD, with respect to a linear ice edge so that WD = 0 represents a parallel-right wind direction, WD = 90 represents directly on-ice winds, etc. Generally the surface (10 m) wind direction, WD_s, is virtually equal to WD_a, while the "turning angle" (the relative geostrophic wind direction at the top of the ABL, WD_g, minus WD_s) is between 15 and 30 degrees. Glendening's [1994] steady-state two-dimensional numerical MIZ ABL simulations using different upper level geostrophic wind directions show that the formation of a secondary "ice-breeze" circulation produces sharp ABL fronts when the relative geostrophic wind direction at the top of the ABL, WD_g, is between 10 and 40 degrees or 170 and 185 degrees. These front-forming wind directions define the parallel-wind regimes. (Table 1).

The relative wind directions from the MIZCLM data were based on surface-layer measurements. The parallel regimes were hard to identify on the MIZCLM data because they represent a narrow and variable wind direction range, they usually last only a few hours because the wind direction changes, and the ice edge orientation is

TABLE 1. Summary of ABL and Ice Movement regimes in MIZs

WD_g[1]	Regimes		Comments
	ABL	Ice Movement	
10 - 50	para.-right	on-ice	thermal and frictional convergence, strong fronts
50 - 130	on-ice	on-ice	warm, ice compaction
130 - 170	on-ice	off-ice	warm, ice destruction regime ($135 < WD_a$[2] < 180)
170 - 185	para.-left	off-ice	thermal convergence, weak fronts
185 - 325	off-ice	off-ice	cold, ice divergence
325 - 10	off-ice	on-ice	cold, brine rejection regime ($315 < WD_a$[2] < 360)

[1]WD_g = geostrophic wind direction relative to the ice edge, see text for details.
[2]WD_a = average boundary layer wind direction relative to the ice edge.

difficult to determine and often highly non-linear. For these reasons, many of the results to be presented are based on a data set with only two unambiguous regimes, off-ice winds and on-ice winds.

Off-ice Wind Regime

The off-ice wind regime, which exists when WD_g is between approximately 185 degrees and 10 degrees (Table 1), is often identifiable on satellite images by the presence of cloud streets aligned with the average ABL wind direction, WD_a; for example see the cloud features just seaward of the FSMIZ in Figure 4. Clear conditions (one-eighth or less total sky cover) iceward of the ice edge are most likely during off-ice winds, occurring during 30% of the MIZCLM spring cases, compared to virtually never for on-ice cases. Longitudinal circulations produce the cloud streets and enhance mixing of momentum and scalars within the ABL. Despite the three-dimensional nature of the longitudinal rolls, their bulk effects can be parameterized with simple one-dimensional numerical ABL models [e.g. Brown, 1990].

A case study of cold season off-ice winds based on observations in the Bering Sea MIZ by Lindsay and Comiskey [1982] was simulated with a one-dimensional slab model by Overland et al. [1983] and Reynolds [1984] and with multi-level, two-dimensional models by Kantha and Mellor [1989a] and Wefelmeier and Etling [1991]. The model results showed that the internal ABL turbulent kinetic energy (TKE) and heat fluxes increased about an order of magnitude across the MIZ due to the different surface roughness and temperature conditions. But the ABL depth only varied by about 10% for this case because ABL growth was limited by an intense capping inversion.

In many off-ice situations, the ABL undergoes substantial increases in thickness over open water. For example, the MIZCLM data set spring Z_i median increases from 155 meters just inside the ice edge to 1240 meters for X greater than 100 km (Figure 10).

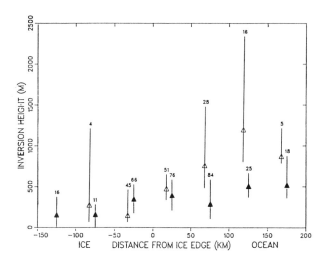

Figure 10. Median inversion base height, Z_i, for the off-ice wind regime during spring (open triangles) and summer (solid triangles) as a function of distance from the ice edge, defined as the 50% concentration isopleth, from MIZCLM data. The vertical bars represent the range between the first and third quartiles, while the numbers indicate the number of soundings in that category.

In the summer, there is a more modest, but still significant increase in Z_i across the MIZ (Figure 10).

Shaw *et al.* [1991] describe aircraft measurements over the BSMIZ near Bear Island on March 24, 1989 when the relative geostrophic wind direction, WD_g, was equal to 345 degrees, producing surface winds that were almost parallel-right, but had some off-ice component. There was no frontal region near the ice edge, in accordance with Table 1. The inversion base, Z_i, increased from about 200 m to 800 m from X = -100 km to X = 100 km, while T_a increased from -20 °C to -10 °C. The horizontal Z_i and T_a characteristics for this case were qualitatively similar to the MIZCLM off-ice medians except that the gradients perpendicular to the ice edge were greater because the ABLs had more time to approach equilibrium with the surface during the nearly-parallel flow. The sloping inversion created large thermal wind effects in the ABL.

An important Z_i scaling parameter we call Z_{max} is the elevation in the ambient (upwind) atmosphere where the potential temperature equals the surface potential temperature, T_o. Z_{max} represents the equilibrium or upper-limit ABL depth due to thermodynamic encroachment and is particularly relevant seaward of the ice edge during off-ice winds when ABL growth is primarily thermodynamic. During the summer, Z_{max}, referenced to O °C, is near or at the surface and mechanical mixing is required to maintain the ABL depth above Z_{max}, while during the spring, Z_{max} is generally between 1200 m and 3000 m elevation and thermodynamic effects alone will create deep ABLs if the air parcels are exposed to open water long enough. The upward vertical fluxes of heat are greater in spring than in summer due to the colder air temperatures (Figure 11) over the virtually constant ocean surface temperatures.

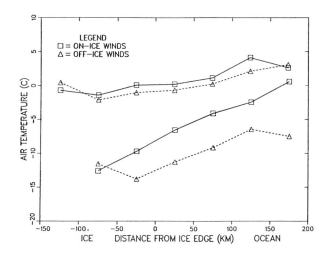

Figure 11. Median 10 m air temperatures during the spring (lower two plots) and summer (upper two plots) periods, from MIZCLM data.

The larger heat fluxes and Z_{max} values cause Z_i to be greater during the spring than during the summer for almost all MIZ locations, especially seaward of the ice edge (Figure 10). This is in contrast to the central Arctic, where radiational cooling of the surface creates the lowest inversions during the cold months [Belmont, 1957; Vowinckel and Orvig, 1970; Sechrist *et al.*, 1989, Appendix E; Serreze *et al.*, 1992].

A crude approximation for the horizontal distance scale, D_T, describing the e-folding distance required for the ABL temperature to adjust to a step change in surface temperature is $D_T = h/C_H$ where h is the downwind depth of the ABL, and C_H is the heat transfer coefficient, referenced to the average ABL temperature, T_a, and wind speed. A first-order approximation for T_a is

$$T_a(X_d) = T_{init} + (T_o - T_{init}) (1 - \exp(-X_d/D_T)) \tag{1}$$

where T_{init} is the initial (over ice) T_a and X_d is the distance downwind from a step change in surface temperature. Applying this to the off-ice ocean regime by setting C_H equal to 1.5×10^{-3} (a typical value reported in the literature for unstable conditions over open ocean, see Hakkinen and Cavalieri [1989]), and h equal to 500 m, gives D_T equal to 333 km, which predicts that relatively large turbulent heat fluxes can be maintained well away from the ice edge during off-ice regimes.

We can test the validity of the D_T scaling parameter by assuming that the median spring MIZCLM near-surface temperatures (Figure 11) represent a single typical case. The median temperature increases from -13 °C (T_{init}) near the ice edge to -7 °C (T_a) at X equals 200 km (assume $X = X_d$) where the sea surface temperature is 1 °C (T_o). The temperature change with distance is consistent with the rate predicted by D_T.

During off-ice winds in the spring, conditions typical of the central Arctic will often occur quite near the edge, if the ice is compacted. In this situation, radiation and wind speed conditions rather than distance from the edge (i.e. advection and surface heat fluxes) may control ABL characteristics. During clear conditions, the surface will cool by radiation, making Z_{max} equal to zero, which means that mechanical mixing controls the depth of the ABL. When surface turbulent heat fluxes are low, which is typical for the central Arctic, the height of the ABL is determined by mechanical production of TKE at the surface, represented by the friction velocity, $u*$, working against a background atmospheric static stability, represented by the Brunt-Vaisala frequency, N. In this situation, $L_N = u*/N$ represents a scaling parameter for ABL depth [Kitaigorodskii, 1988; Kitaigorodskii and Joffre, 1988; Overland and Davidson, 1992]. If one assumes that $u*$ is approximately proportional to wind speed and that N is random, one expects that the median wind speed and inversion height will be linearly-related. The validity of this scaling parameter is demonstrated by a comparison of wind speed vs median Z_i for over-ice (X < 0) conditions (Figure 12). Note the approximately linear relationship between surface wind speed and Z_i during clear skies. When conditions are overcast, the median Z_i values are generally higher and not correlated with wind speed (Figure 12). Mechanical mixing is apparently not an important factor in determining Z_i during overcast conditions.

In the summer, the median ABL temperatures from MIZCLM increase by only about 4 °C from X = -150 km to X = 200 km (Figure 11). This is because the wet ice/snow surfaces keep the ABL temperature over the ice near 0 °C while the ocean temperature is in the -1.7 °C to 4 °C range. Despite these modest temperature gradients, some significant changes in ABL physics occur in summer MIZ regions. For example, Fairall and Markson [1987] and Kellner et al., [1987] used low-level aircraft measurements in the MIZ during the summers of 1983 and 1984, respectively, to find significantly larger TKE, heat flux and momentum fluxes in the unstable ABLs over open ocean areas compared to nearby stable ABLs over ice-covered regions. These results demonstrate that a small change in the surface temperature is significant if the near-surface changes from stable to unstable conditions or vice-versa.

Surface winds in MIZs are dominated by the synoptic and mesoscale features described previously, which generally have horizontal scales larger than the MIZ width. Nevertheless, there is a consistent, albeit noisy, trend of increasing wind speeds in the positive X direction for all the season and wind direction cases in the MIZCLM set (Figure 13). Gorshkov's [1983] atlas also shows a strong average wind speed gradient in the FSMIZ region for all seasons. For example, in April Gorshkov's wind speeds changed from approximately 4.5 ms^{-1} to 7.5 ms^{-1} across the FSMIZ, which is very close to the MIZCLM values. This shows that the increased surface roughness and ABL stability over the ice-covered regions in the MIZ has a significant damping effect on wind speed. The spring off-ice regime has particularly high winds over the open ocean, a result of the unstable conditions.

Because ice in free drift moves at about 45 degrees to the right of the surface wind stress vector, an off-ice surface wind direction between 315 and 360 degrees will move the ice toward the ice pack or "on-ice" (Table 1). During cold seasons, this situation is

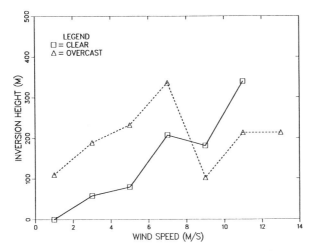

Figure 12 Wind speed vs median inversion base height over ice (X < 0) during clear and overcast conditions for different wind speed bins, from MIZCLM.

potentially favorable for deep convection because the cold off-ice winds cause freezing and brine rejection just off the ice edge, but do not form a solid ice cover because any newly-formed ice is pushed back toward the pack ice by the Ekman drift, leaving the brine behind [Guest and Davidson, 1991a]. Assuming that the pack ice is relatively motionless inside the MIZ, this situation will create a compact MIZ despite the off-ice winds.

The northern part of the FSMIZ is unique among the world's MIZ regions, because the position of the ice edge does not have large seasonal changes. In this region, the ice edge location is primarily controlled by oceanic features. Large melting events occur when winds transport the ice over warm water, allowing oceanic heat fluxes to melt the ice. This creates fresh water which further enhances ice transport toward warm waters by limiting the momentum gained from the wind to a shallow wedge of low-density Arctic water [McPhee *et al.*, 1987].

On-ice Wind Regime

The on-ice regime, which occurs when WD_g is between approximately 50 degrees and 170 degrees (Table 1) is usually associated with one or more layers of stratus clouds which form as warm moist marine air in the ABL is cooled over the MIZ. The solid cloud layers obscure the ABL flow patterns and ice features, and make low-level aircraft operations too dangerous. Therefore, the only direct observations of this regime are ship-based measurements.

If the ice-surface temperature, T_i, is less than the sea-surface temperature, a stable, new ABL may form within the previous near-neutral, marine ABL and prevent TKE

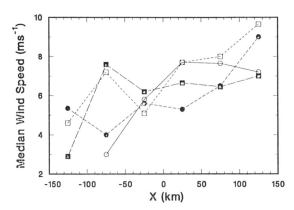

Figure 13. Median 10 m wind speeds for summer (solid symbols) and spring (open symbols) for off-ice (squares) and on-ice (circles) wind regimes, from MIZCLM data. The average standard deviation is 4.7 ms^{-1} and the average standard error is 0.8 ms^{-1}. The is no data for spring, on-ice wind conditions at X < -100 km.

and other surface-generated effects from reaching above the new ABL. This occurs if the local Obukhov length scale, L, [Stull, 1988, p. 181] over the ice is less than the pre-existing mixed-layer depth. L is approximately proportional to the wind stress divided by the temperature difference between the surface and 10 meters elevation; therefore the rougher ice surfaces will tend to increase L and TKE, thus preventing the formation of an internal ABL. However, the stabilizing effect of a relatively cold surface will counteract the increased roughness. The air-surface temperature difference can be quite modest for internal ABLs to form, even over the rough MIZ. For example, using typical MIZ transfer coefficients and setting the wind speed to a higher-than-average 10 ms^{-1} and the pre-existing ABL inversion depth to 500 meters, results in formation of an internal ABL with a 1.2 °C or greater temperature difference.

Kantha and Mellor's [1989a] and Glendening's [1994] MIZ ABL simulations for the on-ice regime assumed that the ice surface temperature was colder than the sea temperature by 10 °C and 4 °C, respectively, using geostrophic wind speeds of 12 ms^{-1} and 5 ms^{-1}. As a result, L was less than 100 meters for their simulations, and shallow, stable, internal ABLs formed over the ice. The temperature structure of the upper part of the old marine ABL and capping inversion persisted over the ice so that two low-level temperature inversions existed, separated by a near-neutral layer, i.e. a "double inversion". The shallow, stable ABLs formed in such cases have a similar effect on ABL dynamics and thermodynamics as the radiation ABLs formed in the central Arctic during winter. Both are characterized by rapid ABL temperature adjustment to surface conditions [Overland and Guest, 1991] and greater surface wind speed reduction (vis-a-vis the geostrophic wind speed) and larger turning angles (WD$_g$ - WD$_s$), compared to typical marine ABLs [Overland and Davidson, 1992]. Assuming a heat transfer coefficient of 1.0 x 10^{-3} and an ABL depth of 100 m results in a temperature adjustment length scale, D$_T$, of 100 km. This length scale is not larger than the MIZ width, which indicates that turbulent surface heat fluxes caused by ABL advection over ice will be small beyond the MIZ when the ABL is shallow, in contrast to the off-ice wind regime.

The above examples suggest that the on-ice wind regime is always characterized by a surface or low-level temperature inversion below a remnant mixed layer and marine inversion. But many factors act to keep the surface temperature very close to the ABL temperature, therefore making L larger than the pre-existing ABL depth and preventing formation of an internal ABL. The foremost of these factors is the presence of clouds within and above the ABL, which are invariably associated with the on-ice wind regime. Our downward longwave radiation measurements in the Nordic Seas MIZ regions indicate that during stratus conditions, the blackbody temperature associated with the downward longwave radiation is usually very close to T_a. This means that the ABL clouds, like ice, snow, and water, are almost radiative blackbodies and will tend to drive the surface temperature toward T_a, while the upward longwave radiation from the surface will drive the cloud bottom temperature (approximately T_a) towards the surface temperature. Further cooling of the upper ABL will occur due to a net upward radiation at the top of the cloud, if broken or no clouds are present above Z_i. If the ice edge is diffuse, there may be significant, concentrations of thin ice or open ocean in regions iceward of the ice edge proper, which makes the effective surface temperature (for determining surface heat fluxes and L) significantly larger than the surface temperature over the thick ice. If light snow is on top of the ice flows, the surface temperature will rapidly adjust to changes in forcing conditions. This means the surface and ABL temperatures will approach equilibrium over a much shorter distance than the D_T length scale, which assumes a non-variable surface temperature.

All of the factors mentioned in the previous paragraph will tend to destabilize the ABL and keep the effective L and TKE large enough to prevent the formation of an internal ABL. Also, Kitaigorodskii and Joffre [1988] suggest that the equilibrium ABL depth can be larger than L by a factor of 30 or more in some cases. Therefore, in many on-ice situations Z_i, will not decrease over the ice and may even increase. For example, Andreas et al. [1984] described an on-ice regime case in the Weddell Sea MIZ (Antarctica) during October, 1981 that had increasing Z_i with increasing ice concentration. This was attributed to the rougher ice surface, which decelerated the flow, resulting in low-level convergence and positive vertical velocities at Z_i. The surface temperature gradient was less than 1 °C across the MIZ and the estimated Obukhov lengths were not small enough for an internal ABL to form. According to a numerical simulation of this case by Bennett and Hunkins [1986] cloud-top cooling and adiabatic lifting significantly cooled the ABL, which would further tend to suppress downward heat fluxes at the surface. When the ABL remains relatively deep, the effects of ABL temperature and moisture advection can extend far into the pack ice [Andreas, 1985].

We have shown that an internal ABL may or may not form during periods of on-ice winds. If internal ABL formation due to advection over a cold surface is significant in

MIZ regions, one would expect that double inversions and/or surface-based inversions would be more common during on-ice winds than off-ice winds. Actually, according to the MIZCLM data, double inversions are more common during off-ice winds than on-ice winds for all location and season categories (Table 2). Surface-based inversions were also more common during off-ice winds than on-ice winds over the ice ($X < 0$). Surface-based inversions were rare seaward of the ice edge during spring for both wind direction regimes. The results summarized in Table 2 indicate that, in the MIZCLM sampling region, surface-based advective inversions are not especially common during on-ice winds. The occurrence of surface-based inversions is correlated with clear skies and light winds, not relative wind direction. This suggests that surface-based inversions in the Nordic Seas MIZs are more often due to radiational surface cooling than ABL advection.

The fact that double inversions are more common for the off-ice regime than the on-ice regime means that the ABL is not cooled from below enough to form detectable internal inversions during a typical on-ice wind regime. By backtracking ABL air parcel trajectories for individual off-ice regime cases from MIZCLM, when a double inversion is present, we find that the air masses over the MIZ had previously been over open water, and then moved over the ice before heading back toward the open water in the local off-ice flow. The longer time over the central ice pack ice for these off-ice cases, compared to the on-ice cases, allows the formation of a stronger inversion between the new ABL and the remnant marine mixed layer. Radiational cooling from the upper inversion base location (which usually corresponds to a cloud top) is able maintain the mixed layer/capping inversion structure at the top of the old marine ABL.

During the spring on-ice regime, the median Z_i from MIZCLM decreases by about 60% going across the ice edge from $X = 50$ km to $X = -50$ km, (Figure 14). The effect on Z_i of increased TKE and convergence over the rough MIZ ice surface is more than counteracted by the increased stability due to the relatively cold ice

TABLE 2. Occurrence of Surface-based Inversions and Double Inversions for Different Seasons and Wind Directions from MIZCLM.

Season	Spring		Summer	
Location[1] (km)	$X < 0$	$X > 0$	$X < 0$	$X > 0$
Wind regime	Surface-Based Inversion (% of total)			
On-ice	10	3	16	8
Off-ice	12	1	20	5
	Double Inversion[2] (% of total)			
On-ice	28	10	26	22
Off-ice	33	39	44	26

[1]Within 200 km of ice edge.
[2]Not mutually exclusive with surface-based inversions.

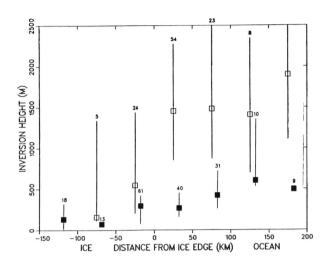

Figure 14 Same as Figure 10 but for the on-ice wind regime. The spring and summer periods are represented by solid and open squares, respectively.

surface, on the average. But the on-ice spring MIZCLM Z_i also have a much larger variability (see quartile ranges) compared to the off-ice spring case (Figure 10), particularly in the location bin just inside the ice edge, -50 km < X < 0 km. The variability indicates that in some cases an inversion forms below the original marine inversion (only the lowest one is counted), while in others the old inversion remains at the top of the surface mixed layer; unlike the off-ice regime, variations in Z_i across the ice edge at any given time are qualitatively different than the median horizontal variations. The median inversions are higher at all locations during on-ice winds, compared to off-ice winds, a result of the initially deeper and warmer ABLs and the lag in ABL adjustment to different surface conditions.

The summer on-ice regime median Z_i values also decrease over the ice (Figure 14); the trend and variability within distance bins is much less than the spring on-ice regime (Figure 14) and somewhat less than the summer off-ice regime. But the effect of over-ice stability dominates over increased surface roughness, even during the summer when horizontal temperature gradients are relatively weak.

Our discussion so far has implicitly assumed that the ABL height, h, defined as the top of the turbulent layer, is the same as the inversion base height, Z_i. In fact, during conditions with surface cooling or a decrease in mechanical TKE production, the ABL height may be much lower than Z_i, but undetectable by the rawinsondes, which do not resolve the turbulent fluctuations. Therefore, shallow internal ABLs during on-ice winds may be more common than implied by the summary of MIZCLM ABL characteristics (Table 2) and the Z_i values in Figure 14.

During the spring, the on-ice regime exhibits a steady cooling trend in the downwind direction, even seaward of the 50% concentration ice edge (0 km < X < 200 km), where the ABL is initially colder than the -1.7 °C to 2 °C sea surface (Figure 11). Some of the downwind decrease in air temperature, T_a, seaward of the ice edge is due to the cooling effect of sea ice which may be present at concentrations below 50%. But most of the decrease toward the ice in the $X > 0$ region is due to the increased probability that the air parcels nearer the ice have a recent over-ice history, despite their current location over the open ocean. Air parcels over a pack ice region can travel across the ice edge into an open ocean region and then may acquire an on-ice component if the parcel trajectory curves back toward the ice edge or if the ice edge curves toward the parcel trajectory. The on-ice T_a values are 5 °C higher than the off-ice regime cases throughout the region -50 km $< X <$ 150 km, indicating the effect on T_a of different initial conditions and the lag in the adjustment of ABL temperature to surface temperature.

In the summer, the on-ice regime median T_a values are about 1.5 °C higher than the off-ice values throughout the MIZ (Figure 11). Otherwise, the T_a characteristics are similar for the two regimes in summer, i.e. temperature gradients exist, but they are weak. The median wind speed increases from the pack ice to the open ocean for spring and summer cases (Figure 13), a result of the horizontal variations in ABL structure and surface conditions.

Usually the ice movement in the on-ice regime is toward the pack ice, but if the surface wind direction, WD_s, is between 135 degrees and 180 degrees, the ice movement will have a seaward component, despite the on-ice winds (Table 1). This wind regime may be associated with intense ice melting because the lower atmosphere is warm due to its marine origin and the ice is transported to regions with warmer waters.

Parallel wind regimes

The parallel-right wind regime, which exists when WD_g is between 10 degrees and 50 degrees, approximately, produces the most dramatic changes in ABL structure (Table 1). According to Glendening's [1994] numerical simulations of the MIZ ABL for different wind directions, the surface temperature gradient across the MIZ generates a secondary circulation or "ice-breeze" that always has a positive X component (off-ice). The ice-breeze opposes the surface wind seaward of the ice edge, which has a small negative X component. At the furthest seaward extent of the ice-breeze circulation, the opposing air masses will cause ABL convergence and deformation. Another factor that contributes to the convergence during the parallel-right regime is the greater wind turning angle over ice-covered regions due to the relatively shallow ABL and rough MIZ ice surface, compared to open ocean regions. This latter effect is similar to the frictional convergence at coastlines modeled by Roeloffzen et al. [1986]. Both the ice-breeze circulation and frictional convergence favor the formation of sharp, quasi-stationary, temperature fronts, discontinuities in ABL depth, wind speed jets and strong vertical velocities near the ice edge, according to models of the

ABL during the parallel-right regime [Glendening, 1994; Kantha and Mellor, 1989a]. During parallel-right winds, the ice will be transported toward the ice pack, creating a compact ice edge and concentrating the surface gradients to a relatively narrow region.

The ice-breeze also enhances convergence in the ABL during parallel-left regimes, when WD_g is between 170 degrees and 185 degrees, approximately, using Glendening's [1994] case (Table 1). But if the horizontal temperature gradient is weak, frictional divergence will be stronger than the ice-breeze convergence and ABL fronts and associated features will be suppressed.

The ABL characteristics are extremely sensitive to small changes in WD_g for both parallel wind cases [Glendening, 1994]. The WD_g ranges in Table 1 are based on cases with the same horizontal temperature change (4 °C) and geostrophic wind speed (5 ms^{-1}) [Glendening, 1994]. The actual WD_g range which induces stationary ABL frontogenesis, and therefore defines the parallel wind regimes, depends on the relative strength of the ice-breeze circulation compared to the background geostrophic wind. Therefore, if the surface temperature gradient is larger and/or the geostrophic wind speed is less than what was used by Glendening [1994], the ice-breeze component will be relatively larger and the geostrophic wind vector must be oriented more on-ice to oppose the ice-breeze. If this is the situation, then the range of WD_g values marking the boundaries of the parallel-right regime will be rotated clockwise (increased WD_g) from the values shown in Table 1 while the parallel-left WD_g range will be rotated counter-clockwise. Conversely, in other situations, the ice-breeze may be relatively weak and the parallel regime WD_g values are rotated in the other directions. These effects cause the actual WD_g values describing the parallel regimes to vary by plus or minus 15 degrees from the Table 1 values, depending on the actual temperature and roughness conditions, with the exception below.

When the geostrophic winds seaward of the ice edge are light (less than approximately 4 ms^{-1}) but with some on-ice component, Table 1 does not apply, because no direction of on-ice geostrophic flow will be able to counteract the ice-breeze in the ice edge region. A front will form over the open ocean at the furthest extent of the ice-breeze circulation. For this reason, weak on-ice geostrophic forcing regimes are in the parallel wind regime, using our "front-forming" definition.

Shapiro et al. [1989] document such a case in the MIZ south of Svalbard. An Arctic front (similar to an ABL front but deeper) with a temperature jump of 6 °C over a 25 km wide strip existed 100 km off the ice edge. Low-level winds were almost directly opposed on different sides of the front. Although this is the only case study of an ice-breeze in a Nordic Sea MIZ in the literature, Glendening's [1994] modeling studies suggest that they are quite common. They have not been observed more often because the required measurements (simultaneous surface and geostrophic wind vectors) are rare and the ice-breeze signal is usually masked by synoptic-scale forcing. A numerical simulation of the Shapiro et al. case is case by Thompson and Burk [1990] demonstrated that frontogenesis was due primarily to horizontal deformation, while tilting caused by vertical wind shear, and differences in surface fluxes tended to destroy the frontal structure. In these and other respects, this front did not differ significantly from cold fronts in the central United States.

There are not enough cases of soundings within each X location bin during unambiguous parallel wind regimes from the MIZCLM data to derive meaningful statistics concerning average variations in ABL structure across the MIZ. But the data available suggest average Z_i and T_a values intermediate to the off-ice and on-ice values. This is consistent with the fact that advection of lower (or higher) inversions or temperature, such as occurs during the off-ice (on-ice) regime, is generally less for the parallel cases, because the ABL is closer to equilibrium with the surface.

During the entire MIZCLM period, continuous surface meteorological measurements documented the passage of several atmospheric fronts, as revealed by sudden changes in temperature, humidity, clouds or fog, and the wind vector. The strongest temperature fronts are associated with synoptic or mesoscale cyclones and/or waves and are characterized by significant surface pressure dips. These types of fronts are not particularly associated with the parallel-wind regime and usually propagate over the MIZ from some other region. For example, cold fronts sometimes form over the pack ice and propagate over the MIZ and open ocean during off-ice winds [e.g. Fett, 1989b, pp. 1D-1 to 1D-25]. Schultz [1987] and Fett (1989a) describe an ABL warm front, in association with an easterly wave, which crosses the FSMIZ during on-ice winds.

But there is a strong association between parallel winds and shallow temperature or humidity fronts near the ice edge which do not have significant surface pressure signatures. An unpublished study of fog formation in MIZs by Guest found that sudden changes from totally clear to foggy conditions or vice-versa virtually always occur when winds are nearly parallel to the ice edge. These events are associated with humidity fronts formed between the dry arctic air and the moist marine air. During these frontal passages in summer, the more moist marine air is sometimes cooler than the dry air, probably a result of fog-top radiational cooling. The parallel wind regime is also associated with modest (3 °C over 20 minutes) ABL temperature fronts associated with the retreat or advance of shallow pools of cold Arctic air. We cannot be sure from the single point measurements whether the fronts described in this paragraph originally form due to the frictional and ice-breeze mechanisms modeled by Glendening [1994] and Kantha and Mellor [1989a] or whether some larger scale deformation field or instability is involved.

If ABL MIZ frontogenesis occurs in nature for any reason, the fronts will not usually remain stationary, because small WD_g changes will cause the front to advect away from the ice edge. It is the transient, but locally-formed, fronts that probably produce the modest temporal jumps in temperature and/or humidity that are often observed from the relatively stationary ship platforms in the MIZ. According to Shapiro and Fedor [1989], ABL fronts originally formed over the ice edge near Spitzbergen can often be traced as far away as the coast of Norway or the United Kingdom.

There are occasional periods when a quasi-stationary ABL front or strong baroclinic zone will remain in the MIZ for longer than a few hours. During a six-day period of

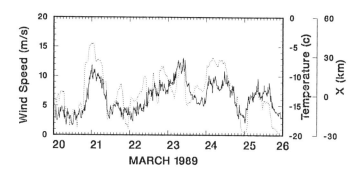

Figure 15. The median 14 meter wind speeds and temperature from the *R/V Polarbjoern* during the March 20 - 25, 1989 period during CEAREX. An approximate X location is shown on the right vertical axis. Note the close correlation between temperature or location and wind speed. This shows that large changes in wind speed can occur across an MIZ even when the geostrophic forcing is constant.

the CEAREX project from March 20 through March 25, 1989 the synoptic geostrophic wind vector was remarkably constant. The magnitude was about 7 ms^{-1} and WD_g was equal to 15 degrees in the part of the FSMIZ that was bisected by the 78.5 degree latitude line [Lackmann *et al.*, 1989]. The surface winds were parallel-right and the temperature of the ABL changed by 16 °C over a 50 km region centered on the ice edge. During this period the ship *R/V Polarbjoern* made 16 transects across the ice edge, as part of an oceanographic survey, while we performed 25 rawinsonde soundings. The soundings revealed a quasi-steady-state situation featuring a 200 m thick wedge of cold (-16 °C) Arctic air opposing a warm (0 °C) 1000 m deep marine ABL. Above 200 m elevation, the structure of the atmosphere was virtually identical on both sides of the ice edge, while below 200 m a strong baroclinic zone was present. Determining the small-scale characteristics of the baroclinic zone was not possible because the isotherms typically changed position by about plus or minus 10 km in the several hours between ship transects and rawinsonde soundings. The changing position of the isotherms may have been caused by horizontal waves along the baroclinic zone or by small changes in WD_g.

The most interesting phenomenon during this period was the close correlation between surface wind speed and temperature (Figure 15). As the ship (and the front) moved with respect to the ice edge, the ship encountered changes in ABL temperature which consistently matched changes in wind speed. The wind speed varied from 4 ms^{-1} in the coldest air to 12 ms^{-1} in the warmest locations. The changes in wind speed were due to the contrast between the shallow stable ABL over the rough ice and the deep marine ABL. This case study demonstrates how ABL fronts and associated features that form during the parallel regimes can have a strong effect on variations in atmospheric forcing (i.e. heat and momentum fluxes) of the ice and ocean. The horizontal changes in atmospheric forcing associated with an ABL front may be stronger than changes in forcing due to the direct effects of a change in surface conditions, such as an ice edge.

Radiative Regime

If the wind speed is below approximately 4 ms^{-1} during cloudy skies, or below 6 ms^{-1} during clear skies, radiation effects, rather than wind direction, will dominate ABL characteristics iceward of the ice edge. The radiative regime is characterized by surface-based inversions or very shallow ABLs in dark seasons and diurnal changes in air temperature and ABL height during the months March to June, and September and October. We observe the radiative regime very close to compact ice edges during light off-ice winds, but never over open water regions. During the summer melt season in July and August, the radiative regime does not exist, i.e. diurnal T_a changes are insignificant, because the surface temperatures are essentially invariant.

Momentum Fluxes

In this section, we examine the transfer of momentum from the atmosphere to the ocean/ice surface, or wind stress, in MIZs. Wind stress provides the crucial dynamic input responsible for currents, ice movement, and mechanical turbulent mixing in the upper ocean. Wind stress is a function of the near-surface wind speed squared, surface roughness and atmospheric stability. MIZs are regions where the changes in surface conditions induce large gradients of wind stress. The resulting wind stress fields may generate various ocean and ice features.

Studies of the effect of wind stress on ocean dynamics in MIZs generally assume that the wind speed is constant everywhere [e.g. Roed and O'Brien, 1983; Roed, 1983, 1984; Smedstad and Roed, 1985; Hakkinen, 1986a, b, 1987; Smith et al., 1988, Kantha and Mellor, 1989b; Ikeda, 1991; Hakkinen et al.; 1992] so that wind stress is simply proportional to surface roughness. However, we demonstrated previously that a significant wind speed gradient generally exists in MIZs. In general, this gradient affects wind stress variability as strongly as changes in surface roughness.

The surface roughness, as parameterized by the neutral drag coefficient, C_{DN}, referenced to 10 m, depends on the type of sea ice which is present (Table 3). The very smooth, smooth, rough, very rough and extremely rough categories in Table 3 represent typical fast ice, central pack ice, inner MIZ ice, outer MIZ ice and central pack intense shear zone ice, respectively [Guest and Davidson, 1991b]. The outer MIZ is defined as the MIZ region adjacent to the ice edge which is directly affected by swell and wave action. If the ice edge is compact, the outer MIZ exists only about 5 km to 10 km into the ice. The region just inside the ice edge has the greatest roughness of any sea ice region except for of intense central pack shear zones. The inner MIZ exists where floes have started to break-up, but are not currently being subjected to significant swell and wave action. The inner MIZ extends into the ice about 100 km and has moderate roughness due to floe edges and rafting, while the central pack is relatively smooth, with intermittent ridges forming the major roughness elements. The drag coefficients of open ocean surfaces are highly variable, but for winds less than 10 ms^{-1}, the average values are similar to central Arctic pack ice.

TABLE 3. The Roughness Length, z_o, and Neutral Drag Coefficient, C_{DN}, for Various Ice and Sea Surfaces.[1]

Ice Type	$z_o \times 10^3$ m median	$C_{DN} \times 10^3$ median	minimum	maximum
Grease	0.0027	0.7	0.6	1.1
Nilas	0.45	1.6	1.4	1.9
Pancake				
< 0.75 m diameter	0.016	0.9	0.7	1.3
0.75 - 1.5 m diameter	0.45	1.6	1.1	2.2
> 1.5 m diameter	2.8	2.4	1.9	2.9
fused	1.0	1.9	1.5	2.6
Young				
smooth	2.4	2.3	1.9	2.7
rough	7.5	3.1	2.6	3.6
First-year				
very smooth[2]	0.33	1.5	1.2	1.9
smooth	1.3	2.0	1.6	2.4
rough	7.5	3.1	2.2	4.0
very rough	21.0	4.2	3.1	5.0
Multiyear				
very smooth[2]	0.33	1.5	1.2	1.9
smooth	2.0	2.2	1.9	2.5
rough	10.0	3.4	2.5	4.1
very rough	27.0	4.6	3.6	5.5
extremely rough	110.0	8.0	6.7	9.1
Greenland Sea ($U < 12$ ms[-1])				
All wind directions	0.80	1.8	0.7	3.0
Ice upwind 2-10 Km	0.23	1.4	1.1	1.8
Steady-state open ocean[3]				
$U = 10$ ms[-1]	0.15	1.3	na	na
$U = 25$ ms[-1]	1.3	2.0	na	na

[1]From Guest and Davidson [1991]. Only from periods when lowest inversion was higher than 150 meters. At least 85% of the stated ice type was upwind.

[2]Includes studies summarized by Overland [1985]

[3]From Smith [1988]. No ranges given.

The drag coefficients values shown in Table 3 are generally consistent with earlier MIZ observational studies summarized by Overland [1985] and more recent results by Anderson [1987], Kellner et al. [1987] and Fariall and Markson [1987]. The FSMIZ has higher drag coefficients than other MIZs such as the Weddell Sea [Martinson and Wamser, 1990; Andreas et al., 1993] because of the higher proportion of thick, multi-year flows which are present.

The ice floe edges in the MIZ represent a significant roughness feature. According to a model of the aerodynamic drag of ice floes by Hansen-Bauer and Gjessing [1988], the maximum drag coefficient for typical FSMIZ floes occurs at 73% ice concentration. At higher concentrations, the form drag effect of the floe edges is diminished due to the sheltering effect of nearby floes. This model is probably most applicable during the summer off-ice wind regimes when floe-to-floe interactions are limited and the predominant multi-year floes have the simplest "boxcar" shape.

When ice that has already been broken-up into small (< 10 m diameter) floes by swell and wind waves is compacted, rafting and flow tipping will create a very rough surface. In this situation, the highest drag coefficients are found with the highest ice concentrations. For example, independent wind stress measurement studies of the summer FSMIZ in 1984, by Anderson [1987] and Guest and Davidson [1987], indicate that the drag coefficient increases with ice coverage, particularly in the range between 70% and 90%, (no measurements were available at higher concentrations). Apparently, the sheltering of the ice flow edges is counteracted by increased roughness due to floe interactions. The highest drag coefficients measured in MIZs by the above studies (4×10^{-3} to 6×10^{-3}) occurred in regions with 90% ice concentration after intense on-ice wind events had crushed and compacted the ice, forming a jumbled mass resembling a jagged boulder field. Ice this rough only occurs in 1 km - 2 km strips inside the edge during strong, ice-compaction wind events, and therefore does not significantly affect the average regional wind stress fields.

During cold seasons, many different types of ice can be present (Table 3). For rough and very rough ice, i.e ice that has been broken-up by swell, the younger types have lower drag coefficients because the ice is relatively thin and the roughness elements cannot protrude too high into the air flow. Grease ice damps ocean capillary waves, thus producing a significantly smoother surface and lower drag coefficient than would occur with normal ocean surfaces (Table 3).

For modeling purposes, an effective drag coefficient [Fiedler and Panofsky, 1972] must be defined for regions where different surface conditions are present within a single gridpoint. This can be accomplished by using areal-weighted averages of the drag coefficients as given in Table 3, or more accurately, by calculating the momentum flux over each surface separately and combining the values to determine an average regional wind stress, similar to the heat flux method described later.

The combined effect of the generally increasing winds from ice to open ocean (Figure 13), and higher drag coefficients in the outer MIZ, result in a typical wind stress field with two maxima, one over the outer MIZ and one out in the open ocean where the wind speed is greatest. In the transition region between the MIZ and the central pack, there can be strong wind stress gradients generated by the combined effect of relatively higher winds and rougher ice over the MIZ. The resulting wind stress curl is often just as significant, but in the opposite direction, as the curl generated by roughness contrasts between the MIZ ice and the open ocean at the ice edge.

Heat Fluxes

Turbulent heat fluxes seaward of the ice edge

The two most important features concerning the energy exchange between the atmosphere and surface in MIZs are the large horizontal gradients of surface temperature and albedo. The former feature generates turbulent heat fluxes as air is advected across surface gradients while the latter feature allows ice-free areas to absorb much more solar radiation than adjacent ice surfaces.

The largest magnitude turbulent sensible and latent heat fluxes occur in the open ocean regions just seaward of the ice edge, according to our estimates of the median turbulent heat fluxes using the bulk method based on the MIZCLM wind speed, humidity, air temperature and sea-surface temperature. We use a fixed temperature and moisture transfer coefficient of 1.5×10^{-3}, which allows comparison with a study of Nordic Sea heat fluxes by Hakkinen and Cavalieri [1989]. The latter study used the gridded Navy Fleet Numerical Oceanography Center wind, temperature and humidity fields for the year of 1979 to determine bulk heat flux statistics in the Nordic Seas, based on this transfer coefficient.

For our study, the median total upward turbulent heat fluxes just seaward of the ice edge ($0 \text{ km} < X < 50 \text{ km}$) during the spring are 178 Wm^{-2} and 101 Wm^{-2} for the off-ice and on-ice wind regimes, respectively. In the location bin $100 \text{ km} < X < 150 \text{ km}$, the turbulent heat flux values for the different regimes are 137 Wm^{-2} and 41 Wm^{-2}. It is interesting that significant upward turbulent heat fluxes occur during on-ice winds despite the air parcels' recent movement toward cooler surfaces. The coolness of the air masses during on-ice wind regimes indicates that the air parcels have previously been over land or ice surfaces. Because of the deep ABLs which form over the open ocean areas of the Nordic Seas, the temperature adjustment length scale, D_T, is several hundred kilometers and heat fluxes remain large well away from coasts or ice edges throughout the Nordic Sea region during cold seasons.

Gorshkov's [1983] atlas shows that a spatial maximum turbulent heat flux (approximately 110 Wm^{-2} during April) occurs in the Fram Strait region just west of Svalbard, with decreasing fluxes toward the MIZ to the west and north. This magnitude is comparable to the MIZCLM results. However, Gorshkov's spatial analysis locates the maximum heat flux approximately two hundred kilometers seaward of the mean ice edge location, probably a smoothing artifact. Hakkinen and Cavalieri's [1989] more detailed April analysis has a maximum turbulent heat flux at the ice edge and a north/south cross-section with the same general pattern as the MIZCLM results. However, Hakkinen and Cavalieri's [1989] magnitudes are approximately twice the MIZCLM values. The tendency for the MIZCLM ships to avoid storms cannot explain this difference. But individual storms do have a significant impact on the total monthly heat flux for an area. The year used by Hakkinen and Cavalieri, 1979, may have had more active synoptic conditions than the years used in the spring MIZCLM set, 1987 and 1989. Considering the importance off these large heat fluxes on processes such as deep convection in the ocean and atmosphere, and deep water formation, it is apparent that further research is required to explain the differences between studies, and to accurately quantify the heat flux climatology of the areas in the Nordic Seas region that are adjacent to ice edges.

Turbulent fluxes over ice-covered regions in the MIZ

The temperature and moisture transfer coefficients used for bulk estimates of heat flux depend on the roughness of the surface. There are few measurements of the heat and moisture transfer coefficients, C_H and C_M, over snow or ice. Andreas' [1987] theoretical study relates scalar transfer coefficients to the roughness Reynolds number. He predicts that the neutral scalar transfer coefficients over sea ice will almost always be somewhere between 1.0×10^{-3} and 1.5×10^{-3}, referenced to 10 meters elevation.

We measured heat fluxes using the profile method from a four-level, six-meter, tower over a large (2 km diameter) multi-year ice floe within the FSMIZ during the summer of 1984. The average near-neutral C_H was 1.5×10^{-3} and C_{DN} was 2.3×10^{-3}; both values had a estimated accuracy of 20%. The difference between this average value of C_H and the Andreas' [1987] formula value is less than the measurement error, suggesting that uncertainty in the value of the transfer coefficients may not be a major problem in determining turbulent heat fluxes from bulk measurements.

Using a numerical model of a surface with intermittent ice strips and leads, Claussen [1991] found that the error due to ignoring local advection effects was, at most, 6% of the total surface heat flux, for typical conditions. What this means is that a reasonable method for determining the areal average turbulent heat flux for a region (such as a model gridpoint) is to calculate the heat flux separately over each type of surface using the same ABL air temperature, but using different surface temperatures, and different roughness and stability-corrected transfer coefficients. The total surface heat flux for that region, which may represent one numerical model gridpoint, is then based on the areal-weighted average of the heat fluxes from each different surface type, e.g. Andreas [1988].

Although the uncertainty in the transfer coefficient is not large, and local advection can be ignored, a major difficulty remains when attempting to estimate heat fluxes over regions with varying ice conditions such as MIZs. The difficulty is that the surface temperature in regions with mixtures of thick ice, thin ice and open water is highly variable spatially, making determination of areal averages difficult. We have made some heat flux and surface temperature measurements over MIZ sea ice, in locations where we were able to place instruments directly on floes, but not enough to determine statistics concerning heat fluxes as a function of distance from the ice edge.

Our profile measurements of sensible heat fluxes over snow-covered floes in MIZs and nearby regions show that the surface of thick ice is virtually always within 3 °C of the air temperature, unless winds are very light, resulting in turbulent sensible fluxes that have magnitudes that are rarely greater than 20 Wm^{-2} [Overland and Guest, 1991]. The flux is both upward and downward so that the average is near zero.

If the ice is "thin", as defined by Guest and Davidson [1992], the surface temperature is affected by conductive heat fluxes through the ice and the upward surface heat flux is intermediate to the open ocean and thick ice values. We do not have direct measurements of heat fluxes over thin ice, but they can be determined quite accurately if the ice thickness is known. Therefore, the crucial parameter for determining turbulent fluxes in ice covered regions, including the MIZ, is the relative areal coverage of the various open ocean and ice thickness surfaces.

In the summer, the ice surface is wet, and remains very close to 0 °C. As a result, the median surface temperature gradients and associated turbulent heat fluxes are much smaller than the spring values, rarely having values over 30 Wm^{-2}, which is similar to the results of Gorshkov [1983] and Hakkinen and Cavalieri [1989] for June and July.

Radiation Fluxes

The difference in albedo between ice surfaces and open water create large gradients in surface heating due to solar or "shortwave" radiation in summertime MIZs. In the summer, average cloudiness in the Fram Strait is about 88% [Warren et al., 1988]. Cloudiness increases slightly the albedo of open water areas and leads, while decreasing slightly the albedo of sea ice compared to clear skies, but there are still substantial differences between surfaces in the FSMIZ. By extrapolating a few albedo measurements by Maykut and Perovich [1985] to the entire 25 day period of the MIZEX84 project, and aided by a radiation model, Francis et al. [1991] determined the average albedo to be approximately 0.6 over solid ice floes, 0.3 over melt ponds and 0.1 over leads or open ocean areas seaward of the ice edge. The total (infrared or "longwave" and shortwave) net (upward minus downward) radiative flux at the surface over open ocean regions (-173 Wm^{-2}) was one-third of the downward shortwave radiation at the 200 mb level, while over sea-ice regions the net radiative flux at the surface (-92 Wm^{-2}) was only one-sixth of the downward shortwave radiation at 200 mb. Considering the entire atmosphere below 200 mb, the atmosphere-open water system has a total net radiative gain of approximately 60 Wm^{-2}, while the atmosphere-ice system is nearly in equilibrium [Francis et al., 1991].

The direct and indirect effects of solar radiation is important in MIZ regions, as evidenced by the annual advance and retreat of ice edges throughout polar regions. Solar radiation directly melts considerable amounts of sea ice in late summer when melt ponds are present [Maykut, 1982; Ebert and Curry, 1993]. Some of the solar radiation absorbed by leads and open areas in MIZs is used to melt the sides of the ice floes, but much goes into bottom melting, which will not affect the areal ice coverage until the ice floes melt completely [Maykut and Perovich, 1987]. Studies which ignore this latter effect seriously over-estimate the feedback between ice coverage and shortwave radiation.

When solar radiation becomes weak, surface cooling by longwave radiation causes the ice edge in the BSMIZ and most other MIZ regions to rapidly advance as new ice is formed. (Ocean heat fluxes prevent large advances of the northern FSMIZ ice edge.) The vertical location of the longwave cooling has an important impact on the ABL dynamics and thermodynamics. Surface cooling during clear skies will stabilize the atmosphere and produce shallow, mechanically-driven ABLs while cooling at the top of cloudy ABLs creates deeper ABLs, whose depth is not a function of surface-generated mechanical mixing (for example see Figure 12).

Downward longwave radiation over sea ice during the CEAREX drift was an average of $75 \, \mathrm{Wm}^{-2}$ greater during low overcast sky conditions than during clear skies [Guest, 1992]. This shows how differences in longwave radiation across boundaries between cloudy and clear areas create horizontal variations in the surface energy balance which are comparable to changes in shortwave and turbulent heat fluxes across an ice edge during moderate winds. But unlike the latter two heat sources, the net surface longwave radiation does not undergo radical changes across MIZs unless cloud conditions also change.

Summary and Conclusions

We examined several of the salient features concerning dynamic and thermodynamic interactions between the atmosphere, ice and ocean in MIZs and surrounding regions in the Nordic Seas. These MIZs are affected by both local forcing conditions, such as the surface temperature and roughness gradients, and global forcing, such as the location of the main polar baroclinic zone and associated cyclonic activity. The local forcing in MIZs causes large variations in ABL structure which, in turn, modify the surface heat and momentum fluxes. Quantifying the radiative and turbulent micro-physical processes in MIZs is essential to understanding the overall air-sea-ice system.

We focused the discussion of synoptic-scale (large) atmospheric processes on cyclones, because they produce the strongest momentum and heat flux events in MIZs. Most of these storms, particularly the more intense ones, form in the open ocean regions south of the MIZs. The dominant cyclone track runs approximately along, or slightly northwest of, the warmest waters in the Norwegian Sea and crosses the MIZ near Svalbard or in the northern Barents Sea. In addition to having a large effect on characteristics and processes within the MIZ, cyclones replace most of the heat lost by radiational cooling in the central Arctic basin.

The Nordic Sea MIZs are not particularly associated with synoptic-scale cyclogenesis, unless a deep baroclinic zone generated by large scale forcing happens to lie over an MIZ region. The local forcing effects from MIZs are apparently not strong enough to induce synoptic-scale cyclogenesis. But synoptic-scale cyclogenesis can occur over MIZs, and even over nearby pack ice, as documented in a case study. In this case, a deep layer of marine air had penetrated into the Arctic basin, creating a marine-like atmospheric environment that was conducive to storm development.

Low-level baroclinicity in MIZ regions can generate mesoscale cyclones (diameters less than 800 km). Generally these mesoscale cyclones are quite weak and short-lived without some other forcing mechanisms such as PVA, CISK or ASII. If one or more of these mechanisms are present, the mesoscale cyclone may develop into a polar low with gale-force surface winds. Although baroclinicity generated at the ice edge is crucial for many polar low developments, the mature storm phase will generally not occur until the baroclinic zone and incipient polar low has been advected over the open ocean several hundred kilometers from the ice edge. For this reason, polar lows rarely affect MIZs directly. An exception occurs in the MIZ between Bear Island and Svalbard, where the unique ice edge geography allows storms, formed within an ABL front to the lee of Svalbard, to travel in a southerly direction with the background flow, but remain in or near the BSMIZ.

Radical changes in ABL characteristics occur across MIZs in response to intense surface gradients of temperature and roughness. If the ABL is shallow, typical of light wind and clear sky conditions over ice, the thermal adjustment of the ABL will occur mostly within the MIZ. But more often, the ABL will be deep enough so that thermal adjustment takes hundreds of kilometers, and ice edge advective effects such as large turbulent heat fluxes can persist far from the MIZ.

Unless the geostrophic forcing is weak, the relative wind direction with respect to the ice edge is a crucial parameter controlling advection, and hence, ABL characteristics in the advection-dominated MIZs. During off-ice winds there may be some drop in the ABL depth a few tens of kilometers off the ice edge due to acceleration over the relatively smooth ocean surface, but, in general, the ABL grows in response to surface heating. During cold seasons, the large upward heat fluxes and deep ABLs over the open ocean cause the surface wind speed to be significantly larger than over the ice. This feature tends to negate the effect on wind stress of any increased roughness over the ice.

During on-ice winds, over the open ocean in the spring, the average air temperature is several degrees below freezing, indicating an ice or land origin for the ABL air mass, despite the recent marine history. The resulting upward heat fluxes are about 60% of the magnitude of the average off-ice heat fluxes, but are still quite large considering that the flow is toward colder surfaces. Once over areas with high ice concentrations, the heat flux becomes downward and a new stable arctic ABL may form within an old marine ABL over the ice. If this occurs, the ABL depth will have a sharp discontinuity where the internal ABL forms. The old upper ABL and capping inversion retain their mean structure, but they are no longer in turbulent contact with the surface. According to theoretical results, a surface only a few degrees colder than the air will create an internal ABL. Although, we often observe the double inversion layers that are associated with internal ABLs, they are not particularly associated with on-ice winds. This is because many factors, the foremost being the ever-present stratus deck during on-ice winds, tend to keep the surface temperature very close to the air temperature, which prevents the stabilization necessary for the formation of an internal ABL.

The largest gradients in ABL characteristics occur when the ABL wind is parallel to, or nearly-parallel to, the ice edge or when the external geostrophic forcing is weakly on-ice. During these situations, the ABL air parcels can approach equilibrium with the surface, resulting in large variations where surface conditions change, such as at the ice edge. A secondary ice-breeze circulation and, for parallel-right winds, a frictional convergence, will further concentrate the ABL gradients into narrow frontal regions. In rare cases, these fronts remain near or in the MIZ for several hours or days, where they strongly affect temperature, humidity and wind variations. Usually, the fronts are advected away from the ice edge, in some cases remaining coherent hundreds of kilometers away. The changes in atmospheric forcing of the surface across an ABL front can be as significant as the changes induced directly by the ice edge.

The MIZs are regions where changes in the wind stress field due to changes in surface roughness have been the basis for theoretical studies of various ocean phenomena. The roughness can be parameterized as a function of ice type and concentration using Table 3. Although roughness is important, the surface wind speed is the most variable and crucial parameter determining the wind stress. Observations and modeling results show that large changes in wind speed, sometimes over short distances, can be more important than roughness changes in affecting variations in wind stress in MIZs. A first-order approximation of wind speed variability to use for ocean and ice MIZ modeling purposes would be to assume the change in wind speed across the MIZ is proportional to the median wind speed variations in Figure 13.

Some of the world's most intense surface turbulent heat fluxes occur in the open ocean regions near MIZs and in leads inside of the ice edge. But values estimated from the bulk method using our open ocean measurements are substantially lower than bulk method estimates from a previous study; the reason for the discrepancy is not clear. In the summer, absorption of shortwave radiation increases in MIZ regions as the open water coverage increases, creating a positive feedback loop. However, the feedback is limited because much of the solar radiation penetrates to a location where the heat melts the bottoms of the flows, instead of the sides.

We examined both how ice and ocean features affect the atmosphere and vice-versa in and near MIZs. Large atmospheric features such as synoptic-scale cyclones and the main polar baroclinic zone have a strong effect on the ice and ocean physical, biological and chemical processes, but these processes do not (at least on short time scales) feed back and significantly affect the large scale atmospheric circulations. Mesoscale atmospheric features, on the other hand, do seem to be at least somewhat affected by the surface MIZ features, as evidenced by the multitude of mesoscale cyclones observed near MIZs and the association of some kinds of polar lows with ABL fronts generated at the ice edge. These mesoscale cyclones, in turn, can strongly affect the ice and ocean. The formation of polar lows requires some kind of instability mechanism in addition to the ABL baroclinicity produced by the ice edge.

The atmospheric heat and momentum forcing in MIZs is affected by both local surface conditions and advection from other regions. Large changes in ABL conditions occur as a result of the surface fluxes, and these ABL modifications feed back to strongly affect the fluxes. For example, even with constant background geostrophic forcing, substantial changes in the surface wind vector, and therefore wind stress, will occur due to changes in ABL height, stability and other atmospheric features. Another example of an important ABL feedback is the relation between ABL depth and horizontal temperature adjustment length scale; surface heat flux affects ABL depth and vice-versa.

Because of the significance of the ABL-induced variations on the surface fluxes in MIZ regions, physical models of ice and ocean processes should include ABL feedback effects in order to have valid simulations of processes involving air-sea interactions. In order to begin to understand the complicated geophysical, chemical and biological processes which occur in MIZs, we must treat the atmosphere, or at least the lowest part, as an integral part of the atmosphere-ice-ocean system. But there are currently no three-dimensional, or even two-dimensional, coupled atmosphere-ice-ocean models of MIZ regions reported in the literature. The ocean response to atmospheric forcing takes so much longer than the scale of synoptic and mesoscale atmospheric changes that a large amount of computer time is required to resolve the important physics in both the atmosphere and ocean. With the advance of computer speed, complete physics models of the MIZs should soon be able to simulate some of the important air-ice-sea feedbacks. But empirical data based on real world observations must play a crucial role in any operational predictive scheme for the atmosphere or ocean in MIZs, much as they do today for local weather forecasts in populated areas. This paper represents an effort at determining the statistics of atmospheric features and forcings in MIZ regions.

List Of Acronyms And Symbols

ABL	Atmospheric Boundary Layer
ASII	Air-Sea Interaction Instability
BSMIZ	Barents Sea Marginal Ice Zone
CEAREX	Coordinated Eastern Arctic Region Experiment
CISK	Conditional Instability of the Second Kind
C_{DN}	Neutral Drag Coefficient
C_H	Heat Transfer Coefficient
C_M	Moisture Transfer Coefficient
D_T	ABL Temperature Adjustment Length Scale
ERS-1	European Research Satellite-1
FSMIZ	Fram Strait/Greenland Sea Marginal Ice Zone
h	ABL Depth
L	Obukhov Length Scale
L_N	ABL Depth Scaling Parameter

MIZ	Marginal Ice Zone
MIZCLM	The Authors' MIZ "Climatology" Data Set
MIZEX84	1984 Summer Marginal Ice Zone Experiment
MIZEX87	1987 Spring Marginal Ice Zone Experiment
N	Brunt-Vaisala Frequency
NSCT	Norwegian Sea Cyclone Track
PVA	Positive Vorticity Advection
SAR	Synthetic Aperture Radar
SB	Serreze and Barry [1988]
T_a	Average ABL Air Temperature
T_i	Surface Interfacial Temperature Over Ice Floes
T_{init}	Initial Upwind T_a
TKE	Turbulent Kinetic Energy
T_s	Surface Interfacial Temperature
$u*$	Friction Velocity
U_s	Wind Speed 10 m Above the Surface
UT	Universal Time
WD	Wind Direction with Respect to Ice Edge Orientation
WD_a	Average ABL WD
WD_g	Geostrophic WD at the Top of the ABL
WD_s	WD 10 m Above the Surface
X	Distance from Ice Edge (Negative Iceward of Edge)
X_d	Downwind Distance from Step Change in T_s
Z_i	Height of Inversion Base
Z_{max}	Height Where Potential Air Temperature Equals T_s

Acknowledgements. The support of the sponsors, the Naval Research Laboratory (Program element 0601153N) and the Office of Naval Research Arctic Program 1125-AR (T. Curtin, G. Geerneart) is gratefully acknowledged. We thank the many individuals involved in the logistics and data collection efforts used for the results presented here.

References

Anderson, R. J., Wind stress measurements over rough ice during the 1984 marginal ice zone experiment, *J. Geophys. Res.*, *92*, 6933-6942, 1987.

Andreas, E. L., Heat and moisture advection over Antarctic sea ice, *Mon. Wea. Rev.*, *113*, 736-746, 1985.

Andreas, E. L., A theory for the scalar roughness and the scalar transfer coefficients over snow and sea ice, *Boundary-Layer Meteorol.*, *38*, 159-184, 1987.

Andreas, E. L., Estimating turbulent surface heat fluxes over polar, marine surfaces, *Proceedings Second AMS Conference on Polar Meteorology and Oceanography*, Madison, Wisconsin, March 29-31, 65-68, 1988.

Andreas, E. L., W. B. Tucker III, and S. F. Ackley, Atmospheric boundary-layer modification, drag coefficient, and surface heat flux in the Antarctic marginal ice zone, *J. Geophys. Res.*, *89*, 649-661, 1984.

Andreas, E. L., M. A. Lange, S. F. Ackley, and P.Wadhams, Roughness of Weddell Sea ice and estimates of the air-ice drag coefficient, *J. Geophys. Res.*, *98*, 12,439-12,452, 1993.

Barry, R. G., Aspects of the meteorology of the seasonal sea ice zone, in *The Geophysics of Sea Ice*, edited by N. Untersteiner, pp. 993-1020, Plenum, New York, 1986.

Belmont, A. D., Lower tropospheric inversions at ice island T-3, *J. Atmos. Terr. Phys.*, Special Supplement: Proceedings of the Polar Atmosphere Symposium, Part I, 215-284, 1957.

Bennett, T. J., and K. Hunkins, Atmospheric boundary layer modification in the marginal ice zone, *J. Geophys. Res.*, *91*, 13,033-13,044, 1986.

Bromwich, D. H., Mesoscale cyclogenesis over the Ross Sea linked to strong katabatic winds, *Mon. Wea. Rev.*, *119*, 1736-1752, 1991.

Bromwich, D. H., J. F. Carrasco, and C. R. Stearns, Satellite observations of katabatic wind propagation for great distances across the Ross ice shelf, *Mon Wea. Rev.*, 120, 1940-1949, 1992a

Bromwich, D. H., J. F. Carrasco, and Z. Liu, Katabatic surges across the Ross Ice Shelf, Antarctica: atmospheric circulation changes and oceanographic impacts, *Preprints Third Conference on Polar Meteorology and Oceanography*, Portland Oregon, September 29 - October 2, 29-32, 1992b.

Brown, R. A., Meteorology, in *Polar Oceanography, Part A, Physical Science*, edited by W. O. Smith, Jr., Academic Press, 1-46, 1990.

Businger, S., Arctic hurricanes, *Am. Scientist*, *79*, 18-33, 1991.

Businger, S., and R. J. Reed, Polar lows, in *Polar and Arctic Lows*, edited by P.F. Twitchell, E. A. Rasmussen and K. L. Davidson, A. Deepak, Hampton, VA, 3-45, 1989.

Carleton, A. M., Synoptic cryosphere-atmosphere interactions in the northern hemisphere from DMSP image analysis, *Int. J. Remote Sensing*, *6*, 239-261, 1985.

Claussen, M., Local advection processes in the surface layer of the marginal ice zone, *Boundary-Layer Meteorol.*, *54*, 1-27, 1991.

Craig, G., and H. Cho, Baroclinic instability and CISK as the driving mechanisms for polar lows and comma clouds, in *Polar and Arctic Lows*, edited by P. F. Twitchell, E. A. Rasmussen and K. L. Davidson, A. Deepak, Hampton, VA, 131-140, 1989.

Ebert, E. E. and J. A. Curry, A intermediate one-dimensional thermodynamic sea ice model for investigating ice-atmosphere interactions, *J. Geophys. Res.*, *98*, 10,085-10,109, 1993.

Emanuel, K. A., An air-sea interaction theory for tropical cyclones, Part I: Steady-state maintenance, *J. Atmos Sci.*, *43*, 585-604, 1986.

Emanuel, K. A., and R. Rotunno, Polar lows as arctic hurricanes, *Tellus*, *41A*, 1-17, 1989.

Fairall, C. W., and R. Markson, Mesoscale variations in surface stress, heat fluxes, and the drag coefficient during the 1983 marginal ice zone experiment, *J. Geophys. Res.*, *92*, 6921-6932, 1987.

Fett, R. W., Polar low development associated with boundary layer fronts in the Greenland, Norwegian and Barents seas, in *Polar and Arctic Lows*, edited by P. F. Twitchell, E. A. Rasmussen and K. L. Davidson, A. Deepak, Hampton, VA, 313-322, 1989a.

Fett, R. W., *Navy Tactical Application Guide, Vol 8, Part 1, Arctic, Greenland/Norwegian/Barents Seas, Weather Analysis and Forecast Applications*, Technical Report 89-07, Naval Environmental Prediction Research Facility (currently Naval Research Laboratory) Monterey, CA, 364 pp., 1989b.

Fett, R. W., *Navy Tactical Application Guide, Vol 8, Part 2, Arctic, East Siberian/Chukchi/Beaufort Seas, Weather Analysis and Forecast Applications*, NRL/PU/7541-92-0005, Naval Research Laboratory, Marine Meteorology Division, Monterey, CA, 388 pp., 1992.

Fiedler, F., and H. A. Panofsky, The geostrophic drag coefficient and the effective roughness length, *Q. J. R. Meteorol. Soc.*, *98*, 213-220, 1972.

Francis, J. A., T. P. Ackerman, K. B. Katsaros, R. J. Lind, and K. L. Davidson, A comparison of radiation budgets in the Fram Strait summer marginal ice zone, *J. Clim.*, *4*, 218-235, 1991.

Frederickson, P. A., P. S. Guest and K. L. Davidson, CEAREX/Haakon Mosby meteorology atlas, *Technical Note 82*, Naval Oceanographic and Atmospheric Research Laboratory (Currently Naval Research Laboratory) Monterey, CA, December, 62 pp., 1990.

Glendening, J., Dependence of boundary-layer structure near an ice-edge coastal front upon geostrophic wind direction, *J. Geophys. Res.*, *99*, 5569-5581, 1994.

Gorshkov, S. G., *World Ocean Atlas, Vol. 3 Arctic Ocean*, Pergamon Press, 184 pp., 1983.

Guest, P. G., A numerical, analytical and observational study of the effect of clouds on surface wind and wind stress during the central Arctic winter, Ph.D. Dissertation, Naval Postgraduate School, 187 pp., 1992.

Guest, P. S., and K. L. Davidson, The effect of observed ice conditions on the drag coefficient in the summer East Greenland Sea marginal ice zone, *J. Geophys. Res.*, *92*, 6943-6954, 1987.

Guest, P. S. and K. L. Davidson, MIZEX 87 meteorology atlas, *Naval Postgraduate School Technical Report, NPS-63-88-004*, Naval Postgraduate School, Monterey, CA, February, 137 pp., 1988.

Guest, P. S., and K. L. Davidson, CEAREX/"O" and "A" camp meteorology atlas, *Naval Postgraduate School Technical Report, NPS-63-89-007*, Naval Postgraduate School, Monterey, CA, September, 70 pp., 1989.

Guest, P. S., and K. L. Davidson, Meteorological triggers for deep convection in the Greenland Sea, in *Deep Convection and Deep Water Formation in the Oceans, Proceedings of the International Monterey Colloquium on Deep Convection and Deep Water Formation in the Oceans,* edited by P.C. Chu and J.C. Gascard, Elsevier Oceanography Series, *57*, 369-375, 1991a.

Guest, P. S. and K. L. Davidson, The aerodynamic roughness of different types of sea ice, *J. Geophys. Res.*, 96, 4709-4721, 1991b.

Guest, P. S., and K. L. Davidson, A study of the factors controlling the value of the near-surface air temperature over sea ice, *Preprints Third Conference on Polar Meteorology and Oceanography, Portland Oregon, September 29 - October 2*, available from the American Meteorological Society, Boston MA, p. 54, 1992.

Hakkinen, S., Coupled ice-ocean dynamics in the marginal ice zones: upwelling/downwelling and eddy generation, *J. Geophys. Res.*, *91*, 819-832, 1986a.

Hakkinen, S., Ice banding as a response of the coupled ice-ocean system to temporally varying winds, *J. Geophys. Res.*, *91*, 5047-5053, 1986b.

Hakkinen, S., A coupled dynamic-thermodynamic model of an ice-ocean system in the marginal ice zone, *J. Geophys. Res.*, *92*, 9469-9478, 1987.

Hakkinen, S., and D. Cavalieri, A study of oceanic surface heat fluxes in the Greenland, Norwegian, and Barents Seas, *J. Geophys. Res.*, *94*, 6145-6157, 1989.

Hakkinen, S., G. L. Mellor and L. H. Kantha, Modeling deep convection in the Greenland Sea, *J. Geophys. Res.*, *97*, 5389-5408, 1992.

Hamilton, S. W., Meteorological features during Phase I of the Coordinated Eastern Arctic Experiment (CEAREX) from 17 September to 7 January 1989, Master's Thesis, Naval Postgraduate School, Monterey, CA, 98 pp., 1991.

Hanssen-Bauer, I., and Y. T. Gjessing, Observations and model calculations of aerodynamic drag on sea ice in the Fram Strait, *Tellus, 40A*, 151-161, 1988.

Ikeda, M., Wind-induced mesoscale features in a coupled ice-ocean system, *J. Geophys. Res.*, *96*, 4623-4629, 1991.

Kantha, L. H. and G. L. Mellor, A numerical model of the atmospheric boundary layer over a marginal ice zone, *J. Geophys. Res.*, *95*, 4959-4970, 1989a.

Kantha, L. H., and G. L. Mellor, A two-dimensional coupled ice-ocean model of the Bering Sea marginal ice zone, *J. Geophys. Res*, *94*, 10,921-10,935, 1989b.

Keegan, T. J., Arctic synoptic activity in winter, *J. Meteor.*, *15*, 513-521, 1958.

Kellner, G., C. Wamser, and R. A. Brown, An observation of the planetary boundary layer in the marginal ice zone, *J. Geophys. Res.*, *92*, 6955-6965, 1987.

Kitaigorodskii, S. A., A note on similarity theory for atmospheric boundary layers in the presence of background stable stratification, *Tellus*, *40A*, 434-438, 1988.

Kitaigorodskii, S. A., and S. M. Joffre, Simple scaling for the height of the stratified atmospheric boundary layer, *Tellus*, *40A*, 419-433, 1988.

Lackmann, G. M., P. S. Guest, K. L. Davidson, R. J. Lind, and J. Gonzalez, CEAREX/Polarbjoern meteorology atlas, *Naval Postgraduate School Technical Report, NPS-63-89-005*, Naval Postgraduate School, Monterey, CA, 550 pp., 1989.

Lindsay, R. W., *MIZEX 84 Integrated Surface Meteorological Data Set and Meteorological Atlas, Second Edition*, Polar Science Center, Univ. of Washington, November 1, 128 pp., 1985.

Lindsay, R. W., and A. L. Comiskey, Surface and upper-air observations in the eastern Bering Sea, February and March, 1981, *Tech. Memo., ERL-PMEL-35*, NOAA, Washington D.C., 90 pp., 1982.

Martinson, D. G., and C. Wamser, Ice drift and momentum exchange in winter antarctic pack ice, *J. Geophys. Res.*, *95*, 1741-1755, 1990.

Maykut, G. A., Large-scale heat exchange and ice production in the central Arctic, *J. Geophys. Res.*, *87*, 7971-7984, 1982.

Maykut, G. A., and D. K. Perovich, MIZEX 84 heat and mass balance data, *Scientific Report*, Applied Physics Lab., University of Washington, Seattle, WA, 73 pp., 1985.

Maykut, G. A., and D. K. Perovich, The role of shortwave radiation in the summer decay of sea ice, *Geophys. Res.*, *92*, 7032-7044, 1987.

McPhee, M. G., G. A. Maykut and J. H. Morison, Dynamics and thermodynamics of the ice/upper ocean system in the marginal ice zone of the Greenland Sea, *J. Geophys. Res.*, *92*, 7016-7031, 1987.

Nakamura, N., and A. H. Oort, Atmospheric heat budgets of the polar regions, *J. Geophys. Res*, *93*, 9510-9524, 1988.

Overland, J. E., Atmospheric boundary layer structure and drag coefficients over sea ice, *J. Geophys. Res.*, *90*, 9029-9049, 1985.

Overland, J. E., and K. L. Davidson, Geostrophic drag coefficients over sea ice, *Tellus*, *44A*, 54-66, 1992.

Overland, J. E., and P. S. Guest, The Arctic snow and air temperature budget over sea ice during winter, *J. Geophys. Res.*, *96*, 4651-4662, 1991.

Overland, J. E., M. Reynolds, and C. Pease, A model of the atmospheric planetary boundary layer over the marginal ice zone, *J. Geophys. Res*, *88*, 2836-2840, 1983.

Rasmussen, E. A., The polar low as an extratropical CISK disturbance, *Quart. J. Roy. Meteor. Soc.*, *105*, 531-549, 1979.

Rasmussen, E. A., A comparative study of tropical cyclones and polar lows, in *Polar and Arctic Lows*, edited by P. F. Twitchell, E. A. Rasmussen and K. L. Davidson, A. Deepak, Hampton, VA, 47-80, 1989.

Rasmussen, E. A., *On mesoscale disturbances in arctic regions*, Polar Low Workshop, Hvanneyri, Iceland, June 23-26, 1992a.

Rasmussen, E. A., *On the definition and classification of polar lows*, Polar Low Workshop, Hvanneyri, Iceland, June 23-26, 1992b.

Reed, R. J., and B. A. Kunkel, The Arctic circulation in summer, *J. Meteor.*, *17*, 489-506, 1960.

Reynolds, M., On the local meteorology at the marginal ice zone of the Bering Sea, *J. Geophys. Res*, *89*, 6515-6524, 1984.

Roed, L., Sensitivity studies with a coupled ice-ocean model of the marginal ice zone, *J. Geophys. Res.*, *88*, 6039-6042, 1983.

Roed, L., A thermodynamic coupled ice-ocean model of the marginal ice zone, *J. Phys. Oceanogr.*, *14*, 1921-1929, 1984.

Roed, L., and J. O'Brien, A coupled ice-ocean model of upwelling in the marginal ice zone, *J. Geophys. Res*, *88*, 2863-2872, 1983.

Roeloffzen J. C., W. D. Van den Berg, and J. Oerlemans, Frictional convergence at coastlines, *Tellus*, *38A*, 397-411, 1986.

Rutherford, S. J., Arctic cyclones and marginal ice zone (MIZ) variability, Master's Thesis, Naval Postgraduate School, Monterey, CA, 90 pp., 1993.

Sater, J. E., A. G. Ronhovde, and L. C. Van Allen, *Arctic Environment and Resources*, The Arctic Institute of North America, Washington DC, 310 pp., 1971.

Schultz, R. R., Meteorological Features During the Marginal Ice Zone Experiment from 20 March to 10 April 1987, Master's Thesis, Naval Postgraduate School, Monterey, CA, 346 pp., 1987.

Sechrist, F. S., R. W. Fett, and D. C. Perryman, *Forecasters handbook for the Arctic*, Technical Report TR 89-12, Naval Environmental Prediction Research Facility (currently Naval Research Laboratory), Monterey, CA, 364 pp., 1989.

Serreze, M. C., and R. G. Barry, Synoptic activity in the arctic basin, 1979-85, *J. Clim.*, *1*, 1276-1295, 1988.

Serreze, M. C., J. D. Kahl, and R. C. Schnell, Low-level temperature inversions of the Eurasian Arctic and comparisons with Soviet drifting station data, *J. Clim.*, *5*, 9411-9422, 1992.

Shapiro, M. A. and L. S. Fedor, A case study of an ice-edge boundary layer front and polar low development over the Norwegian and Barents Seas, in *Polar and Arctic Lows,* edited by P. F. Twitchell, E. A. Rasmussen and K. L. Davidson, A. Deepak, Hampton, VA, 257-258, 1989.

Shapiro, M. A., T. Hampel, and L. S. Fedor, Research aircraft observations of an arctic front over the Barents sea, in *Polar and Arctic Lows,* edited by P. F. Twitchell, E. A. Rasmussen and K. L. Davidson, A. Deepak, Hampton, VA, 279-289, 1989.

Shaw, W. J., R. L. Pauley, T. M. Gobel, and L. F. Radke, A case study of atmospheric boundary layer mean structure for flow parallel to the ice edge: aircraft observations from CEAREX, *J. Geophys. Res.*, *96*, 4691-4708, 1991.

Smedstad, O. M. and L. P. Roed, A coupled ice-ocean model of ice break-up and banding in the marginal ice zone, *J. Geophys. Res*, *90*, 876-882, 1985.

Smith, D. C.,IV, A. A. Bird, and W. P. Budgell, A numerical study of mesoscale ocean eddy interaction with a marginal ice zone, *J. Geophys. Res*, *93*, 12,461-12,473, 1988.

Smith, S. D., Coefficients of sea surface wind stress, heat flux, and wind profiles as a function of wind speed and temperature, *J. Geophys. Res*, *93*, 15,467-15,472, 1988.

Stringer, W. J., D. G. Barnett, and R. H. Godin, *Handbook for sea ice analysis, NEPRF CR 84-03*, Naval Environmental Prediction Research Facility (currently Naval Research Laboratory), Monterey, CA, 1984.

Stull, R. B., *An Introduction to Boundary Layer Meteorology*, Kluwer Academic Publishers, 666 pp., 1988.

Thompson, W. T. and S. D. Burk, An investigation of an Arctic front with a vertically nested mesoscale model, *Mon. Wea. Rev.*, *119*, 233-261, 1991.

Twitchell, P. F., E. A. Rasmussen and K. L. Davidson, eds., *Polar and Arctic Lows*, A. Deepak, Hampton, VA, 436 pp., 1989.

U.S. Navy, *U.S. Navy Marine Climatic Atlas of the World: Volume VI, NAVWEPS 50-1C-553*, 69-87, 1963.

Vowinckel, E., and S. Orvig, The climate of the north polar basin, *World Survey of Climatology, Vol. 14, Climates of Polar Regions*, edited by S. Orvig, Elsevier, 129-252, 370 pp., 1970.

Walsh, J. W. and Chapman, W. L., Short-term climatic variability of the arctic. *J. Clim.*, *3*, 237-250, 1990.

Warren, S.G., C. J. Hahn, J. London, R. M. Chervin, and R. L. Jenne, Global distribution of total cloud cover and cloud type amounts over the ocean, *NCAR/TN-317 + STR*, National Center for Atmospheric Research, Boulder, CO, 200 pp., 1988.

Wefelmeier, C., and Etling, D., The influence of sea ice distribution on the atmospheric boundary layer, *Z. Meterol.*, *41*, 333-342, 1991.

Weldon, R., *Satellite training course notes, part IV, cloud patterns and the upper air wind field*, Applications Division, National Environmental Satellite Service, NOAA, Washington, DC, 79 pp., 1979.

Wilson, C., Climatology of the cold regions, part a, *Report 1-A3a*, U.S. Army CRREL, Hanover, N.H., 141 pp., 1967.

Wos, K. A., A climatology of polar low occurrences in the Nordic seas and an examination of katabatic winds as a triggering mechanism, Master's Thesis, Naval Postgraduate School, Monterey, CA, 154 pp., 1992.

3

Small-Scale Physical Processes in the Arctic Ocean

Laurie Padman

Abstract

Small-scale physical processes, defined here as those having time and/or space scales that are smaller than the resolution of basin-scale circulation models, are responsible for water mass production and modification. Several processes are known to be relevant in the Arctic Ocean, both in the central basin and in the extensive marginal shelf seas. In this chapter I discuss the relevance of small-scale processes to attempts to model the Arctic air/sea/ice system and it's relationship to global ocean/atmosphere models, summarize what is presently known about these processes and their spatial variability, and suggest areas of further research.

Introduction

This chapter is intended to complement reviews by Carmack [1990] (large-scale physical processes), Muench [1990] (mesoscale phenomena), Häkkinen [1990] (modeling), and McPhee [1990] (small-scale processes), which all appeared in the volume, "*Polar Oceanography*." Recent progress in modeling the Arctic sea-ice climate system, particularly the characteristics, dynamics and thermodynamics of the sea ice, has been summarized by Barry et al. [1993].

I will first review the principal features of the Arctic Ocean. I then discuss some basic concepts of mixing, and summarize small-scale processes that are relevant to the Arctic. A discussion of the spatial variability of mixing rates follows. Many of the critical links between small and large-scale processes are still very poorly understood, and there is

Arctic Oceanography: Marginal Ice Zones and Continental Shelves
Coastal and Estuarine Studies, Volume 49, Pages 97–129
Copyright 1995 by the American Geophysical Union

considerable scope for differences of opinion concerning which processes are most relevant. The reader should therefore keep in mind that this review reflects the author's own assessment of physical processes that are most worthy of study, rather than suggesting a consensus among the Arctic research community. I reiterate that the review is intended to be complementary to those cited above: the reader should look to the references to provide a detailed understanding of the complex Arctic air/sea/ice (ASI) system.

Background

There is a growing realization that the Arctic can respond rapidly to changes in atmospheric conditions, and also play an important role in regional and global climate [ARCUS report, 1993]. Numerical and simple analytical models predict that the stability of the Arctic ASI system is very sensitive to small perturbations in the atmosphere and ocean properties. The relatively weak density stratification in many polar regions, particularly the Greenland Sea, implies that deep, convectively-driven mixing can be initiated by small changes in atmospheric or oceanic conditions. Deep convection may play a significant role in long time-scale (decades to centuries) modeled climate variability [Mysak et al., 1990; Häkkinen, 1993].

One key ingredient in model sensitivity is the sea-ice distribution: the recent review by Barry et al. [1993] discusses the relevant characteristics of sea-ice dynamics, thermodynamics, and their inclusion in climate models. Sea-ice effectively insulates the Arctic surface waters from the cold polar atmosphere. Multi-year ice, which is typically 2-6 m thick [Bourke and Garrett, 1987], can reduce the heat flux in winter by two orders of magnitude compared with open water. Ice has a feedback role in modifying the radiative transfers at the air/ocean interface because of the significantly higher albedo of ice and snow relative to open water. A coupled ocean/atmosphere GCM is therefore intimately affected by its ability to adequately model the ice response to the ocean, atmosphere, and radiation balance.

Ice concentration and thickness is affected by the upward flux of oceanic heat, as well as by stress-driven ice advection and divergence. Throughout most of the Arctic, the upper-ocean hydrography resembles Figure 1 (EUBEX station 308: see Perkin and Lewis [1984]). A thin, well-mixed surface layer, near the freezing point for the local salinity, overlies a relatively warm ($T > 0°C$) and saline ($S > 34.4$ psu) layer of Atlantic Intermediate Water (AIW) [Coachman and Aagaard, 1974; Bourke et al., 1988]. A comprehensive water mass classification is provided in Figure 4.11 of Carmack [1990]. Away from the West Spitsbergen Current (WSC), which is the original source of AIW, the surface layer is usually separated from the AIW by a cold halocline [Coachman and Aagaard, 1974; Aagaard et al., 1981]. Local variations occur due to proximity to sources of shelf-modified water, river influx, and the oceanic inflows through the Bering and Fram Straits [Coachman and Aagaard, 1974]. Below the AIW, there is a cold, weakly-stratified layer (Deep Water) that extends to the seabed. Deep circulation and the production of Deep Water are discussed by Aagaard [1981], Aagaard et al. [1985], and Carmack [1990].

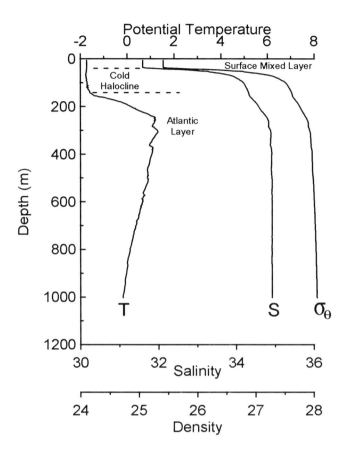

Figure 1. Representative upper-ocean vertical profiles of potential temperature (*T*), salinity (*S*) and potential density (σ_θ) in the Arctic Basin, cast 308 (84.485°N, 17.212°E) from *EUBEX* (Perkin and Lewis, 1984). The surface mixed layer, cold halocline, and Atlantic layer are indicated.

Over the central basin, the density stratification of the cold halocline provides an effective barrier to the upward flux of heat from the AIW towards the ice base, since entrainment of halocline water by surface stress or buoyancy fluxes involves very little upward heat transport. The halocline cannot simply be a mixture of local surface water and AIW, but instead must be maintained by lateral advection of water formed on the adjoining continental shelves. At least three possible formation mechanisms exist: salinization of cold surface water during ice formation [Aagaard et al., 1981]; mixing of upwelled AIW with cold, fresh summer shelf water [Aagaard et al., 1981]; and melting of ice in the MIZ as the Atlantic Water first enters the ice-covered Arctic basin as a surface current [Steele et al., 1994]. The shelf seas, which account for about 1/3 of the total area of the Arctic, undergo a pronounced seasonal cycle. In summer they are largely ice-free [Parkinson,

1991], and have low salinity due to a combination of riverine inputs and melting of the previous winter's sea ice [Coachman and Aagaard, 1974; Hanzlick and Aagaard, 1980; Carmack, 1990]. In fall as the air temperatures drop, rapid freezing occurs with consequent re-salinization of the shelf water through brine rejection. Salinization rates drop rapidly as the ice thickness increases. In regions where the newly-created ice is exported off the shelves by wind stress divergence, however, the salinity can continue to increase during the water's residence time on the shelf.

The second method of halocline formation, mixing of upwelled AIW with cold, fresh shelf water, was suggested by Coachman and Barnes [1962] and has been observed by Hufford [1974] and Aagaard [1977] on the Beaufort Sea shelf. Garrison and Becker [1976] and Mountain et al. [1976] document upwelling of AIW in Barrow Canyon, which appears to act as both a drain for shelf-modified water off the shelf and a preferred route for AIW injection onto the shelf. Hanzlick and Aagaard [1980] found upwelled AIW on the northern slope of the Kara Sea, particularly in the St. Anna Trough. They noted that a region extending northeast from the northern tip of Novaya Zemlya is often the last to freeze and has a small mean ice thickness, implicating the upwelling of AIW in determining local sea ice conditions.

Formation of halocline water by melting at the MIZ as Atlantic Water first enters the ice-covered central Arctic has been investigated by Steele et al. [1994]. Melting of sea-ice freshens and rapidly cools the Atlantic Water. Moore and Wallace [1988] note that the large latent heat of fusion for ice can be modeled as a mixing end-member with a temperature of about -64°C for a bulk sea-ice salinity of about 7 psu. Hence, near-freezing halocline water can be produced by incorporation of a relatively small amount of melted ice into the Atlantic Water at the MIZ.

Neither the total rate of halocline water formation, nor the rates due to each of the above processes, are well known. Aagaard et al. [1981] estimated total mean production of halocline water at between 2.5 and 5 Sverdrups (Sv). Martin and Cavalieri [1989] suggested that freezing under the eastern Arctic coastal polynyas was responsible for about 1.2 Sv, with a later estimate for the entire Arctic polynya system being about 2 Sv (see Steele et al. [1994]). Steele et al. estimated that ice melting at the MIZ contributed about 0.4 Sv to the total, which is small in a basin-averaged sense but may be significant near the Barents Sea. If, as seems probable, shelf formation of halocline water is indeed critical to the overall stability of the ASI system, understanding the various processes that influence this production balance is essential. The present level of uncertainty is illustrated by a comparison of analyses of the Arctic response to a proposed diversion of major Siberian rivers away from the Arctic for agricultural use. Much of the fresh-water input to the Arctic Ocean occurs through summer river runoff into the shelf seas. Of a total riverine freshwater influx of about 3500 km^3yr^{-1}, the three major Russian rivers (the Yenisei, Ob, and Lena) contribute a total of about 1650 km^3yr^{-1} [Treshnikov, 1985]. The Bering Strait inflow of low-salinity North Pacific water is equivalent to a freshwater flux of about 1700 km^3yr^{-1} [Aagaard and Coachman, 1975; Aagaard and Carmack, 1989]. Aagaard and Coachman [1975] suggested that the diversion of the Siberian rivers could produce, in the absence of feedback mechanisms, a permanently ice-free Arctic Ocean by reducing the production of intermediate-density, cold halocline water over the shelves to the point where the surface layer was no longer buffered from the warm AIW. Once this state was reached, the upward oceanic heat flux could prevent re-formation of the insulating ice cover. This is similar in spirit to Gordon's [1991] proposal that the deep-convecting Weddell polynya of the mid-1970s represents a second stable equilibrium state for the Southern Ocean. Rudels [1989] argued, however, that reduced (even zero) river input into the Arctic could be compensated by the increased importance of the low-salinity inflow from the North Pacific through Bering Strait. Yet another possibility is that the seasonal freeze/melt cycle over the central basin could maintain a stable, low-salinity surface layer, independent of external fresh water sources [Aagaard et al., 1981]. In this scenario, the upward heat flux to the ice actually *stabilizes* the ice cover by melting sufficient ice to maintain the low salinity of the surface layer.

The effect of modifying the distribution of fresh water within the Arctic extends well beyond the central basin. As Aagaard and Carmack [1989] noted, the amount of ice and reduced-salinity water exported as the East Greenland Current into the Greenland and Iceland Seas can be critical to the probability of deep convection occurring in these regions. The same paper proposed that the North Atlantic salinity anomaly of the 1960s and 1970s was due to increased fresh water discharge from the Arctic Ocean. Häkkinen [1993] also found this response in a GCM forced by observed daily winds for the period 1955-1975. Manak and Mysak [1989] and Mysak et al. [1990] found evidence in their numerical model for a positive feedback between Arctic sea ice cover and fluctuations in northern Canadian atmospheric conditions (see also: Walsh and Chapman [1990]; Yang and Neelin [1993]). In essence, a positive ice concentration anomaly in the western Arctic is advected through Fram Strait into the Greenland Sea, where it is seen upon melting as a negative surface salinity anomaly that stabilizes the vertical density profile. The likelihood of deep convection and therefore the amount of oceanic heat made available to the atmosphere is therefore reduced. Finally, the changed atmospheric conditions over the Greenland Sea affect the atmospheric circulation over northern Canada, creating the necessary conditions for the next western Arctic ice concentration anomaly. The feedback loop imposes an inter-decadal time scale on both northern hemisphere climate and Arctic ice conditions.

As the above discussion indicates, significant progress has been made in our understanding of the ASI system without considering the details of the small-scale processes that will be described below. Methods such as compilation and interpretation of freshwater and heat budgets provide a relatively clear image of how the ASI system, as a steady-state system, works. However, as the various river-diversion scenarios discussed above indicate, there are areas of significant uncertainty which could be clarified by understanding the actual mechanisms for water mass modification. Assessing the future state of the Arctic ASI system, as a component of an evolving global ocean/atmosphere system, will require an understanding of the processes that actually bring about conversions of one water type to another, and those that determine the upward flux of oceanic heat to the sea ice and atmosphere. Ultimately, water mass modification occurs at the small scales at which molecular diffusivities of heat and salt become important

($O(1)$) cm). Linking these scales and the those that can presently be modeled are the small-scale physical processes that are the focus of this chapter.

Basic Concepts of Mixing

Mixing processes that are relevant to the Arctic processes discussed above can be divided, approximately, into two categories: (i) those that occur due to direct contact with an oceanic boundary; and (ii) those that occur in the pycnocline away from the direct influence of boundaries (see Figure 2). Category (i) applied to the Arctic includes buoyancy flux at the surface due to salt rejection during freezing, and radiative transfers of heat. The category also includes surface-layer turbulence driven by stress at the air/sea or ice/sea interface, and friction at the seabed. Category (ii) includes the pycnocline mixing processes of internal wave (IW) instabilities and double-diffusion. The relative importance of each process varies throughout the Arctic. For example, under multi-year ice the surface heat and salt fluxes are typically negligible, although regionally-averaged surface fluxes might still be significant due to the ubiquitous presence of leads [Morison et al., 1992, 1993; Martinson, 1990]. In regions near sources of internal waves (see Padman and Dillon [1991] and D'Asaro and Morison [1992]), heat from the AIW can be actively mixed upwards through the halocline and into the surface mixed layer. In the central basins, however, diapycnal fluxes through the pycnocline may be dominated by double-diffusion [Padman and Dillon, 1987]. Over the shelf during the fall freeze and in regions of strongly divergent ice motion, surface buoyancy fluxes might easily dominate

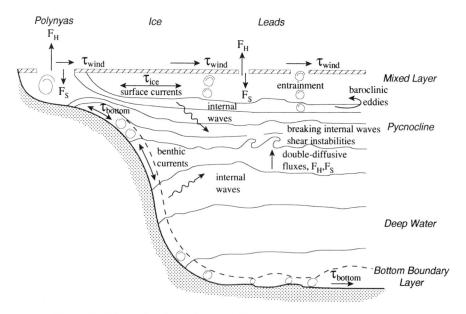

Figure 2. Schematic of significant mixing processes in the Arctic Ocean.

the rates of water mass conversion [Martin and Cavalieri, 1989]. In shallow water, strong currents due to either wind forcing or tides (see below), can create enough mixing through benthic friction to maintain a well-mixed water column.

Despite the large-scale ramifications of mixing processes, the actual transfer of fluid properties from one water parcel to another occurs by molecular diffusion across spatial property gradients on scales of order 0.01 m. Because of this mismatch between scales, oceanographers have adopted the convention of "effective diffusivity", which mimics the Fickian processes responsible for molecular diffusivity but can be applied to much larger spatial scales. An effective diffusivity K_r for property C in direction r is defined by

$$F_r = K_r \langle \partial C / \partial r \rangle \qquad (1)$$

where F_r is the resultant flux and the angle brackets denote some spatial and temporal averaging scheme. One can think of K_r as a bulk parameterization of all the processes within the averaging space that ultimately contribute to the *molecular* diffusion within that space. In the pycnocline, the processes that we hope to parameterize in this way include shear and advective instabilities associated with internal gravity wave motions, and double-diffusive convection [Gregg, 1987]. At boundaries we are interested in processes such as vertical transport by stress-driven turbulence and buoyancy-driven convection [McPhee, 1990; Martinson, 1990].

Effective diffusivities vary significantly both in time and space. The lower limits are the molecular values found in non-turbulent (laminar) flows, about 10^{-6}, 10^{-7} and 10^{-9} m^2s^{-1} for momentum, heat and salinity respectively. Within the main mid-latitude thermocline, a typical value of vertical diffusivity K_v due to instabilities in the internal wave field is about 10^{-5} m^2s^{-1} [Gregg, 1989; Ledwell et al., 1993]. That is, turbulence increases the vertical flux of heat by about two orders of magnitude over what would occur in a laminar stratified flow, and the flux of salt by about four orders of magnitude. Several measurement programs now show that K_v in the pycnocline can vary significantly [Toole et al., 1994; Wijesekera et al., 1993a]. In the under-ice, stress-driven mixing layer, maximum values of K_v can exceed 10^{-2} m^2s^{-1} [McPhee and Martinson, 1993]. Horizontal diffusivity K_h might be significantly greater than this, $O(1)$ m^2s^{-1} or larger (see Ledwell et al. [1993]), because density gradients no longer constrain the scales of the motions responsible for the mixing.

Despite evidence for spatial variations in K_v and K_h, until recently computational limitations prevented inclusion of prognostic models of turbulence into GCMs. Indeed, K_v and K_h have frequently been specified simply to ensure computational stability, given a particular model's requirements for spatial resolution and time stepping. Recently, however, Häkkinen and Mellor [1992] incorporated a turbulence closure scheme [Mellor and Yamada, 1974, 1982] for K_v into their Arctic GCM, instead of specifying a constant value. This scheme determines diffusivity based on the static and dynamic stability of the resolved velocity and hydrographic fields. To satisfy the model's need for both computational stability and realistic parameterization of sub-gridscale processes, a minimum diffusivity of $10^{-5} m^2 s^{-1}$ was imposed. This model represents the best effort to date to parameterize turbulence within an Arctic GCM. Nevertheless, as will be demonstrated below, there is much more variability in K_v than can be generated by the Mellor-Yamada turbulence closure applied solely to the resolved velocity and hydrographic fields. We are *not* yet able to suggest a greatly improved parameterization of K_v, but our understanding of the small-scale processes that need to be considered has greatly improved in the last decade.

Small-scale Physical Oceanographic Processes

Several recent Arctic oceanographic programs have included the collection of data designed specifically to investigate the mechanisms by which heat, salt, and momentum are transported vertically in the upper ocean (Table 1). Based largely on these studies, the small-scale processes discussed below appear to be the most relevant to modification of the larger-scale oceanic environment.

TABLE 1. Recent Arctic experimental programs emphasizing small-scale oceanic processes.

Project	Location	Time
Marginal Ice Zone Experiment 1983/84(**MIZEX**) [†]	Northern Fram Strait and Yermak Plateau	Spring/Summer 1983 & 1984
Arctic Internal Wave Experiment (**AIWEX**)	Canada Basin (western Arctic)	Spring, 1985
Coordinated Eastern Arctic Experiment (**CEAREX**) "O" Camp	Yermak Plateau (eastern Arctic)	Spring, 1989
Arctic Leads Experiment (**LeadEx**) [‡]	Canada Basin (western Arctic)	Spring, 1992

[†] See *Johannessen* [1987] and associated papers in the same volume of *J. Geophys. Res.* for an overview.

[‡] See Morison et al. [1993] for an overview.

Benthic Stress

The turbulent stress at the seabed, τ_B, due to a benthic current $u_{B,r}$ can be approximated by a quadratic drag law,

$$\tau_B = \rho_w c_{B,r} u_{B,r}^2 \tag{2}$$

where ρ_w is the fluid density and $c_{B,r}$ is the quadratic drag coefficient for the reference height r. Benthic stress is a sink of kinetic energy, and limits the velocities that result, for example, from wind and tidal forcing. It is particularly important in shallow water such as over the shallow Arctic shelf seas. For example, Huthnance [1981] noted that the Arctic tides, which are largely driven by tidal energy flux from the North Atlantic rather than by local astronomical forcing, are significantly attenuated by benthic friction as the incoming wave propagates across the broad, shallow Barents Sea.

The quadratic description of stress in (2) indicates that total stress in a multiple-component flow field cannot be obtained by simple addition of the single-component stresses. An example that is particularly relevant to the eastern Arctic shelf seas is the addition of tides to "mean" currents resulting from wind or ice stress at the upper surface. In several locations, notably near Bear Island and in the southern Barents Sea, the tidal current exceeds 1 ms^{-1} [Huthnance, 1981; Kowalik and Proshutinsky, 1993], which is significantly greater than the mean circulation in these regions [Harms, 1992]. The averaged boundary layer thickness and mixing rates will therefore be set primarily by the oscillatory tidal currents rather than by the mean flow. The mean flow forced by pressure gradients and surface stress then experiences a much higher benthic friction than would otherwise be present. One result of this interaction is the generation of a "residual" mean circulation: Harms [1992] obtained maximum residual currents around Svalbard and Bear Island of about 0.01 ms^{-1} due to the addition of the M_2 tide to the barotropic modeled Barents Sea circulation. Harms further noted that a relationship exists between the residual circulation and the position of the Polar Front, which separates cold, southward-flowing water near Svalbard coast and the northeastward inflow of Atlantic Water.

Near regions of strong tidal motion and sloping bathymetry, the combination of bottom friction and planetary vorticity as the tidal current flows across the slope creates an alongslope mean circulation. This process, called tidal rectification, is described by Loder [1980] and has been applied to the Yermak Plateau region by Padman et al. [1992]. Typically, at the upper edge of the continental slope, a rectified flow of about 15% of the tidal current amplitude can be expected [D. Boyer, pers. comm., 1992]. Given the generally sluggish mean flows in the Arctic, tidal rectification may be important in the dynamics of alongslope advection of Atlantic Water in the eastern Arctic, where the tidal currents are greatest (Figure 3). Padman et al. [1992] note that this mechanism might help explain the presence of the thin filament of Atlantic Water found along the western slope of the Plateau during CEAREX [Muench et al., 1992]. In the shallow water of the western Barents Sea, the effect might be even more important. A recent, extremely high-resolution

(0.75 km) model of this region [Kowalik and Proshutinsky, 1994] indicates that residual circulation around Bear Island due to the presence of both semi-diurnal and diurnal tidal components is about 0.1 ms^{-1}.

Tidal currents also increase the potential for sediment resuspension. The importance of this in recent Arctic research can be shown with one example. As is now known, the

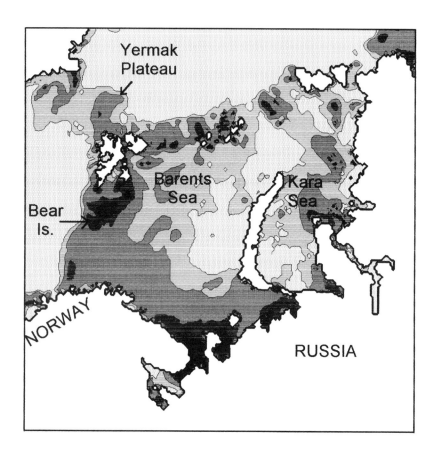

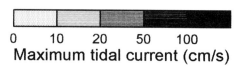

Figure 3. Maximum value of current (cm.s^{-1}) from summation of principal tidal components [Z. Kowalik, pers., comm., 1993].

eastern Barents and Kara Seas around Novaya Zemlya have been used as dumping grounds for radio-active wastes [Molnia, 1993]. In addition, radionuclides are carried into the Kara Sea from contamination sites along the Siberian rivers. Will such contaminants remain locked in the sediments, or be diffused into the central Arctic Basin? Strong tidal currents in these regions can resuspend sediments and the associated contaminants, which can then be transported elsewhere in the Arctic by "mean" currents due to wind forcing and thermohaline circulations.

Surface Stress

In the open ocean, surface stress is applied directly by the wind. As for the benthic layer case above, the stress τ_S can be approximated by a quadratic drag law:

$$\tau_s = \rho_a c_a W_{10}^2 \tag{3}$$

where ρ_a is the air density (about 1.2 kg m^{-3}), c_a is a drag coefficient, and W_{10} is the wind speed at 10 m above the surface. A typical value of c_a is about 0.0015. The drag coefficient depends strongly, however, on both the wind speed and the temperature contrast between the ocean and atmosphere [Smith, 1988]. The transfer of energy from the wind field to the upper ocean is further complicated by the development of secondary circulations in the mixed layer, such as Langmuir cells [Weller et al., 1985], and the possibility for energy transfer into the underlying pycnocline through internal wave generation [Wijesekera and Dillon, 1991]. Nevertheless, (3) is presumably a reasonable estimate of surface stress for open water, for example in the ice-free coastal seas in summer and in polynyas.

Under sea ice, stress at the ice/water interface is related to the ice-relative current, u_{rel}. In the same manner as for benthic and open-water surface stresses, we approximate the ice/water stress τ_S by a quadratic drag law,

$$\tau_s = \rho_w c_{S,r} u_{rel,r}^2 \tag{4}$$

where $c_{S,r}$ is a drag coefficient for a reference height r which is usually taken as 1 m. This drag coefficient varies from about 0.001 for smooth first-year ice [Langleben, 1982] to about 0.006 for multi-year Arctic sea ice [Johannessen, 1970; Langleben, 1980].

When the ice is accelerated by the wind, u_{rel} may simply reflect the wind-forced motion of the ice relative to an almost-stationary water column. This case has been studied in considerable detail by McPhee and Smith [1976] and McPhee [1986, 1990, 1992]. Another case that has been shown to be relevant to the Arctic, however, is the presence of an oceanic current to which the motion of the pack ice is *not* strongly coupled. A clear example of this is tidal flow under land-fast sea ice, such as in the straits of the Canadian Archipelago [Cota et al., 1987]. In this case the mixing due to the ice-relative tidal flow can cause upward transport of nutrients from the pycnocline to the algal community on the ice base [Cota et al., 1987; Cota and Horne, 1989; Conover et al., 1990; Prinsenberg and Bennett, 1987]. Less obvious, however, is the presence of ice-relative tidal flows in the central Arctic Basin, such as those described by Padman et al. [1992] near the Yermak Plateau. Here, u_{rel} arises due to the small horizontal spatial scales of topographically-trapped, tidal-band vorticity waves in a region and season (winter) where the ice cover is consolidated. The ice sheet therefore responds to an area-average of the tidal stress over a region in which the tidal currents actually vary significantly. When the ice is weaker, as in summer and in the MIZ, the ice more easily couples with the local tidal currents. High diurnal ice velocities have been observed over the Yermak Plateau from satellite tracking of surface buoys [Hoffman, 1990; Prazuck, 1991] and the drift of ships positioned in the summer MIZ [McPhee, 1984; Morison et al., 1987]. Other currents with small horizontal scales can also generate ice-relative flows, e.g. submesoscale eddies and internal gravity waves [Morison et al., 1985].

Ice-relative currents can create extremely high surface mixing rates [Padman and Dillon, 1991; McPhee and Martinson, 1993]. Energetic turbulence in the mixed layer increases the entrainment rate, i.e. the rate at which pycnocline fluid is incorporated into the mixed layer. If there is no cold halocline, surface mixing entrains the Atlantic Water, providing a supply of heat to the base of the ice. If the cold halocline is present, the same amount of entrainment (i.e. the same change in total potential energy) simply raises the bulk salinity of the mixed layer, with very little upward flux of heat. Surface-generated turbulence, both under the ice and in the seasonally-open water, determines the mixing of the low-salinity spring ice melt water through the upper ocean, and hence the preconditioning of the upper ocean hydrography prior to the next autumn's freeze. Both the spatial distribution and the timing of surface-stress events within the seasonal cycle can therefore be important to the evolution or stability of the upper ocean. Steele and Morison [1993] noted that much of the upward heat flux from the Atlantic Water to the surface, either open or ice-covered, occurs in response to intermittent atmospheric cyclones. They further postulated that the seasonal cycle of heat input to the Arctic Ocean in the WSC may be weak because the highest loss rates to the atmosphere should occur, in response to variations in cyclone strength and occurrence frequency, at the same time of year (autumn and early winter) as the maximum in WSC heat transport.

The spatial variation of the surface stress-driven mixing rates causes horizontal variability in surface layer properties, e.g. mixed layer depth (Figure 4) and temperature (Figure 5). Padman and Dillon [1991] also comment that the spatial variation of ice-relative current imposes stress and strain on the ice sheet, and could presumably play a role in modifying ice concentration through the sheet's response (lead formation and ridging) to internal stresses. This topic is presently being explored by Z. Kowalik [pers. com., 1994], using his high-resolution barotropic tidal model.

Surface Buoyancy Fluxes

Both heat and salt contribute to the surface buoyancy fluxes in the Arctic Ocean. Aagaard and Greisman [1975] estimate that a basin-averaged, annual-average surface heat loss of about 11 Wm^{-2} is required to balance the Arctic heat budget, after the heat transports through the Fram and Bering Straits and the Canadian Archipelago, and river runoff, have been considered. The freshwater budget is balanced by freshwater input from river runoff and exchanges through the Straits and Archipelago of both water and ice. Evaporation and

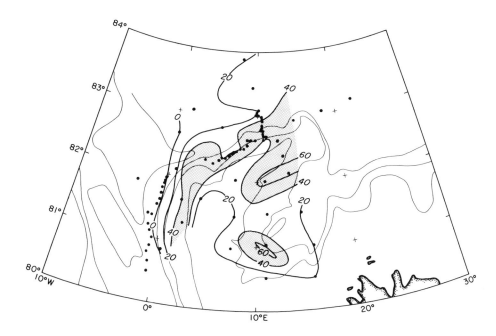

Figure 4. Regional variation of mixed layer depth (in meters) during CEAREX [after Padman et al., 1992]. Regions where the mixed layer is deeper than 40 m are shaded. Dots represent station locations [see Muench et al., 1992].

precipitation over the Arctic Basin are a negligible component of the freshwater budget [Coachman and Aagaard, 1974].

The heat loss is highly variable in both space and time. Sea ice effectively insulates the ocean from the atmosphere: the heat loss through open water in winter can be two orders of magnitude higher than under multi-year ice. A crucial issue in understanding the Arctic heat budget is therefore the spatial variability of sea ice concentration and thickness. Maykut [1978], for example, has suggested that the total heat flux through open leads in the central Arctic is about the same as the total flux through the multi-year ice, even though leads constitute only a small percentage of the total area. A discussion of the oceanography of winter leads is provided by Morison et al. [1992]. Coastal polynyas, where the open water is maintained by offshore advection of newly-formed ice, are other important regions for ventilation of oceanic heat to the atmosphere. It is likely that the total heat loss from the Arctic Ocean is dominated by fluxes through polynyas and leads, as well as in the MIZ [Steele et al., 1994].

Surface cooling at the surface obviously is a significant component of the Arctic heat balance. Nevertheless, cooling plays a negligible role in modifying a fluid parcel's buoyancy (and hence it's equilibrium depth in the basin), because the thermal expansion coefficient is quite small for seawater at temperatures near the freezing point. Much more

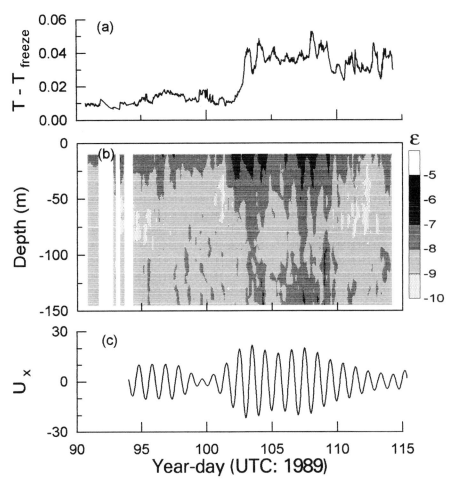

Figure 5. Transects of: (a) surface mixed layer temperature relative to the freezing point , $T\text{-}T_{freeze}$; (b) dissipation rate of turbulent kinetic energy, $\log_{10}\varepsilon$ ($m^2 s^{-3}$); and (c) cross-slope diurnal tidal current ($cm.s^{-1}$) at the CEAREX O Camp [Padman and Dillon, 1991].

significant is the associated injection of salt into the upper ocean during ice formation. Ice production is highest in open water since that is where the atmospheric cooling is greatest, particularly in winter. Salinization by surface cooling under coastal polynyas is a major source of halocline water formation over the shelves [Martin and Cavalieri, 1989]. Steele et al. [1994] suggest also that a combination of direct surface cooling of the Atlantic Water plus ice melting in the Fram Strait MIZ can contribute significantly to halocline water formation.

The ultimate fate of water that has been made dense by the surface flux of salt during ice formation depends on its final density. Numerical studies are presently being undertaken on the dynamics and thermodynamics of dense water formation under coastal polynyas

[Gawarkiewicz and Chapman, 1994; Chapman and Gawarkiewicz, 1994], and under leads [Smith and Morison, 1993]. Properties such as initial density (upwelled AIW compared with diluted river runoff, for example), and residence time for the water under the open lead/polynya are clearly important. The residence time for water under a polynya is in part set by the time scale for onset of the baroclinic instability at the polynya's margin [Gawarkiewicz and Chapman, 1994], but will also depend on advective "flushing" by the local geostrophic and wind-driven currents, and by tidal advection.

Internal Gravity Waves

Internal waves (IWs) are ubiquitous in the oceanic pycnocline. Instabilities in the IW wave field lead to turbulent diffusion of momentum, and scalars such heat, salt and buoyancy (see Gregg [1987] for a detailed review). Recent measurements in the main oceanic thermocline suggest that K_v is typically about $10^{-5}m^2s^{-1}$ [Gregg, 1989; Ledwell et al., 1993; Toole et al., 1994]. This is significantly smaller than the value of $10^{-4}m^2s^{-1}$ required in the deep ocean for a steady state balance in basin-scale advection-diffusion models [Munk, 1966]. One possible way to reconcile the mid-ocean measurements with the model requirements is to assume that most mixing actually occurs in "hot spots" near steep or rough topography. Toole et al. [1994] found significantly larger values of K_v in the deep ocean near a seamount than elsewhere over the abyssal plain. Wijesekera et al. [1993a] found, from data collected at the CEAREX O Camp, a strong correlation between mixing rates and the proximity to the steep topography of the Yermak Plateau in the eastern Arctic. These and other recent measurements support the view that mixing hot spots, perhaps related to topography but also to mesoscale structures such as fronts and eddies, might dominate basin-scale mixing rates.

Significant progress is now being made in relating microstructure mixing rates to the more-easily measured IW spectral properties [Gregg, 1989; Wijesekera et al., 1993a]. Gregg's model was developed for wave fields described by the Garrett-Munk ["G-M": Garrett and Munk, 1972a; Munk, 1981] "universal" IW spectrum, which is a reasonable approximation to the energy density and shape of the internal wave spectrum in many locations in the oceanic permanent thermocline. This model is structurally consistent with earlier theoretical studies of non-linear wave-wave interactions in a G-M wave field [Henyey et al., 1986; McComas and Müller, 1981], although the predicted magnitudes differ for each of these studies. Much interest is now focussed, however, on the behavior of anomalous, particularly high-energy, wave fields, because of their association with regions of energetic turbulence [Wijesekera et al., 1993a,b; Toole et al., 1994]. The generalization of the existing models to non-G-M wave fields is an ongoing study being addressed by theoretical, numerical, and empirical modeling.

The Arctic is known to be a region where anomalous (non-G-M) wave fields are, in fact, the norm [Levine et al., 1985, 1987; Levine, 1990; D'Asaro and Morehead, 1991]. In the central Beaufort Sea, for example, the average IW energy density is extremely small, the slope of the frequency spectrum is flatter than G-M, and the waves are significantly less coherent in the vertical than in the canonical G-M wave field. The reasons for the anomalous wave fields are not clear, although the presence of ice cover is obviously important. In early spring, when both the AIWEX and CEAREX programs were carried out, the ice cover is quite rigid. The quasi-rigid ice sheet affects the surface generation of internal waves by changing the mechanisms by which atmospheric stress is transmitted to the upper ocean. Since the spatial structure and translation velocity of the wind field are important to the efficiency of near-inertial IW generation by atmospheric forcing [Gill, 1982, Chapter 9], and since near-inertial IWs appear to be a significant source of oceanic turbulence because of the high associated velocity shears [Garrett and Munk, 1972b; Munk, 1981; Gregg, 1987], the changed boundary stress condition imposed by compact sea-ice is a likely candidate for the reducing surface forcing of IWs. Furthermore, as McPhee and Kantha [1989] described, IWs can be generated by the motion of rough ice moving across the mixed layer. Waves are generated that have horizontal wavenumbers determined by the horizontal spatial scales of ice roughness. When the keel heights are comparable to the mixed layer depth, significant energy can be transferred to the IW field in the upper pycnocline. Wijesekera and Dillon [1991] discuss a complementary model in which internal waves are generated by the pycnocline's response to organized mixed layer convective structures ("eddies") in the sheared, upper-ocean equatorial Pacific.

The reduction of near-inertial forcing and the possible increase in production rate for higher-frequency waves through wavenumber-coupling with the ice roughness may explain many of the differences between typical Arctic IW spectra and those from lower latitudes. It is also possible that the dissipation mechanisms for IWs are sufficiently different that they play a role in the observed spectra. For example, sea ice increases the dissipation rate of any IWs that are generated, since the friction at the ice/water interface is higher than at the air/water interface in the open ocean [Morison et al., 1985]. Mixing in the double-diffusive staircases that are a common feature in the Arctic pycnocline (see below) might also attenuate IWs as they propa-gate vertically from their source, at either the surface or seabed [Padman and Dillon, 1989; Padman, 1994].

Models of turbulence due to IWs rely on non-linear wave-wave interactions to move energy to smaller vertical scales, however the final stage is usually a shear instability (see Turner [1973] and Gregg [1987]). For shear-driven turbulence in the stably-stratified pycnocline, Osborn [1980] noted that it was possible to approximate the effective diffusivity by measuring either the density or velocity microstructure. This technique has been widely used, with the most common approach being to first calculate the rate of dissipation of turbulent kinetic energy, ε, from the measured microscale velocity shears, $\partial u/\partial z$ and $\partial v/\partial z$. Shears are usually measured with "airfoil" shear sensors [Osborn and Crawford, 1980; Padman and Dillon, 1987, 1991]. The equation relating ε to the microscale shear measurements is

$$\varepsilon = \frac{15}{2}\nu\,\frac{\langle \partial u/\partial z^2\rangle + \langle \partial v/\partial z^2\rangle}{2} \tag{5}$$

where ν is the kinematic viscosity and $\langle\,\rangle$ indicates some vertical averaging [Osborn, 1980]. The factor, 15/2, arises from assuming that the turbulence is isotropic. The shears must be measured to vertical scales of a few centimeters to accurately estimate ε, however this now a fairly mature technology.

Once ε is known, the vertical scalar diffusivity can be approximated from

$$K_v \approx \frac{0.2\varepsilon}{N^2} \tag{6}$$

where N is the buoyancy frequency averaged over some time and space scales. Three Arctic experiments, AIWEX, CEAREX, and LEADEX (see Table 1) have included collection of simultaneous profiles of ε and N suitable for calculating K_v from (6).

A commonly-used parameter for assessing the dynamic stability of a stratified shear flow (i.e. the likelihood of shear instabilities being generated) is the gradient Richardson number, Ri, defined as

$$Ri = \frac{g}{\rho} \frac{\partial \rho / \partial z}{(\partial U / \partial z)^2} = \frac{N^2}{(\partial U / \partial z)^2} \tag{7}$$

where g is the accelation due to gravity, $\rho(z)$ is the fluid density, and $U(z)$ is the horizontal component of fluid velocity. The Richardson number can therefore be regarded as the ratio of the stabilizing buoyancy gradient to the destabilizing velocity field. The flow is dynamically stable for large values of Ri and may be unstable when Ri is small. Both observations and theory suggest that the transition from stable to unstable occurs for $Ri \sim 1/4$ for Kelvin-Helmholtz instabilities [Turner, 1973], and $Ri \sim 1$ for advective instabilities in a finite-amplitude internal wave field [Orlanski and Bryan, 1969]. Turbulence driven by shear instabilities is a highly intermittent process [Baker and Gibson, 1987], implying that many realizations of ε are required in order to calculate an averaged value of ε (and hence K_v) that could be useful to basin-scale modeling efforts.

Turbulence closure schemes for numerical models, e.g. Mellor and Yamada [1974, 1982], include a Richardson-number dependence for determining dynamic stability. Even if the scheme fairly represents the physics of turbulence generation, however, it is limited by the model's resolution of the density and velocity fields that determine the local values of Ri. If the variation of $\partial \rho / \partial z$ or $\partial U / \partial z$ that initially creates the low Richardson number instability occurs on time and/or space scales smaller than those resolved by the model, the resultant increased mixing rates and higher diffusivity will not be modeled. This is clearly the case for internal waves which, as we have discussed above, explain most of the mixing that is found in the pycnocline.

At the CEAREX O Camp [Padman and Dillon, 1991], extremely energetic mixing events were found when the diurnal tidal current amplitude was large. The peak cross-slope diurnal tidal currents at O Camp were about 0.3 ms^{-1} compared with a mean alongslope flow of about 0.02 ms^{-1} [Muench et al., 1992] and a mean southward ice velocity of about 0.05 ms^{-1} in the Transpolar Drift [Moritz and Colony, 1987]. That is, the diurnal tide was the dominant source of kinetic energy during this experiment. As was discussed above, velocity shear is generally assumed to be a prerequisite for turbulence generation in a stably-stratified ocean. During CEAREX, the shear was not, however, provided directly by the diurnal tidal currents, which were quite uniform with depth [Padman et al., 1992]. Instead, what was found was a quasi-diurnal occurrence of large-amplitude wave packets, which were propagating towards the north and northwest, i.e. away from the Yermak Plateau [Czipott et al., 1991]. The wave packets have large velocity shears associated with them, and also large strain rates, i.e. Ri (Eq. 7) is locally reduced by simultaneously decreasing $\partial\rho/\partial z$ and increasing $\partial U/\partial z$. This is in contrast to the open ocean where most mixing appears to be correlated with the near-inertial frequency waves, in which changes in $\partial\rho/\partial z$ are insignificant compared with variability of $\partial U/\partial z$ [Garrett and Munk, 1972b; Gregg, 1987].

While the generation mechanism for these high-frequency wave packets is not yet understood, the diurnal periodicity and relationship to the mean cross-slope current magnitude [Padman et al., 1991; Wijesekera et al., 1993a] undoubtedly implicates tidal currents. Padman et al. [1991] also found that the second most significant time scale of variability in dissipation rate was near six hours, and noted that the highest turbulence levels occurred when IWs with periods near six hours were propagating upwards. In a separate program in which current profiles in the upper 300 m were measured from a drifting buoy, Plueddemann [1992] found evidence over the Yermak Plateau of upward propagation of near-inertial (semi-diurnal) wave energy. With these observations in mind, the following two scenarios for IW generation in this region are plausible. First, diurnal tidal currents flowing across the bottom topography might generate 6-hour waves (the fourth harmonic) that can then propagate upwards into the high stratification of the main pycnocline, with a subsequent non-linear cascade of energy to the higher observed frequencies (near N) of the intermittent wave packets. Alternatively, shear associated with the upward-propagating, near-inertial IWs observed by Plueddemann [1992] might provide the necessary background shear for higher frequency IWs to become unstable.

Regardless of the actual mechanism for internal wave and turbulence generation, this region of the eastern Arctic clearly differs from the open ocean in that energy radiating upwards from the seabed is more important than surface forcing, at least during the winter and early spring when these measurements were made. Significant further research in this area is justified by the importance of the mechanisms for the dispersion of the AIW heat content within the Arctic.

Double Diffusion

Double-diffusion is a mechanism in which the differing molecular diffusivities of heat and salt drive significant vertical fluxes of these scalar properties [Turner, 1973]. For double-diffusion to occur, the mean vertical gradients of T and S must have the same sign. If T and S both increase with depth, the process is called "diffusive convection": if T and S both decrease with depth, it is denoted "salt-fingering". Schmitt [1994] reviews both instability types. Examples of double diffusion in the Arctic are given by Neshyba et al.

[1971], Melling et al. [1984], Perkin and Lewis [1984], Padman and Dillon [1987, 1988, 1989], and Padman [1994].

In regions with strong intrusions, both diffusive-convection and salt-fingering can be found on the upper and lower edges of warm intrusions respectively. Intrusions are frequently found at the edges of the West Spitsbergen Current (WSC) [Perkin and Lewis, 1984; Padman and Dillon, 1991], near the ice edge (e.g. Paquette and Bourke [1979]), and at shelf fronts, for example at the locations where shelf-modified water first interacts with waters that are characteristic of the deep basin. The horizontal scales of intrusions are typically a few kilometers [Paquette and Bourke, 1979; Coachman, 1986], and the vertical scales are between 1 and 100 m (see Perkin and Lewis [1984]; Toole and Georgi [1981]; and Coachman [1986]). An example of intrusions can be seen in Figure 1 at the core of the AIW near 300 m depth. The dynamics of intrusions and the associated isopycnal mixing and double-diffusive vertical fluxes are clearly important in understanding the final properties of the shelf-modified water entering the main basin halocline, and would benefit from further study, but is beyond the scope of this review.

In the central Arctic under sea-ice and away from fronts in water mass properties, the upper-ocean stratification typically consists of cold, fresh surface water overlying the warmer, more saline AIW, and diffusive convection dominates. A vertical profile of T, S, and potential density, σ_θ, obtained using a microstructure profiler while the CEAREX O Camp was located over a submesoscale eddy, shows the sequence of extremely sharp interfaces separating well-mixed convective layers that is characteristic of diffusive-convection (Figure 6). The ratio of density gradients due to salt and heat is parameterized by the density ratio, R_ρ, defined by

$$R_\rho = \frac{\beta \langle S_z \rangle}{\alpha \langle T_z \rangle} \tag{8}$$

where α and β are, respectively, the thermal expansion and saline contraction coefficients, and $\langle T_z \rangle$ and $\langle S_z \rangle$ are the vertical gradients of temperature and salinity. The density stratification corresponding to the thermal gradient is intrinsically unstable: static stability is maintained by $\langle S_z \rangle$. It follows from (8) that for a diffusive-convective staircase, $R_\rho > 1$. For $R_\rho \approx 1$, the gradient of T almost cancels that of S in the equation of state. At large values of R_ρ, salinity dominates in the calculation of density.

Laboratory measurements demonstrate than heat (and salt) fluxes are much larger for $R_\rho \approx 1$ than for $R_\rho \gg 1$ [Marmorino and Caldwell, 1976; Taylor, 1988]. Furthermore, the effective diffusivity for heat in a staircase is typically much greater than that for salt, except for $R_\rho \approx 1$ [Kelley, 1984]. In the Canada Basin above the warm core of the Atlantic Layer, $R_\rho \approx 4$-5. In this case, fluxes due to diffusive-convection will be extremely small: Padman and Dillon [1987] estimated vertical heat fluxes due to diffusive-convection of about 0.04 Wm^{-2}. The mean vertical diffusivities of heat and salt are therefore almost certainly dominated by other processes, for example the vertical component of isopycnal mixing and advection. As Padman [1994] showed, however, double-diffusive heat and salt

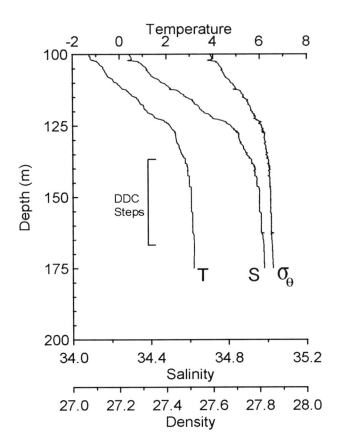

Figure 6.   Vertical profiles of temperature (T), salinity (S), and potential density (σ_θ) in the CEAREX O Camp eddy. Well-developed diffusive-convective steps are visible near 150 m.

fluxes can be substantially higher above the warm cores of submesoscale eddies centered on the Atlantic Layer, if R_ρ is small. For the eddy that Padman [1994] studied the estimated upward heat flux due to diffusive-convection was about 4 Wm^{-2}. Double-diffusive transport of heat and salt may also be an important process as intrusions of shelf-modified water propagate into the basin interior to form the cold halocline and, as noted above, at the lateral boundaries of the AIW [Perkin and Lewis, 1984]. Unlike shear-driven mixing, double-diffusion is known to have different diffusivities for heat and salt [Marmorino and Caldwell, 1976; Kelley, 1984], providing a mechanism for modification of the T-S relationship.

It is important to note two significant differences between fluxes driven by double-diffusion and those due to internal waves: (i) the buoyancy flux due to shear instabilities is down-gradient, whereas it is up-gradient in diffusive-convective systems because it is then driven by the intrinsically-unstable thermal gradient; and (ii) the effective vertical diffusivities for heat and salt ($K_{v,T}$ and $K_{v,S}$ respectively) are assumed to be equal in the case of shear instabilities. In environments where diffusive-convection dominates, $K_{v,T}$ is greater than $K_{v,S}$ [Kelley, 1984].

Submesoscale Eddies

From hydrographic and current data from the AIDJEX pilot and main programs in the 1970s [Newton et al., 1974; Manley and Hunkins, 1985], it is known that the western Arctic is densely populated by small baroclinic eddies, also known as submesoscale coherent vortices (SCVs). Eddies have also been observed along ice edges and near the West Spitsbergen current [Johannessen et al., 1987]. McWilliams [1985] reviews some general properties of oceanic SCVs, including: their stability; interactions with topography, horizontally-sheared flows and other SCVs; and ability to transport anomalous properties long distances from their source. The subsurface vorticity maxima of many SCVs are probably due to reduction in the surface angular momentum by frictional stress at the ice/water interface [Ou and Gordon, 1986]. Padman et al. [1990] suggested, however, that the subsurface vorticity might actually be increased by critical layer absorption of internal gravity wave momentum. Wave momentum can be absorbed when the wave phase velocity matches the local mean fluid velocity relative to the wave source. Therefore, even in a horizontally isotropic wave field, only those waves travelling in the same direction as the rotating vortex can be absorbed, applying a torque to the eddy. A fascinating aspect of this hypothesis is that the eddy vorticity might be retained even as its anomalous hydrographic properties are being erased by vertical and horizontal mixing. Larger-scale processes such as interactions with topography, mean-sheared currents, or other SCVs [McWilliams, 1985] possibly dominate over internal wave absorption in the evolution of these features. There is a need to better understand the dynamics of these SCVs and their role in large-scale dispersion and transport within the Arctic.

Manley and Hunkins [1985] found that SCVs occupied roughly 25% of the Canada Basin surveyed in AIDJEX. Example of eddy structure are presented by Hunkins [1974], Manley and Hunkins [1985], Padman et al. [1990], and D'Asaro [1988a]. SCVs usually have anomalous temperature and salinity properties relative to the background, suggesting a remote generation site. Typically, the SCVs also have subsurface vorticity maxima centered on either the Bering Strait inflow or the Atlantic Layer. For these reasons, it has been proposed that the eddies originate through instabilities in the boundary currents flowing into the Arctic Basin [D'Asaro, 1988b]. Then, based on the mean circulation velocities in the Arctic and the location of the observed SCVs relative to the apparent generation sites, they appear to be very long-lived. One important consequence of these observations is that the SCVs may contribute significantly to the horizontal dispersion of heat, salt, and kinetic energy within the Arctic. Manley and Hunkins [1985] suggest that SCVs are responsible for more than 30% of the total kinetic energy in the upper 200 m of the Beaufort Sea and Canada Basin. From most studies to date of the western Arctic SCVs, they appear to be primarily relevant to the hydrography and circulation of the deeper basin. However, recent modeling efforts aimed at studying the dynamics of dense, shelf-modified water over the continental shelf and slope indicate that the offshore transport of water formed by cooling under a coastal polynya occurs primarily through offshore advection of small eddies formed at the polynya's edge by baroclinic instability [Gawarkiewicz and Chapman, 1994]. A companion study by Chapman and Gawarkiewicz [1994] further suggests that the eddy flux will be preferentially channeled offshore by submarine canyons, however the *total* offshore flux does not change significantly due to the presence of canyons.

Small mesoscale eddies are also frequently observed in the marginal ice zone (MIZ) in the eastern Arctic and Greenland Sea. These eddies can be clearly seen in satellite infrared images in ice-free regions, and occasionally in the surface circulation of unconsolidated ice in the MIZ [Johannessen et al., 1987]. Two generation sources are implicated: topographic trapping; and combined barotropic and baroclinic instability in the boundary currents in Fram Strait. An example of the former is an eddy near the Molloy Deep, which has been observed several times [Johannessen et al., 1987; Bourke et al., 1987]. The propagating eddies formed from boundary current instability are presumably the source for the deep eddies centered on the Atlantic Layer in the Canada Basin [D'Asaro, 1988a]. Häkkinen [1986] discussed the generation of ice-edge eddies in a coupled ice/ocean model. The ice-edge eddies, by providing a mechanism for periodically advecting warm water under the ice, play a significant role in ice edge thermodynamics. Johannessen et al. [1987] suggested that ice edge retreat due to eddies could be about 1-2 km day^{-1}. A similar response can be postulated for the periodic advection of warm water at the ice edge due to tides.

Regardless of whether eddies form from boundary current instabilities, localized convective cooling under coastal polynyas, or in the MIZ, they clearly represent a dispersion mechanism for water that is characteristic of the thermocline (AIW) and western Arctic halocline (Bering Strait inflow). Furthermore, from studies elsewhere in the world ocean, eddies are known to interact with topography at shelf breaks, leading to incursions of eddy water onto the shelf [e.g. McClean-Padman and Padman, 1991], as well as enhanced benthic stirring and, under consolidated pack-ice, local surface layer mixing. Since it is likely that much of the lateral dispersion of shelf-modified and boundary current water types throughout the Arctic is due to these features, they deserve considerable further research.

Spatial Variability of Mixing Rates

From the above discussion, is it obvious that mixing rates in the oceanic boundaries will vary significantly, both regionally and through the seasonal cycle of sea ice concentration. For example, buoyancy-forced mixing associated with ice production can vary by over two orders of magnitude between open water in winter, and under thick, multi-year pack ice. Mixing at the seabed will vary significantly, depending on such processes as wind-driven and geostrophic flow, tides, eddies and internal waves. Most of these processes have been shown to be most important in the shallow water of the shelf seas, where halocline water is first formed, and at the shelf break. In the surface layer, stress-driven mixing rates will vary with the wind speed (in open water), or the ice-relative surface current (under sea ice) [McPhee, 1990; McPhee and Martinson, 1993] which is again related to processes such as wind-forcing of the ice sheet, tides, eddies, and internal waves.

What is less well-known is the degree of spatial and temporal variability in the pycnocline. In the present context, this is particularly relevant for several reasons, including: provision of AIW heat through the cold halocline to the base of the surface mixed layer and ultimately to the sea ice; coupling through the pycnocline between surface and benthic layers on the shelves; and the depth to which the shelf-modified water, having mixed to some extent with the resident water of the deep basin, will intrude into the basin. The clearest indication of spatial variability comes from comparison of microstructure measurements collected from ice camps during AIWEX [Padman and Dillon, 1987] and CEAREX [Padman and Dillon, 1991]. The vertical heat flux through the pycnocline above the temperature maximum (Figure 7a) and the effective vertical diffusivity for heat (Figure 7b) vary by 2-3 orders of magnitude between these experiments. During AIWEX the ice camp remained over the deep water of the Canada Basin for several weeks and, below the mixed layer, very few patches of turbulence were found [Padman and Dillon, 1989; Padman et al., 1990]. During CEAREX, the oceanography ice camp (O Camp) drifted from the deep water of the Nansen Basin over the slope of the Yermak Plateau, then into northern Fram Strait. Background oceanic conditions varied substantially throughout the drift, and the subdivisions in Figure 7 reflect these changes. The mean diffusivity was substantially higher in the pycnocline during CEAREX than in the mid-latitude permanent thermocline [Gregg, 1989], and some reasons have been proposed to explain this [Wijesekera et al., 1993a, b].

There is no *a priori* reason to suppose that these two experiments represent the extreme values of heat flux and diffusivity in the Arctic pycnocline. Our present best hypothesis to explain the high diffusivities near the Yermak Plateau is that strong, topographically-amplified diurnal tides [Padman et al., 1992] interact with the steep and rough topography of the plateau slope to produce large-amplitude, high-frequency wave packets [Czipott et al., 1991], that then dissipate strongly near the plateau. Recent tidal modeling by Kowalik [1993] and Kowalik and Proshutinsky [1993] indicate that there are several other sites where tidal currents are stronger than over the Yermak Plateau (see Figure 3). Some of these sites are in shallower water than the plateau, so that the near-bottom buoyancy frequency is higher and the prospects for internal wave generation are correspondingly greater. Work in progress is addressing the potential for strong mixing in these locations.

The diffusivity at AIWEX in the Canada Basin, while it is small, is substantially higher than molecular. Padman and Dillon [1987; 1989] determined that double-diffusive convection was the primary cause of diapycnal fluxes in the central Canada Basin, except when submesoscale eddies (SCVs) were present [Padman et al., 1990]. Evaluating the true mean vertical flux in the Canada Basin is complicated however, by the apparent high density of SCVs. SCVs have substantial vertical velocity shears associated with their baroclinicity, and may also be regions of anomalous internal wave (IW) activity [Padman

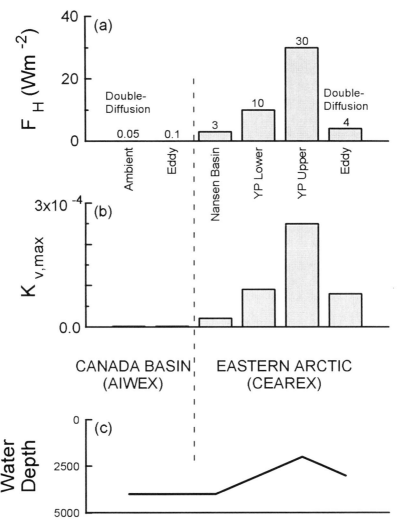

Figure 7. Variability of (a) maximum upward heat flux from the Atlantic Layer, F_H, and (b) the associated effective vertical diffusivity, $K_{v,max}$, evaluated from microstructure profiles obtained at AIWEX [Padman and Dillon, 1987, 1989; Padman et al., 1990] and CEAREX O Camp [Padman and Dillon, 1991]. Approximate water depth for each experiment is shown in (c).

et al., 1990; D'Asaro and Morison, 1992; R. Pinkel, pers. comm, 1990]. Given that these features are quite common, the mixing rates in eddies may have considerable impact on the basin-averaged vertical transports.

Double-diffusive convection is implicated in the high thermal diffusivity found above the vorticity maximum of a submesoscale eddy that was sampled during CEAREX [Padman, 1994]. In this case, the heat flux was about 4 Wm^{-2} (see Figure 7), which is significant compared with the mean heat loss of Atlantic Water heat averaged over the central Arctic basin [Coachman and Aagaard, 1974].

Other studies indirectly confirm this view of strong spatial variability of mixing rates in the Arctic pycnocline. Melling et al. [1984] found from a regional survey of the Beaufort Sea and Canadian Archipelago that double-diffusive steps were only present well away from the shelf and slope. In the Beaufort Sea and Canada Basin, double-diffusive staircases will be formed very slowly because the heat fluxes at high density ratio (Eq. 7) are extremely small. We interpret the absence of steps as an indication that some other mixing process, presumably related to mean and/or internal wave shear, is responsible for disrupting the development of double-diffusive steps. The fluxes associated with this other stirring mechanism are presumably greater than the diffusive-convective fluxes, which attempt to regenerate the staircase structure after perturbations caused by shear-induced mixing events. Another example is the current data collected from the Arctic Environmental Drifting Buoy (AEDB: Plueddemann [1992]). The AEDB drifted over the deep Nansen Basin, across the center of the Yermak Plateau, then into the Greenland Sea *via* Fram Strait. As Plueddemann [1992] showed, the IW energy density was much greater over the plateau than in the other regions. Using the arguments of Gregg [1989] and Wijesekera et al. [1993a], we expect that turbulence levels were higher over the plateau than elsewhere during the AEDB's drift. More evidence for this hypothesis is provided by D'Asaro and Morison [1992], who used current shears from expendable current profilers deployed over the Nansen Basin and Yermak Plateau. Figure 8, adapted from D'Asaro and Morison, indicates that diffusivities are much greater over the plateau and the Nansen-Gakkel ridge than over the relatively featureless and deep Nansen Basin.

Diurnal (tidal), and semi-diurnal (tidal or near-inertial), motions are significant sources for much of the mixing, both in the pycnocline and at the surface and benthic stress boundaries. This is particularly true in the shallow shelf seas and near abrupt topography such as the shelf break. While the latest Arctic tidal model is still barotropic [Kowalik and Proshutinsky, 1993], it nevertheless provides a guide to likely regions of strong benthic and under-ice friction, and enhanced pycnocline mixing by tide-generated internal gravity waves. With this dependence in mind, a variety of complementary approaches that can be taken towards the goal of parameterizing mixing in GCMs include:

- continued development and validation of tidal models, and the inclusion of modeled tidal currents into GCMs in order to improve surface and benthicfriction parameterizations;
- extension of tidal modeling to the baroclinic case, particularly to investigate the flux of energy from the barotropic tides into the upward-propagating near-inertial waves;
- improved mapping of the internal wave spectrum, particularly with respect to major topographic features and regions of energetic tidal currents as identified by numerical tidal models;
- theoretical development of internal wave generation mechanisms by tidal flow over steep and rough topography in an ocean with a realistic vertical structure of N;

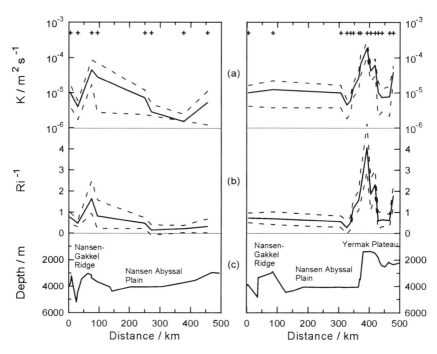

Figure 8. Spatial variability of (a) vertical diffusivity (K_v), (b) inverse Richardson number $(Ri^{-1}=S_{10}^2/N^2$, where S_{10}^2 is the internal wave velocity shear variance on 10 m scales), and (c) water depth [after D'Asaro and Morison, 1992]. Diffusivity is estimated from internal wave parameters using the Gregg [1989] model. Dashed lines in (a) and (b) indicate 95% confidence limits. Symbols (+) in top panel indicate locations of XCP profiles. For station locations, see *D'Asaro and Morison*.

- modeling of the energy transfer to dissipative scales by wave-wave interactions in the observed wave fields (i.e. extending existing Garrett-Munk based models to the non-GM spectra found in the Arctic); and
- developing a clearer understanding of the role of a quasi-rigid, rough upper boundary (the pack ice) on the generation, propagation and decay of internal waves.

Concluding Remarks

Carmack [1990], in his review of the large-scale oceanography of polar oceans, concluded that "Despite ongoing work ... there is much to learn about the large-scale physical oceanography of (polar) regions." Carmack's caution is even more appropriate to small-scale physical oceanography. We are really just at the stage of defining the critical small-scale processes that need to be considered in modeling polar oceans, and have only poorly-formed ideas about relating these processes to modeled, larger-scale phenomena. Figure 9 shows a flow chart of selected oceanographic processes and their interactions,

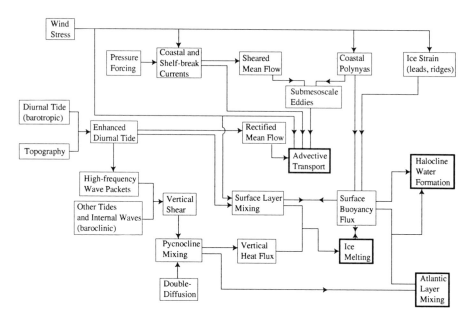

Figure 9. Draft flow chart showing the relationships between various physical processes in the Arctic Ocean.

and reflects in large measure my own interpretation of the issues most deserving of further research. Each reader might wish to add favorite processes, and perhaps in time the true complexity of small-scale processes in the Arctic would emerge. Nevertheless, the flow chart represents something quite fundamental: small-scale physical processes have an extremely complex and significant impact on all aspects of Arctic hydrography and circulation.

Despite the complexity of Figure 9, I hope to have shown in this review that progress is being made in our understanding of Arctic small-scale physical oceanographic processes that have relevance to large-scale modeling efforts. In the next decade we can expect to see significant progress in numerical modeling of Arctic processes such halocline water formation under coastal polynyas, and heat loss from the surface layer through leads and polynyas, and tidal currents. Observational techniques continue to improve, ranging from satellite determination of ice concentration distribution [e.g. Massom and Comiso, 1994] to sensors capable of recording the extremely small ice strain and tilt associated with underlying oceanic processes such as internal waves [Czipott et al., 1991]. The combination of modeling and improved observations of small-scale processes should lead to a greatly improved understanding of the Arctic and it's role in global climate.

Acknowledgements. Conversations with Tom Dillon, Murray Levine, Miles McPhee, Jamie Morison, Robin Muench, Rob Pinkel, Al Plueddemann, and Hemantha Wijesekera, among others, have had a significant impact on my understanding of Arctic small-scale processes. Bjørn Gjevik and Zygmunt Kowalik have developed my appreciation of the role of tidal currents in the Arctic. Support was provided by the Office of Naval Research, contract N00014-90-J-1042.

References

Aagaard, K., STD measurements in possible dispersal regions in the Beaufort Sea, *Annual Report, Contract 03-5-022-67, TO1*, Department of Oceanography, Univ. Washington, 34 pp., 1977.

Aagaard, K., On the deep circulation in the Arctic Ocean, *Deep-Sea Res.*, 28A, 251-268, 1981.

Aagaard, K., A synthesis of the Arctic Ocean circulation, *Rapp. P.v. Réun. Cons. int. Explor. Mer*, 188, 11-22, 1989.

Aagaard, K., and E.C. Carmack, The role of sea ice and other fresh water in the Arctic circulation, *J. Geophys. Res.*, 94, 14,485-14,498, 1989.

Aagaard, K., and L.K. Coachman, Towards an ice-free Arctic Ocean, *EOS*, 56, 484-486, 1975.

Aagaard, K., L.K. Coachman, and E.C. Carmack, On the halocline of the Arctic Ocean, *Deep-Sea Res.*, 28A, 529-545, 1981.

Aagaard, K, and P. Greisman, Toward new mass and heat budgets for the Arctic Ocean, *J. Geophys. Res.*, 80, 3821-3827, 1975.

Aagaard, K., J.H. Swift, and E.C. Carmack, Thermohaline circulation in the Arctic mediterranean seas, *J. Geophys. Res.*, 90, 4833-4846, 1985.

ARCUS (Arctic Research Consortium of the United States) Report, *A Plan for Integration*, Fairbanks, AK, 1993.

Barry, R.G., M.C. Serreze, J.A. Maslanik, and R.H. Preller, The Arctic sea ice-climate system: observations and modeling, *Rev. Geophys.*, 31, 397-422, 1993.

Baker, M.A., and C.H. Gibson, Sampling turbulence in the stratified ocean: statistical consequences of strong intermittency, *J. Phys. Oceanogr.*, 17, 1817-1836, 1987.

Bourke, R.H., and R.P. Garrett, Sea ice thickness in the Arctic Ocean, *Cold Reg. Sci. Technol.*, 13, 259-280, 1987.

Bourke, R.H., M.D. Tunnicliffe, J.L. Newton, R.G. Paquette, and T.O. Manley, Eddy near the Molloy Deep revisited, *J.Geophys. Res.*, 92, 6773-6776, 1987.

Bourke, R.H., A.M. Weigel, and R.G. Paquette, On the westward turning branch of the West Spitsbergen Current, *J. Geophys. Res.*, 93, 14,065-14,077, 1988.

Carmack, E.C., Large-scale physical oceanography of polar oceans, in *Polar Oceanography (Part A)*, edited by W.O. Smith, Jr, pp 171-222, Academic Press, San Diego, Calif, 1990.

Chapman, D.C., and G. Gawarkiewicz, Formation and offshore transport of dense shelf water in the presence of a submarine canyon, Abstract, *EOS Transactions*, 75(3), p 74, 1994.

Coachman, L.K., Observations of finestructure formed in a continental shelf front (Southeastern Bering Sea), In: *Marine Interfaces Hydrodynamics*, Ed. J.C.J. Nihoul, 215-255, 1986.

Coachman, L.K., and K. Aagaard, Physical oceanography of Arctic and subarctic seas, in *Marine Geology and Oceanography of the Arctic Seas,* edited by Y. Herman, pp 1-72, Springer-Verlag, Berlin, 1974.

Coachman, L.K., and C.A. Barnes, Surface water in the Eurasian basin of the Arctic Ocean, *Arctic*, 15, 251-277, 1962.

Conover, R.J., G.F. Cota, G. Harrison, E.P.W. Horne, and R.E.H. Smith, Ice/water interactions and their effect on biological oceanography in the Arctic Archipelago. In: *Canada's Missing Dimension: Science and History in the Canadian Arctic Islands*, Canadian Museum, Ottawa, 1990.

Cota, G.F., and E.P.W. Horne, Physical control of arctic ice algal production, *Mar. Ecol. Prog. Ser.*, 52, 111-121, 1989.

Cota, G.F., S.J. Prinsenberg, E.B. Bennett, J.W. Loder, M.R. Lewis, J.L. Anning, N.H.F. Watson, and L.R. Harris, Nutrient fluxes during extended blooms of Arctic ice algae, *J. Geophys. Res.*, 92, 1951-1962, 1987.

Czipott, P.V., M.D. Levine, C.A. Paulson, D. Menemenlis, D.M. Farmer, and R.G. Williams, Ice flexure forced by internal wave packets in the Arctic Ocean, *Science*, 254, 832-835, 1991.

D'Asaro, E.A., Observations of small eddies in the Beaufort Sea, *J. Geophys. Res.*, 93, 6669-6684, 1988a.

D'Asaro, E.A., Generation of submesoscale vortices: a new mechanism, *J. Geophys. Res.*, 93, 6685-6694, 1988b.

D'Asaro, E.A., and M.D. Morehead, Internal waves and velocity finestructure in the Arctic Ocean, *J. Geophys. Res.*, 96, 12,725-12,738, 1991.

D'Asaro, E.A., and J.H. Morison, Internal waves and mixing in the Arctic Ocean, *Deep-Sea Res.*, 39, S459-S484, 1992.

Garrett, C.J.R., and W.H. Munk, Space-time scales of internal waves, *Geophys. Fluid Dyn.*, 2, 225-264, 1972a.

Garrett, C.J.R., and W.H. Munk, Oceanic mixing by breaking internal waves, *Deep-Sea Res.*, 19, 823-832, 1972b.

Garrison, G.R., and P. Becker, The Barrow submarine canyon: a drain for the Chukchi Sea, *J. Geophys. Res.*, 4445-4453, 1976.

Gawarkiewicz, G., and D.C. Chapman, A numerical study of dense water formation and transport on a shallow, sloping continental shelf, Abstract, *EOS Transactions*, 75(3), p 74, 1994.

Gill, A.E., *Atmosphere-Ocean Dynamics*, 662 pp, Academic Press, San Diego, Calif, 1982.

Gordon, A.L., Two stable modes of Southern Ocean winter stratification, in *Deep Convection and Deep Water Formation in the Oceans*, edited by P.C. Chu and J.C. Gascard, pp 17-35, Elsevier Press, 1991.

Gregg, M.C., Diapycnal mixing in the thermocline: a review, *J. Geophys. Res.*, 92, 5249-5286, 1987.

Gregg, M.C., Scaling turbulent dissipation in the thermocline, *J. Geophys. Res.*, 94, 9686-9698, 1989.

Häkkinen, S., Coupled ice-ocean dynamics in the marginal ice zones: upwelling/downwelling and eddy generation, *J. Geophys. Res.*, 91, 819-832, 1986.

Häkkinen, S., Models and their applications to polar oceanography, in *Polar Oceanography (Part A)*, edited by W.O. Smith, Jr, pp 335-384, Academic Press, San Diego, Calif, 1990.

Häkkinen, S., An Arctic source for the Great Salinity Anomaly: a simulation of the Arctic ice-ocean system for 1955-1975, *J. Geophys. Res.*, 98, 16,397-16,410, 1993.

Häkkinen, S., and G.L. Mellor, Modeling the seasonal variability of a coupled Arctic ice-ocean system, *J. Geophys. Res.*, 97, 20,285-20,304, 1992.

Hanzlick, D., and K. Aagaard, Freshwater and Atlantic Water in the Kara Sea, *J. Geophys. Res.*, 85, 4937-4942, 1980.

Harms, I.H., A numerical study of the barotropic circulation in the Barents and Kara seas, *Continental Shelf Res.*, 12, 1043-1058, 1992.

Henyey, F.S., J. Wright, and S.M. Flatté, Energy and action flow through the internal wave field: an eikonal approach, *J. Geophys. Res.*, 91, 8487-8495, 1986.

Hoffman, P.J., Transpolar sea ice drift in the vicinity of the Yermak Plateau as observed by Arctemiz 86 buoys, M.S. thesis, 99 pp., Nav. Postgrad. Sch., Monterey, Calif., 1990.

Hufford, G.L., On apparent upwelling in the southern Beaufort Sea, *J. Geophys. Res.*, 79, 1305-1306, 1974.

Hunkins, K., Subsurface eddies in the Arctic Ocean, *Deep-Sea Res.*, 21, 1071-1033, 1974.

Huthnance, J.M., Large tidal currents near Bear Island and related tidal energy losses from the North Atlantic, *Deep-Sea Res.*, 28, 51-70, 1981.

Johannessen, O.M., Note on some vertical profiles below ice floes in the Gulf. of St. Lawrence and near the North Pole, *J. Geophys. Res.*, 75, 2857-2862, 1970.

Johannessen, O.M., Introduction: summer marginal ice zone experiments during 1983 and 1984 in Fram Strait and Greenland Sea, *J. Geophys. Res.*, 92, 6716-6718, 1987.

Johannessen, J.A., O.M. Johannessen, E. Svendsen, R. Scuchman, T. Manley, W.J. Campbell, E.G. Josberger, S. Sandven, J.C. Gascard, T. Olaussen, K. Davidson, and J. Van Leer, Mesoscale eddies in the Fram Strait marginal ice zone during the 1983 and 1984 Marginal Ice Zone Experiments, *J. Geophys. Res.,* 92, 6754-6772, 1987.

Kelley, D.E., Effective diffusivities within oceanic thermohaline staircases, *J. Geophys. Res.*, 89, 10,484-10,488, 1984.

Kowalik, Z., Modeling of topographically-amplified diurnal tides in the Nordic Seas, *J. Phys. Oceanogr.*, (in press) 1993.

Kowalik, Z., and A.Y. Proshutinsky, Diurnal tides in the Arctic Ocean, *J. Geophys. Res.*, 98, 16,449-16,468, 1993.

Kowalik, Z., and A.Y. Proshutinsky, Topographic enhancement of tidal currents in the Arctic Ocean, Abstract, *EOS Transactions*, 75(3), p 126, 1994.

Langleben, M.P., Water drag coefficient at AIDJEX, Station Caribou, in *Sea Ice Processes and Models,* edited by R.S. Pritchard, pp 464-471, Univ. Of Washington, Seattle, 1980.

Langleben, M.P., Water drag coefficient of first-year sea ice, *J. Geophys. Res.*, 87, 573-578, 1982.

Ledwell, J.R., A.J. Watson, and C.S. Law, Evidence for slow mixing across the pycnocline from an open-ocean tracer-release experiment, *Nature,* 364, 701-703, 1993.

Levine, M.D., Internal waves under the pack ice during the Arctic internal wave experiment: the coherence structure, *J. Geophys. Res.*, 95, 7347-7357, 1990.

Levine, M.D., C.A. Paulson, and J.H. Morison, Internal waves in the Arctic Ocean: comparison with lower-latitude observations, *J. Phys. Oceanogr.*, 15, 800-809, 1985.

Levine, M.D., C.A. Paulson, and J.H. Morison, Observations of internal gravity waves under the Arctic pack ice, *J. Geophys. Res.*, 92, 779-782, 1987.

Loder, J.W., Topographic rectification of tidal currents on the sides of Georges Bank, *J. Phys. Oceanogr.*, 10, 1399-1416, 1980.

Manak, D.K., and L.A. Mysak, On the relationship between Arctic sea ice anomalies and fluctuations in northern Canadian air temperature and river discharge, *Atmos.-Ocean.*, 27, 682-691, 1989.

Manley, T.O., and K. Hunkins, Mesoscale eddies of the Arctic Ocean, *J. Geophys. Res.*, 90, 4911-4930, 1985.

Marmorino, G.O., and D.R. Caldwell, Heat and salt transport through a diffusive thermohaline interface, *Deep-Sea Res.*, 23, 59-67, 1976.

Martin, S., and D.J. Cavalieri, Contributions of the Siberian shelf polynyas to the Arctic Ocean intermediate and deep water, *J. Geophys. Res.*, 94, 12,725-12,738, 1989.

Martinson, D.G., Evolution of the Southern Ocean winter mixed layer and sea ice: open deepwater formation and ventilation, *J. Geophys. Res.*, 95, 11,641-11,654, 1990.

Massom, R., and J.C. Comiso, The classification of Arctic Sea ice types and the determination of surface temperature using very high resolution radiometer data, *J. Geophys. Res.*, 99, 5201-5218, 1994.

Maykut, G.A., Energy exchange over young sea ice in the central Arctic, *J. Geophys. Res.*, 83, 3646-3658, 1978.

McClean-Padman, J.L., and L. Padman, Summer upwelling on the Sydney inner continental shelf: the relative roles of wind forcing and mesoscale eddy encroachment, *Continental Shelf Res.*, 11, 321-345, 1991.

McComas, C.H., and P. Müller, The dynamic balance of internal waves, *J. Phys. Oceanogr.*, 11, 970-986, 1981.

McPhee, M.G., Drift velocity during the drift-station phase of MIZEX 83, *MIZEX Bull. IV, CRREL Spec. Rep. 84-28*, U.S. Army Cold Reg. Res. and Eng. Lab., Hanover, N.H., 1984.

McPhee, M.G., The upper ocean, in *The Geophysics of Sea Ice*, edited by N. Untersteiner, pp 339-394, Plenum Press, New York, 1986.

McPhee, M.G., Small-scale processes, in *Polar Oceanography (Part A)*, edited by W.O. Smith, Jr, pp 287-334, Academic Press, San Diego, Calif, 1990.

McPhee, M.G., Turbulent heat flux in the upper ocean under sea ice, *J. Geophys. Res.*, 97, 5365-5379, 1992.

McPhee, M.G., and L.H. Kantha, Generation of internal waves by sea ice, *J. Geophys. Res.*, 94, 3287-3302, 1989.

McPhee, M.G., and D.G. Martinson, Turbulent mixing under drifting pack ice in the Weddell Sea, *Science*, 263, 218-221, 1993.

McPhee, M.G., and J.D. Smith, Measurements of the turbulent boundary layer under sea ice, *J. Phys. Oceanogr.*, 6, 696-711, 1976.

McWilliams, J.C., Submesoscale, coherent vortices in the ocean, *Rev. Geophys.*, 23, 165-182, 1985.

Melling, H., R.A. Lake, D.R. Topham, and D.B. Fissel, Oceanic thermal structure in the western Canadian Arctic, *Continental Shelf Res.*, 3, 233-258, 1984.

Mellor, G.L., and T. Yamada, A hierarchy of turbulence closure models for planetary boundary layers, *J. Atmos. Sci.*, 31, 1791-1806, 1974.

Mellor, G.L., and T. Yamada, Development of a turbulence closure model for geophysical fluid problems, *Rev. Geophys.*, 20, 851-875, 1982.

Molnia, B., Alarming reports motivate international meeting on Arctic contaminants, *Witness the Arctic*, 1(2), 11, 1993.

Moore, R.M., and D.W.R. Wallace, A relationship between heat transfer to sea ice and temperature-salinity properties of Arctic Ocean waters, *J. Geophys. Res.*, 93, 565-571, 1988.

Morison, J.H., C.E. Long, and M.D. Levine, Internal wave dissipation under sea ice, *J. Geophys. Res.*, 90, 11,959-11,966, 1985.

Morison, J.H., M.G. McPhee, T.B. Curtin, and C.A. Paulson, The oceanography of winter leads, *J. Geophys. Res.*, 97, 11,199-11,218, 1992.

Morison, J.H., M.G. McPhee, and G.A. Maykut, Boundary layer, upper ocean, and ice observations in the Greenland Sea marginal ice zone, *J. Geophys. Res.*, 92, 6987-7011, 1987.

Morison, J.H., and the LeadEx group, The LeadEx experiment, *EOS*, 74, 393-397, 1993.

Moritz, R.E., and R. Colony, Statistics of sea ice motion, Fram Strait to the North Pole, *Proc. (4) Int. Conf. Offshore Mech. Arct. Eng., 7th*, 75-82, 1987.

Mountain, D.G., L.K. Coachman, and K. Aagaard, On the flow through Barrow Canyon, *J. Phys. Oceanogr.*, 6, 461-470, 1976.

Muench, R.D., Mesoscale phenomena in the polar oceans, in *Polar Oceanography (Part A)*, edited by W.O. Smith, Jr, pp 223-286, Academic Press, San Diego, Calif, 1990.

Muench, R.D., M.G. McPhee, C.A. Paulson, and J.H. Morison, Winter oceanographic conditions in the Fram Strait–Yermak Plateau region, *J. Geophys. Res.*, 97, 3469-3483, 1992.

Munk, W.H., Abyssal recipes, *Deep-Sea Res.*, 13, 707-730, 1966.

Munk, W.H., Internal waves and small-scale processes, in *Evolution of Physical Oceanography,* edited by B.A. Warren and C. Wunsch, MIT Press, pp 264-291, 1981.

Mysak, L.A., D.K. Manak, and R.F. Marsden, Sea-ice anomalies observed in the Greenland and Labrador Seas during 1901-1984 and their relation to an interdecadal Arctic climate cycle, *Clim. Dyn.*, 5, 111-133, 1990.

Neshyba, S., V.T. Neal, and W. Denner, Temperature and conductivity measurements under ice-island T-3, *J. Geophys. Res.*, 76, 8107-8120, 1971.

Newton, J.L., K. Aagaard, and L.K. Coachman, Baroclinic eddies in the Arctic Ocean, *Deep-Sea Res.*, 21, 707-719, 1974.

Orlanski, I., and K. Bryan, Formation of the thermocline step structure by large-amplitude internal gravity waves, *J. Geophys. Res.*, 74, 6975-6983, 1969.

Osborn, T.R., Estimates of the local rate of vertical diffusion from dissipation measurements, *J. Phys. Oceanogr.*, 10, 83-89, 1980.

Osborn, T.R., and W.R. Crawford, An airfoil probe for measuring turbulent velocity fluctuations in water, in *Air-Sea Interactions: Instruments and Methods*, edited by F. Dobson, L. Hasse, and R. Davies, pp 369-386, Plenum Press, 1980.

Ou, H.W., and A.L. Gordon, Spin-down of baroclinic eddies under sea ice, *J. Geophys. Res.*, 91, 7623-7630, 1986.

Padman, L., Momentum fluxes through sheared oceanic diffusive-convective steps, *J. Geophys. Res.*, 1994 (submitted).

Padman, L., and T.M. Dillon, Vertical heat fluxes through the Beaufort Sea thermohaline staircase, *J. Geophys. Res.*, 92, 10,799-10,806, 1987.

Padman, L., and T.M. Dillon, On the horizontal extent of the Canada Basin thermohaline steps, *J. Phys. Oceanogr.*, 18, 1458-1462, 1988.

Padman, L., and T.M. Dillon, Thermal microstructure and internal waves in an oceanic diffusive staircase, *Deep-Sea Res.*, 36, 531-542, 1989.

Padman, L., and T.M. Dillon, Turbulent mixing near the Yermak Plateau during the Coordinated Eastern Arctic Experiment, *J. Geophys. Res.*, 96, 4769-4782, 1991.

Padman, L., T.M. Dillon, H.W. Wijesekera, M.D. Levine, C.A. Paulson, and R. Pinkel, Internal wave dissipation in a non-Garrett-Munk ocean, *Proceedings of the 'Aha Huliko'a Hawaiian Winter Workshop*, University of Hawaii at Manoa, Honolulu, 1991.

Padman, L., and M.D. Levine, Turbulence and thermal finestructure in the western Weddell Sea, *Antarctic Journal of the United States*, XXVII (5), 105-107, 1992

Padman, L., M.D. Levine, T. Dillon, J. Morison, and R. Pinkel, Hydrography and microstructure of an Arctic cyclonic eddy, *J. Geophys. Res.*, 95, 9411-9420, 1990.

Padman, L., A.J. Plueddemann, R.D. Muench, and R. Pinkel, Diurnal tides near the Yermak Plateau, *J. Geophys. Res.*, 97, 12,639-12,652, 1992.

Paquette, R.G., and R.H. Bourke, Temperature fine structure near the sea-ice margin of the Chukchi Sea, *J. Geophys. Res.*, 84, 1155-1164, 1979.

Parkinson, C.L., Interannual variability of the spatial distribution of sea ice in the north polar region, *J. Geophys. Res.*, 96, 4791-4801, 1991.

Perkin, R.G., and E.L. Lewis, Mixing in the West Spitsbergen Current, *J. Phys. Oceanogr.*, 14, 1315-1325, 1984.

Plueddemann, A.J., Internal wave observations from the Arctic Environmental Drifting Buoy, *J. Geophys. Res.*, 97, 12,619-12,638, 1992.

Polzin, K.L., J.M. Toole, and R.W. Schmitt, Finescale parameterization of turbulent dissipation, *J. Phys. Oceanogr.* (in press), 1994.

Prazuck, C., Anomalous diurnal currents in the vicinity of the Yermak Plateau, Ph.D. dissertation, 125 pp., Nav. Postgrad. Sch., Monterey, Calif., 1991.

Prinsenberg, S.J., and E.B. Bennett, Mixing and transports in Barrow Strait, the central part of the Northwest Passage, *Continental Shelf Res.,* 7, 913-935, 1987.

Rudels, B., The formation of Polar Surface Water, the ice export and the exchanges through the Fram Strait, *Prog. Oceanog.*, 22, 205-248, 1989.

Schmitt, R.W., Double diffusion in oceanography, *Annu. Rev. Fluid Mech.*, 26, 255-285, 1994.

Smith, D.C., and J.H. Morison, A numerical study of haline convection beneath leads in sea ice, *J. Geophys. Res.*, 98, 10,069-10,083, 1993.

Smith, S.D., Coefficients for sea surface wind stress, heat flux, and wind profiles as a function of wind speed and temperature, *J. Geophys. Res.*, 93, 15,467-15,472, 1988.

Steele, M., J.H. Morison, and T.B. Curtin, Halocline water formation in the Barents Sea, *J. Geophys. Res.*, 1994 (submitted).

Taylor, J., The fluxes across a diffusive interface at low values of the density ratio, *Deep-Sea Res.*, 35, 555-567, 1988.

Toole, J.M., and D.T. Georgi, On the dynamics and effects of double-diffusively driven intrusions, *Prog. Oceanogr.*, 10, 123-145.

Toole, J.M., K.L. Polzin, and R.W. Schmitt, New estimates of diapycnal mixing in the abyssal ocean, *Science* (submitted), 1994.

Treshnikov, A.F., *Atlas Arktiki. Arkt.-Antarkt.*, Nauchno-Issled. Inst., Moscow, 1985.

Turner, J.S., *Buoyancy Effects in Fluids*, Cambridge University Press, New York, 1973.

Walsh, J.E., and W.L. Chapman, Arctic contribution to upper-ocean variability in the North Atlantic, *J. Climate*, 3, 1462-1473, 1990.

Weller, R.A., J.P. Dean, J. Marra, J.F. Price, E.A. Francis, and D.C. Boardman, Three-dimensional flow in the upper ocean, *Science*, 227, 1552-1556, 1985.

Wijesekera, H.W., and T.M. Dillon, Internal waves and mixing in the upper equatorial Pacific Ocean, *J. Geophys. Res.*, 96, 7155-7125, 1991.

Wijesekera, H.W., L. Padman, T.M. Dillon, M.D. Levine, C.A. Paulson, and R. Pinkel, The application of internal wave dissipation models to a region of strong mixing. *J. Phys. Oceanogr.*, 23, 269-286, 1993a.

Wijesekera, H.W, T.M. Dillon, and L. Padman, Some dynamical and statistical properties of turbulence in the oceanic pycnocline. *J. Geophys. Res.*, 98, 22,665-22,679, 1993b.

Yang, J., and J.D. Neelin, Sea-ice interaction with the thermocline circulation, *Geophys. Res. Letters*, 20, 217-220, 1993.

4

New Insights on Large-Scale Oceanography in Fram Strait: The West Spitsbergen Current

Jean-Claude Gascard, Claude Richez and Catherine Rouault

Abstract

Following a review of selected relevant papers published during the last decade - starting with the MIZEX 84 experiment - we present new information based on hydrological and neutrally buoyant floats data which shows an undocumented pattern of warm and salty waters embedded in the West Spitsbergen Current at its inception into the Arctic Ocean. We emphasize the role of topography, in particular the Yermak Plateau, on this new pattern of circulation. We then strengthen the importance of the circulation of these waters on the ice edge, maintained beyond 80°N North of Svalbard year round. Finally, we present new information about the recirculation in the Fram Strait and Greenland Sea of warm, salty West Spitsbergen water masses escaping from the Arctic Ocean just before and/or after having been transformed in Arctic Intermediate Waters.

Introduction

The Fram Strait is a key feature for the large-scale world ocean circulation. It controls the main exchanges of heat and salt between the Arctic Ocean and the North Atlantic Ocean. It also has a strong impact on global climate, particularly in the northern hemisphere. It interacts strongly with the North Atlantic Subpolar gyre at seasonal, interannual and decadal time scales. For these reasons, many polar scientists have dedicated important work in the recent past to a better understanding of large-scale oceanic circulation in the Fram Strait.

The transports of ice and water across the Fram Strait had long been recognized as important factors for the World Ocean Circulation and Climate related problems. The

Arctic Oceanography: Marginal Ice Zones and Continental Shelves
Coastal and Estuarine Studies, Volume 49, Pages 131–182
Copyright 1995 by the American Geophysical Union

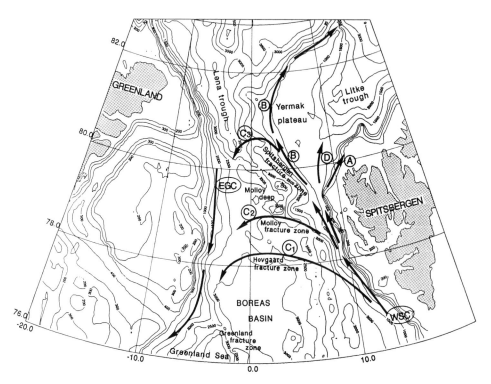

Figure 1. Schematic circulation showing the West Spitsbergen multi-path current system: the coastal branch A, the West Yermak branch B, the recirculating branches C1 on the Hovgaard Fracture Zone, C2 on the Molloy Fracture Zone, C3 on the Spitsbergen Fracture Zone and the Litke Trough Branch D. Bottom Topography in meters

"Great Salinity Anomaly" identified in the North Atlantic Ocean during the 1960s and 1970s by Dickson et al. [1988], more likely resulting from an excess fresh water discharge (including sea-ice) from the Arctic Ocean [Aagaard and Carmack, 1989; Häkkinen, 1993], is a good example. A general description of sea ice in the Fram Strait has been given by Wadhams [1983], Vinje and Finnekasa [1986] and completed by Gascard, et al. [1988], and documents in great detail summer conditions during Mizex 1984. The Transpolar Ice Drift approaching Fram Strait, is composed of two main streams: *one* coming from the North, entering the Lena Trough at about 81°N, merges with the East Greenland Current (EGC) at about 80°N slightly east of 5°W, then following the East Greenland Continental Slope all the way down to the southern tip of Greenland at 60°N; *the other* approaches from the North-East, crossing the Yermak Plateau and merging with the northern stream between 80°N and 79°N. The northern ice stream is primarily composed of thick and sometimes large multiyear ice floes, rather than younger, thinner and smaller first and second year ice floes found in the North-Eastern ice stream. The ice edge in Fram Strait and North of Svalbard is mainly populated with north-eastern ice floes. The annual total flux of sea-ice through Fram Strait had been evaluated by Vinje and Finnekasa [1986] at around 5000 km³. Most of it is composed of northern ice floes, since a large amount of the north-eastern ice floes cannot escape the melting induced by the northward flow of warm and salty waters carried on by the West Spitsbergen Current (WSC).

The WSC has a very complex filamentary structure, much more complex than the EGC. We will use "branch" instead of "filament" [Perkin and Lewis, 1984], to define the WSC multi-path current system. Four types of branches (Figure 1) have been described in the literature associated with the WSC in Fram Strait: (1) *a north-eastward coastal branch A*, following the shelf break North of Svalbard and described by Aagaard, et al. [1987]. (2) *a northward branch B* following the continental slope West of the Yermak Plateau, mentioned by Perkin and Lewis [1984] and more recently by Muench et al. [1992]. (3) *westward recirculating branches C* following topographic fracture zones, crossing Fram Strait in southeast-northwest direction at different latitudes from 78°N to 81°N, described by Quadfasel et al. [1987], Gascard et al. [1988] and Bourke et al. [1988]. (4) *a fourth type of branch D* has recently been proposed by Manley et al. [1992], and concerns primarily the upper layer. According to the authors, this branch enters the Litke trough from the South and recirculates cyclonically around the periphery of the Yermak Plateau. A general upper cyclonic circulation around the Yermak Plateau has also been proposed by Muench et al. [1992]

The coastal branch A has been studied in great detail by Aagaard et al. [1987] from data collected in 1977. Bourke et al. [1988] analyzed hydrological data collected in 1985 and presented a quite complete description of the EGC and WSC baroclinic transports in Fram Strait. They estimated the baroclinic transports for both currents to be close to 1.5 Sverdrups, if the WSC recirculating part merging in the EGC is excluded. This value has to be compared with 3 to 6 Sverdrups total averaged transport for WSC, including the barotropic component, proposed by different authors (extreme values are from 0 to 9 Sverdrups [Hanzlick, 1983]. Regarding EGC, Foldvik et al. [1988] estimated yearly mean total transport is 3 Sv. for the first 700 m beneath the surface, with no seasonal variability. All authors emphasized the role of topography in Fram Strait on the WSC multi-path current system, which is not surprising given the strength of the barotropic component. According to Bourke et al. [1988], 20% of WSC is trapped in the coastal branch A and represents the only part of the WSC warm and salty waters penetrating into the Arctic Ocean. The 80% remaining of the WSC would be composed entirely of WSC C branches recirculating westwards and then southwards as soon as they merge with the EGC. Most of this recirculation should occur between 78°N and 81°N. No WSC branch B was found progressing northwards beyond 81°N. This result was somewhat in contradiction with Perkin and Lewis [1984] and Muench et al. [1992], who indicated warm and salty lenses at 82°- 83°N on the northern edge of the Yermak Plateau, unless these lenses structures were related to interruptions of the WSC branch B. Bourke et al. [1988] observations might have well corresponded to one of these interruptions. Rudels [1987] also found that a large part of the Atlantic Water in the WSC recirculated in the northern vicinity of the Strait.

In this paper, based on new data collected primarily during the Arctemiz 88 experiment, we will document and analyze in great and unprecedented detail the northward WSC branch B and the westward-southward recirculating WSC branches C from Fram Strait to the Greenland Sea.

In the *first and second sections* we will present two data sets : the first, based on 111 CTDO$_2$ hydrological stations taken West and North of Spitsbergen in August-September 1988 during Arctemiz 88, will define water masses characteristics. We will then compute

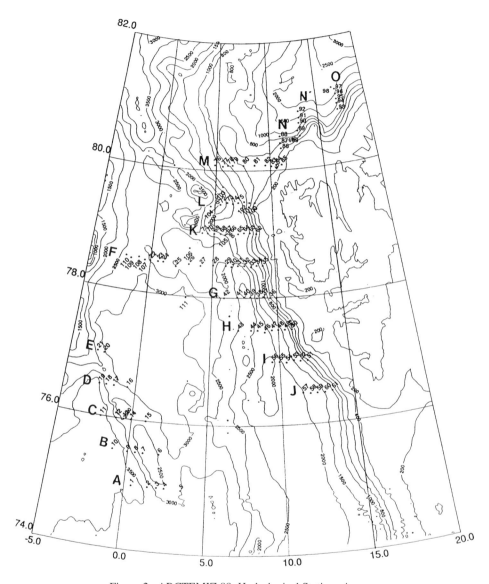

Figure 2. ARCTEMIZ 88: Hydrological Stations Array.

geostrophic currents and baroclinic transports; the second is based on the trajectories of 12 neutrally buoyant drifting floats launched late August-early September 1988 (ARCTEMIZ 88) in Fram Strait. *8 floats, ballasted for 300 m depth*, were deployed in the core of the WSC, along the continental slope West of Spitsbergen from 76°30N up to 79°30N and between 500 m and 1500 m isobaths. *4 floats, ballasted for 1000 m depth*, were launched in the Molloy Deep area. In addition, we will show the trajectory of other floats launched in Fram Strait during spring 1989 (CEAREX 89) and will compare them with float trajectories obtained during the summer experiment MIZEX 84 [Manley et al.,

1987; Quadfasel et al., 1987; Gascard et al., 1988]. All these trajectories range from a period of a few months to over one year. In a *third section*, we will discuss three main topics raised by these new observations : the first topic is related to the importance of the Yermak Plateau topography. The float trajectories reveal an undiscovered path for the circulation of WSC warm and salty waters (**branch B'**) entering into the Arctic Ocean and strongly influenced by the topography of the Plateau. The second topic deals with the "Return Atlantic Current" (RAC), which corresponds to the recirculating WSC branches C escaping from the Arctic Ocean just before and/or after having been transformed into Arctic Intermediate Waters. These recirculating waters play a major role in the central Greenland Sea, where deep convection occurs in winter to form the new Greenland Sea Deep Water, a basic component of the North Atlantic Deep Water. They also play a major role as a main driving force for the global thermohaline circulation. The third topic emphasizes the role of oceanic heat and salt fluxes primarily due to the northward propagating WSC branch B' constraining the ice edge North of Spitsbergen. We will compare our results with recent estimations made by Aagaard et al. [1987], Quadfasel et al. [1987] and Untersteiner [1988] of the ice and heat balance in Fram Strait. This is the ice edge northernmost latitude, although paleoclimatologists recognise it as the ice edge having undergone the most dramatic variations between glacial-interglacial periods, most likely because it is closely related to dramatic changes in large scale ocean circulation. Finally, a *last section* will develop our conclusions.

Hydrological Data

Water Masses

111 hydrological stations were conducted between August 15 and September 6, 1988 with the French Research Vessel Cryos (Figure 2). 5 short EW sections (21 stations) were first taken across the Greenland Fracture zone, followed by 8 EW sections (86 stations) across the continental slope West of Spitsbergen and 3 NS short sections (14 stations) across the northern slope completed the network. For all these stations, we used a CTDO$_2$ Neil Brown Mark III. 700 water samples were taken for tracers analysis and to calibrate the CTDO$_2$ sensors. The pressure sensor was calibrated in the laboratory before and after the cruise. The hysteresis and non linearity for the pressure sensor have been compensated to give an accuracy of 1 decibar between 0 and 6000 decibars. The temperature sensor, after laboratory calibration, still had a bias at low temperatures of nearly -0.01°C, as indicated by in situ calibration using reversing thermometers. This bias has been corrected and all the temperatures are given within an accuracy of 0.005°C and a precision of 0.002°C. The conductivity sensor was calibrated using more than 700 samples collected during the experiment and the final accuracy is equivalent to 0.005 psu with a standard deviation of 0.004 psu. The oxygen sensor was calibrated to a final accuracy of 0.05 ml/l and a standard deviation of 0.05 ml/l.

A global θ-S diagram (Figure 3), using data between the surface and 1500 m depth, taking into account all the hydrological stations (except the 21 first ones, conducted across the Greenland Fracture Zone and not considered in this paper) shows the characteristic water masses involved in the West Spitsbergen Current: following Swift and Aagaard [1981] and Aagaard et al. [1987] nomenclature, the *Atlantic Water* (AW:

THETA-0.0 / SALINITY DIAGRAM

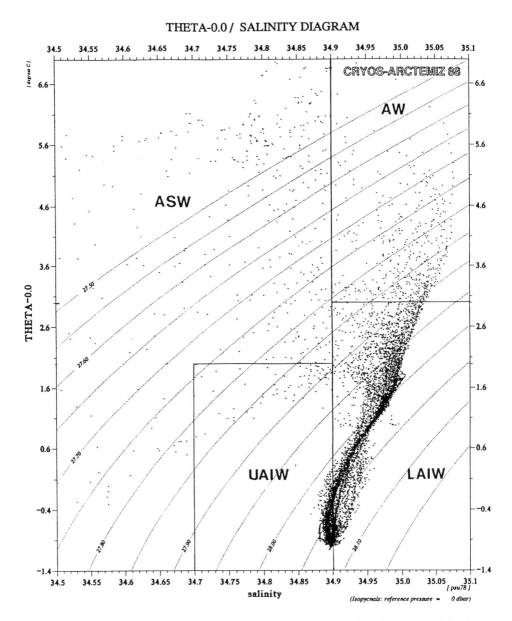

Figure 3. Potential Temperature - Salinity Diagram for the ARCTEMIZ 88 Hydrological Stations 22 to 111, from surface to 1500 decibars, by 10 dbars. AW: Atlantic Water; ASW: Arctic Surface Water, UAIW and LAIW: Upper and Lower Arctic Intermediate Water.

Temperature> 3°C and Salinity> 34.9 psu), the *Lower Arctic Intermediate Water* (LAIW: Temperature< 3°C and Salinity> 34.9 psu), the *Upper Arctic Intermediate Water* (UAIW: Temperature< 2°C and 34.7< Salinity< 34.9 psu), and the *Arctic Surface Water* (ASW: Temperature> 2°C, with Salinity< 34.9 psu and Temperature< 2°C with Salinity < 34.7).

The diagram shows that the most abundant water masses in the eastern Fram Strait are the AIW and AW, between 3 and 6 °C, mainly located along the West Spitsbergen continental slope. ASW is only present in westernmost stations, while *Polar Water* (PW: Temperature< 0°C and Salinity< 34.4 psu) - not represented in this diagram - is found at the surface only in the northernmost sections M, N, N' and O.

The Southern Sections

From sections taken across the continental slope West of Spitsbergen, one can identify the core of the WSC characterized by warm and salty Atlantic waters (> 4°C and > 35.00 psu) in the eastern part of the different sections. All along the short southernmost section J, the salinity maximum extends between 80 to 300 m depth (Figure 4.a). In section I, extending more to the West (Figure 4.b), a much thinner second maximum appears between 100 to 160 m, while the main core extends between 80 and 400 m depth, as in section J, in the eastern part of the section. In these two sections, AW is associated with an oxygen minimum (Figure 5). In Section H (Figure 4.c), at 77°30N, and also in section G (Figure 4.d), at 78°N, water with salinity > 35 psu is only present on the eastern side of the section, between 100 and 350 to 400 m depth. In these two sections, as in section K (Figure 6.a), at 79°N, a second core of slightly less salty and much colder high oxygen water (S > 34.98 psu, θ > 1.5°C, O_2 <7.3 ml/l) is found on the western side. Surprisingly, section F (Figure 7), at 78° 30'N, shows, only in the eastern part of the section, two or three small patches of salinity > 35.0 psu at around 200 to 250 m depth, as if the major part of the WSC disappeared or was deviated to the east of the section !

The geostrophic velocity calculations along sections often show, side by side, quite intense northward and southward flows, indicative of strong shear. In Section H (Figure 8.a) the baroclinic part of the WSC main core, east of the section, corresponds to a northward baroclinic transport of about 0.5 Sv, while, offshore of this main core, one can see a very intense anticyclonic eddy. This eddy, 50 km in diameter, extends down to 1000 m, at least for the baroclinic part, and corresponds to a maximum current of 14 cm.s^{-1} between 100 and 400 m depth. The eddy structure is very symmetrical; the integrated eddy baroclinic transport is about 2 Sv., much more than the baroclinic transport associated with the WSC main core. The water masses characterizing this anticyclonic eddy are mostly LAIW, very dense (close to 28), with salinity up to 34.98 psu (Figure 4.c) and very cold temperature , 1.5°C in the core and extending to 0°C at 1000 m (Figure 8.b).8.a).

Section K (Figure 6.a) also plainly shows the WSC main core, between stations 62 and 65, with a maximum current of 12 cm.s^{-1} (Figure 6.b), located over 1200 m isobath, and with a baroclinic transport of about 0.5 Sv, which is coherent with Bourke et al. [1988] at this latitude. Offshore, one can observe a strong southward current more likely corresponding to the eastern part of an other intense anticyclonic eddy implying a baroclinic transport of about 1.5 Sv (as strong as the eddy at 77°30N (section H: Figure 8.a).

Section F (Figure 9), the longest among our sections, is puzzling. As previously mentioned, the WSC core does not seem to be visible in this section. It could exist east of stations 34 and 35 over 800 to 1000 m depth. However, one can observe, in the center of section F at 5°E, a gigantic meander or a cyclonic eddy 100 km in diameter extending

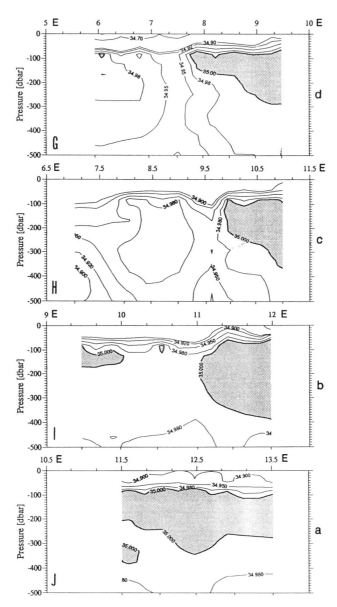

Figure 4. Vertical distribution, from South to North, of Salinity between 0 and 500 decibars. *Dark grey*: Salinity > 35 psu. **(a)** Section J at 76° 30'N; **(b)** Section I at 77°N; **(e)** Section H at 77° 30'N; **(d)** Section G at 78°N.

from 3°E to 7°E and from top to bottom (2000 m depth). The integrated eddy baroclinic transport associated with this feature is about 2 Sverdrups. In Figure 7, one notes that the center part of this feature is characterised by lower salinity (34.90 instead of 34.95 psu) and colder temperatures (1°C colder in the center than on the edge). The 0°C and 1°C isotherms are lifted up respectively from 600 m and 400 m depth outside the eddy to 400

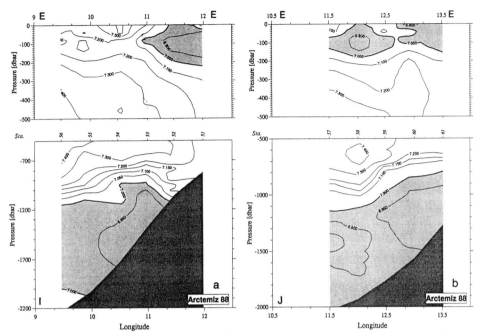

Figure 5. Vertical distribution of Oxygen (ml/l) at (a) Section I (b) Section J. - *Light grey:* Oxygen < 7 ml/l.

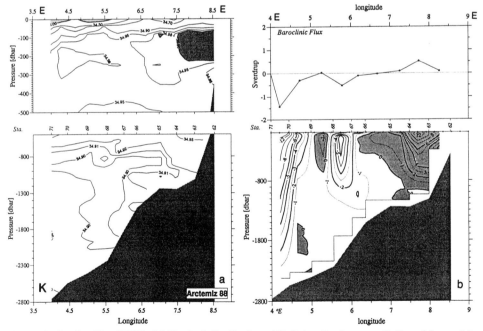

Figure 6. Section K at 79°N: (a) Vertical distribution of Salinity. *Dark grey*: salinity > 35 psu. (b) Baroclinic Geostrophic Velocity and Baroclinic Transport. Reference surface, taken at the deepest common value between 2 successive stations, is indicated along the section by a light line. *Dark grey*: Northward positive velocity values.

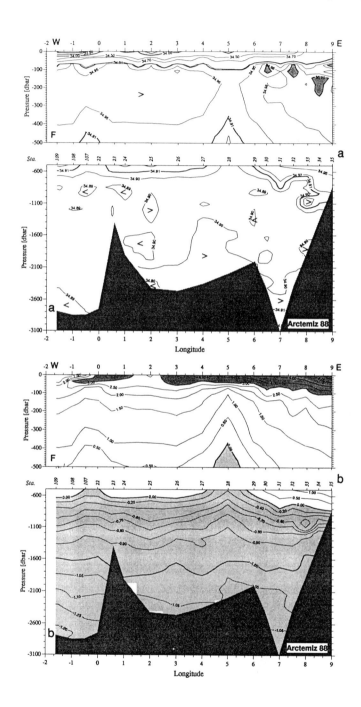

Figure 7. Section F at 78°30'N: (a) Vertical distribution of Salinity. *Dark grey*: salinity > 35 psu. (b) Vertical distribution of Potential Temperature : *Dark grey*: Temperature > 3°C - *Light grey:* Temperature < 0°C

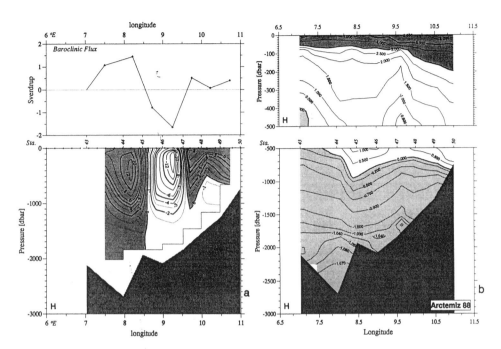

Figure 8. Section H at 77°30'N: (a) Baroclinic Geostrophic Velocity (cm.s⁻¹) and Baroclinic Transport. Reference surface, taken at the deepest common value between 2 successive stations, is indicated along the section by a light line. *Dark grey*: Northward positive velocity values. (b) Vertical distribution of Potential Temperature : *Dark grey*: Temperature > 3°C - *Light grey*: Temperature < 0°C

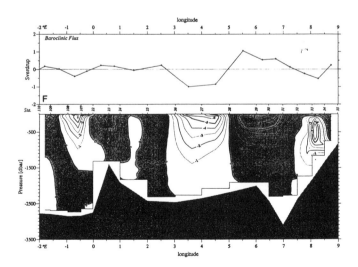

Figure 9. Section F at 78° 30'N: Baroclinic Geostrophic Velocity (cm.s⁻¹) and Baroclinic Transport. Reference surface, taken at the deepest common value between 2 successive stations, is indicated along the section by a light line.*Dark grey*: Northward positive velocity values.

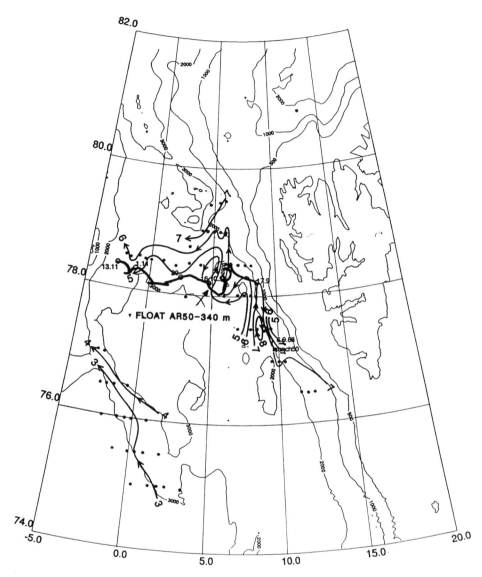

Figure 10. Dynamic Topography at 300 decibars relative to 1500. *Superimposed:* Trajectory of float 50 from September 17, 1988 to November 13.

m and 200 m depth in the center, intruding respectively in the 1°C and 2°C domain. These figures confirm the *unsteadiness of the WSC, and its propensity to form eddies.*

This feature also appears on the dynamic topography at 300 dbar relative to 1500 (Figure 10). Superimposed on this topography, the westward undulating trajectory of float 50, launched on September 6 around 77°15N, suggests the float drift was strongly influenced by these eddies (cf. § "The Hovgaard Fracture Zone eddy. The Return Atlantic Current

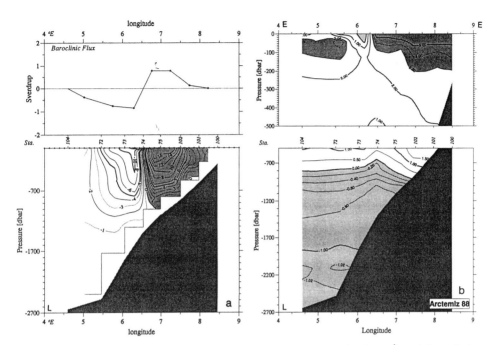

Figure 11. Section L at 79°30'N: (a) Baroclinic Geostrophic Velocity (cm.s⁻¹) and Baroclinic Transport. Reference surface, taken at the deepest common value between 2 successive stations, is indicated along the section by a light line. *Dark grey*: Northward positive velocity values. (b) Vertical distribution of Potential Temperature: *Dark grey*: temperature > 3°C - *Light grey*: temperature < 0°C

(RAC) at 78°N"). However, one should not forget that in doing such a comparison, there is a time lag (1 to 2 months) between the float drift and the time the execution of section F.

The geostrophic velocity along section L (Figure 11.a) highlights the presence of a *cyclonic* eddy at 79°30N, mainly composed of AIW. It is interesting to compare it with the *anticyclonic* eddy observed in section H, at 77°30N. Both are basically composed of the same water masses and extend to a comparable depth (1000 m), but their environment is quite different. In section H (Figure 9.b), deep isotherms sink under the anticyclonic eddy core, while in section L, at 79° (Figure 11.b), the same isotherms rise up towards the cyclonic eddy core. The cyclonic eddy is also surface, rather than subsurface, intensified for the anticyclonic eddy.

WSC branching around 79°N

In section L (Figure 12.a), at 79°30N, and M (Figure 12.b), at 80°N, the two separate cores of WSC, one to the East and one to the West of the section, show the well-known split of the WSC, North of Spitsbergen, into two branches : branch **A** turning East and

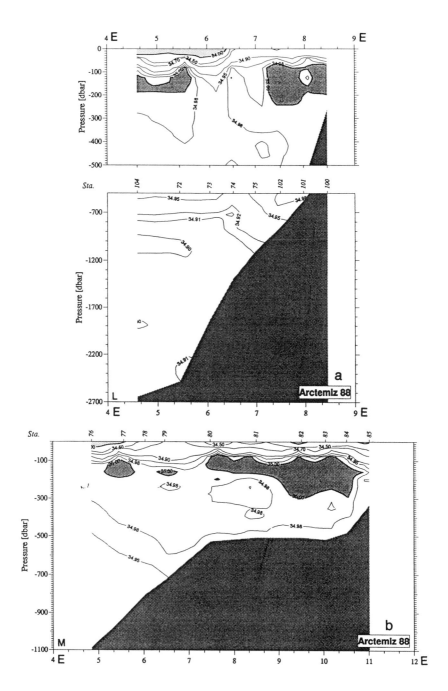

Figure 12. Vertical distribution of Salinity. *Dark grey*: Salinity > 35 psu - *Light grey:*   salinity < 34 psu. (a) Section L at 79°30'N. (b) Section M at 80°05'N

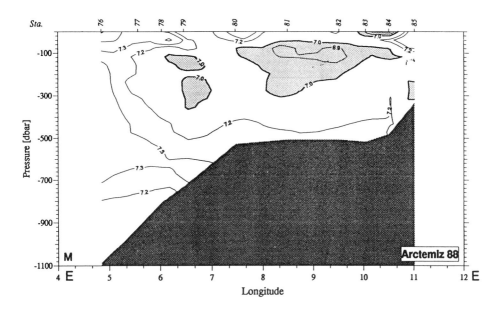

Figure 13. Section M at 80°05'N: Vertical Distribution of Oxygen (ml/l).:*Light grey:* Oxygen < 7 ml/l.

flowing along the northern coast of the Svalbard along isobaths less than 500 m deep, and branch **B**, centered above isobaths 800 m, going northwards following the western flank of the Yermak Plateau [Perkin and Lewis, 1984; Aagaard et al., 1987]. The O_2 minimum associated with the WSC, is more pronounced for the coastal branch than for the north going branch (Figure 13), indicative of a less active mixing with ambient water for the coastal branch. The geostrophic baroclinic velocity section M (Figure 14) confirms the branching of the WSC in this area. The baroclinic transport to the North, on the western side of the section M, is about 1 Sv, between stations 79 and 76, while the transport associated with the coastal branch is quite low in comparison (about 0.1 Sv), in agreement with Bourke et al. [1988] estimations.

The Northern Sections

The South-North Section N' (Figures 15.a and .b), at 13°E, shows the offshore part of the WSC coastal branch. Salinity maxima reach 35 psu for a corresponding temperature of 2.5°C at 300m depth. A very distinct surface intensified cyclonic eddy appears, about 37 kilometers in diameter, centered above the 1200 m isobath (Figure 15.c), on the offshore side of the WSC coastal branch. It is characterised by Arctic Intermediate Waters (< 2°C) and seems very similar to the cyclonic eddy observed at 79°30N (Figure 11).

The easternmost section O (Figure 16) reveals two cores, composed of remnants of Atlantic Waters (S>35 psu). The offshore core is very shallow, extending from the surface

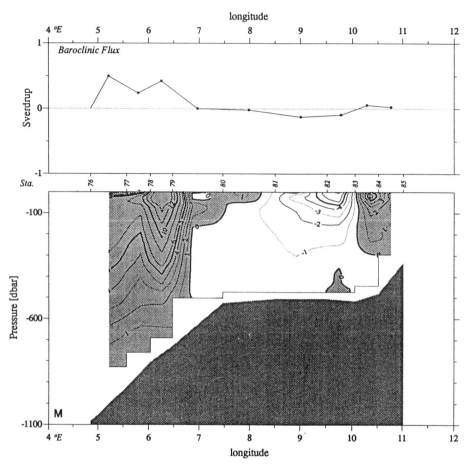

Figure 14. Section M at 80°05'N: Baroclinic Geostrophic Velocity (cm.s^{-1}) and Baroclinic Transport. Reference surface, taken at the deepest common value between 2 successive stations, is indicated along the section by a light line. *Dark grey*: Northward positive velocity values.

down to 150 m, under a thin layer of ASW (T < 0°C, at station 98). The Atlantic Water core is surprisingly warm (> 4°C), and causes much ice melting as shown by an AVHRR image (Figure 17). The northern coast of Nordauslandlet (Island of Svalbard) was completely free of ice and the melting zone extended to 81°N and 20°E. By comparison, another AVHRR image (Figure 33), taken in Summer 89, shows the ice covering the whole area between 80°-81°N and 15°-20°E.

Regarding *heat flux*, it is interesting to compare our 1988 observations with Aagaard et al. [1987] (fig. 7 p. 3783), based on 1977 observations in the same area (stations AA-x). Station AA-6 at 79°N and 9°E in 1977, and station 62 on section L indicate the same mean temperature (4°C) in the 100-200 m layer, but salinity at station 62 (35.06 psu) is much higher than at station AA-6 (34.94 psu). At 80°N, station AA-94 indicated 3°C and

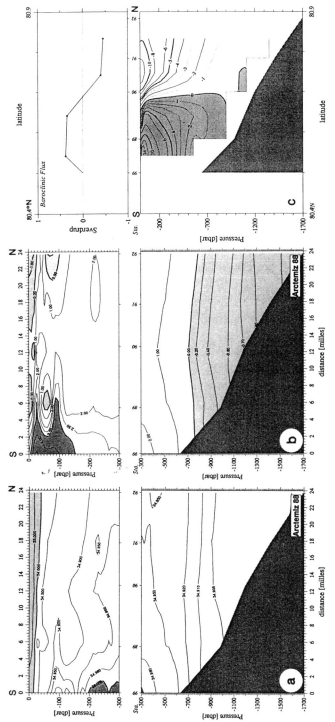

Figure 15. South-North Section N' along 13°E: (a) Vertical distribution of Salinity. *Dark grey:* salinity > 35 psu - *Light grey:* salinity < 34 psu. (b) Vertical distribution of Potential Temperature: *Dark grey:* temperature > 3°C - *Light grey:* temperature < 0°C. Baroclinic Geostrophic Velocity (cm.s⁻¹) and Baroclinic Transport. Reference surface, taken at the deepest common value between 2 successive stations, is indicated along the section by a light line. *Dark grey:* Eastward positive velocity values.

34.94 psu as mean values for the 100-200 m layer, compared to 3°C and 35.03 psu at our stations 82-83 (section M). Farther North, station AA-37 (81°10N and 18°E) indicated 3°C and 34.92 psu compared to 2°C and 34.96 psu at our station 93 (section O). The temperatures are quite similar in 1977 and 1988 in the WSC coastal branch, confirming the heat flux calculations of Aagaard et al. [1987]. However, salinities are much higher in 1988 (> 0.1 psu) and more melting occurs, which explains colder temperatures in 1988 farther East. One can conclude, referring to this thorough inspection of the WSC hydrological structure, to:

- *the unsteadiness characteristic of the WSC*: The mean baroclinic transport associated with the WSC is of the order of 0.5 Sverdrups in agreement with Bourke et al. [1988] estimations. The maximum baroclinic geostrophic velocities are 14 cm.s^{-1}.

- *the high eddy activity*: Eddies are both cyclonic and anticyclonic, mainly composed of Arctic Intermediate Waters strongly interacting with Atlantic Waters which are the main component of WSC. Cyclonic eddies are surface intensified (max. baroclinic velocities 15 cm.s^{-1} at surface) and anticyclonic eddies are subsurface intensified (max. baroclinic geostrophic velocities 14 cms^{-1} at depth 200-500 m). Typical baroclinic transports associated with eddies are much larger than typical baroclinic transport associated with WSC suggesting eddies are more baroclinic and WSC more barotropic.

- *the presence of two cores in the WSC North of 79°N*, confirming the WCS branching at this latitude. Active mixing is occurring in the offshore branch, more exposed to interactions with AIW than the coastal branch. Heat losses associated with the coastal branch North of Spitsbergen are in agreement with earlier observations [Aagaard et al., 1987], but WSC Atlantic Waters were 0.1 psu saltier in 1988 than in 1977, causing more ice melting.

Floats Data

During the second part of August and the first week of September 1988, 12 neutrally buoyant drifting floats with SOFAR acoustic transmitters (260 Hz) and 6 subsurface moorings equipped with autonomous listening stations (ALS) were deployed West and North of Svalbard (Figure 18). Floats were expendable and ALSs had to be recovered. 4 floats (46 to 49) were ballasted for 1000 m depth and 8 (50 to 57) for 300 m depth. Each float weighted over 300 kgs in air and the ballast was determined to within a few grams by double hydrostatic weighting. Floats were made out of a 4m long aluminium tube, 30 cm in diameter. A 260 Hz resonant aluminium tube (like an organ pipe), about 2 m long and 30 cm in diameter, was fixed at one end and the total length of the instrument was about 6 m. A transducer activated the resonance at 260 Hz. Floats were about twice less compressible than sea water, but, because of some uncertainties in compressibility coefficient and thermal expansion, it was difficult to compute the necessary ballast to better than 5 to 10%. The "300 m" floats sank initially to depths ranging from 315 to 345 m and the "1000 m" floats to 1055 and 1065 m depth.

The 6 ALS moorings were recovered the following year, the 3 southernmost (M1, M3, M4) by the German Research Vessel Valdivia between August 19 to 21, 1989, and 2 out of the 3 northern moorings by the Norwegian Coast Guard Andenes on September 2 and 3, 1989. The 6th mooring (M7) could not be recovered because of heavy ice conditions and has never been retrieved, even after another try, one year later. 3 out of the 5 ALSs, recovered in August and September 1989 worked well (M1, M5 and M6), but data were

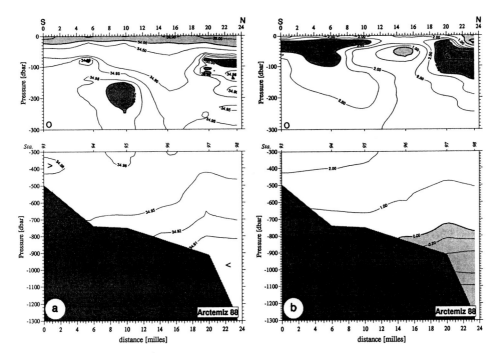

Figure 16. South-North Section O along 17°E: (a) Vertical distribution of Salinity. *Dark grey*: salinity > 35 psu - *Light grey:* salinity < 34 psu. (b) Vertical distribution of Potential Temperature: *Dark grey*: temperature > 3°C - *Light grey:* temperature < 0°C.

never produced by the two last ones. Floats sent a SOFAR signal 3 times a day at 8 hours intervals with great precision (accuracy: 0.1 second). The ALS measured the arrival times with the same accuracy, corresponding to a 150 m uncertainty, with a sound speed of 1460 m.s^{-1}. After recovery, the ALSs clock drift was measured by comparison to a reference clock; drift was corrected linearly over one year. The clock drift of the floats could also be determined so long as 3 ALSs at a time received their acoustic signals. The accuracy for positioning float also depended on the accuracy of the ALS mooring location which we estimated to be about 1/2 km given mooring motion and the uncertainty on mooring anchor location. The total accuracy for float location is estimated at about +/- 1 km in the case float signal could be received by 3 ALSs.

The results obtained during the first 2 months after deployment (September-October 1988), are excellent; the receiving conditions were at their best and the ice cover remained at minimum extent. As soon as freezing started late October, acoustic propagation became more limited and only a few floats could be located later in the year until December 1988. During winter, January-March 1989, none of the floats were located although some float signals were still received by one ALS (the nearest), indicating that at least some floats were still transmitting. Surface ice conditions are obviously the limiting factor due to a *decorrelation of the acoustic signal each time acoustic rays are reflected by the ice*. The last float, located early December 1988, was received again by the ALSs early April 1989, when pack-ice started to retreat. This float was then continuously located until June 1989 when the ALSs stopped recording

Figure 17. NOOA 7 AVHRR satellite image on September 4, 1988. A light grey mask is applied to clouds. In red, sea surface temperature > 4°C. *Superimposed:* Trajectories of ARCTEMIZ 88 floats 52 (purple-pink), 53 (green), 54 (white) and 57 (red).

prematurely, due to an error in the tape length. Additional floats were launched by Manley and Owens in Fram Strait in Spring 1989 during CEAREX 89, and a second array of 4 ALSs was deployed for another year in the Greenland Sea in September 1989, by the US Vessel "Bartlett". It was recovered by the "Bartlett" in August 1990, allowing the completion of 2-year SOFAR float tracking in the Arctic Ocean, Fram Strait and Greenland Sea, from 72°N up to 82°N and between 10°E to 10°W.

Next, we will first describe, in detail, the 3-month period (Sept-Oct-Nov 1988), following the initial deployment, late August 1988, of 12 floats. Then we will comment on the large scale drift of some floats tracked during 2 years: August 1988-August 1990. We will

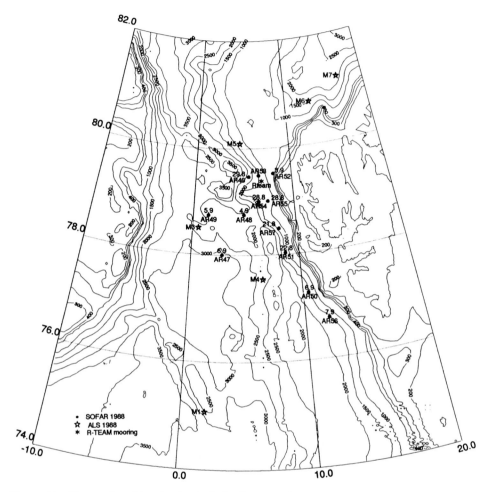

Figure 18. Time and position for Arctic Sofar floats AR 46 to AR 57 deployments (dots) during Arctemiz 88 (August-September 1988). Positions of moorings M1 to M6 (stars) with Autonomous Listening Stations (ALS) and currentmeters (R-TEAM - double cross). Bottom Topography in meters

consider the topographic effect and compare the 1988 results with those obtained in summer during Mizex 84 (June-July 1984), and will show the privileged paths of the Return Atlantic Current, and the Ocean-Ice interactions North of Spitsbergen.

3-month period : September-November 1988

Figures 19 and 20 represent, from September to November 1988 (Arctemiz 88), the overall trajectories of the 12 floats launched in Fram Strait, late August-early September

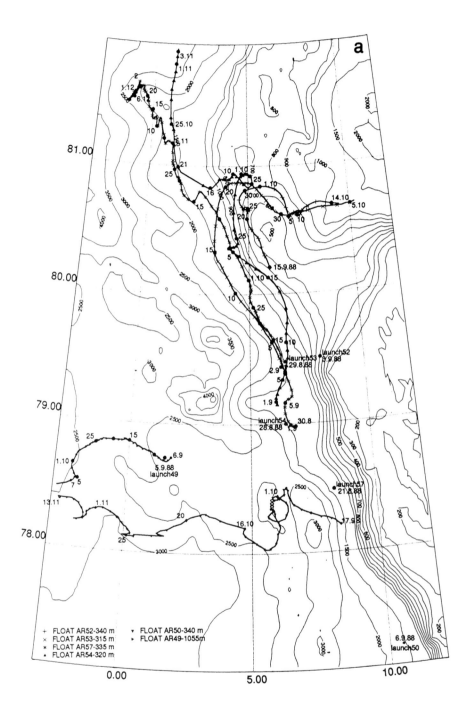

Figure 19.a.   Trajectories of 5 shallow (about 300 m) and 1 deep (49 - about 1000 m) Sofar floats from August 1988 until December 1988. Bottom Topography in meters

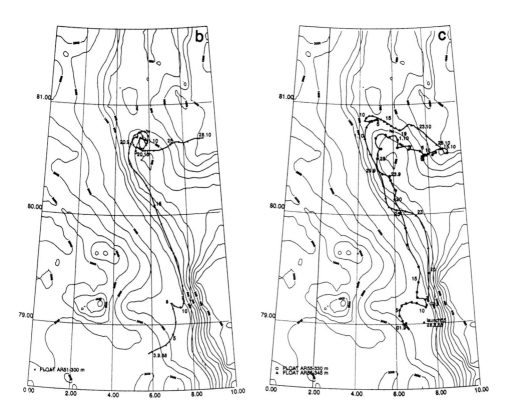

Figure 19.b and .c. Trajectories of shallow (about 300 m) Sofar floats from August 1988 until December 1988. Bottom Topography in meters

1988: None of the floats has been dragged along in the coastal WSC North of Spitsbergen (WSC branch A). Even Float 52, deployed closer to the continental shelf, remained in the main stream (WSC branch B). All the trajectories have been strongly influenced by topographic features: the continental slope West of Spitsbergen and West of the Yermak Plateau, the Yermak shallows, the Hovgaard, Molloy and Spitsbergen Fracture Zones.

In the following, we will highlight the informative elements that these observations provide concerning the West Spitsbergen Current (WSC), in particular: the *WSC eddies* following topographic Fracture Zones (Hovgaard, Molloy) and the *Return Atlantic Current* (RAC), the WSC circulation *around and across the Yermak Plateau* (WSC branches B and B'), and the ice edge north of Spitsbergen .

The West Spitsbergen Current (76°45-79°30N)

The Figure 21 and Table 1 represent the mean initial displacement for each "300 m" float between its deployment and its first localization, calculated at the time an acoustic signal

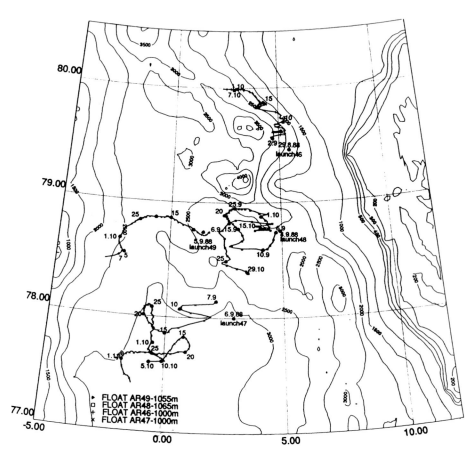

Figure 20. Trajectories of 4 deep (1000 m) Sofar floats from August 1988 until October 1988. Bottom Topography in meters

from the float has been received by at least two ALSs, and the mean speed (in cm.s^{-1}) associated with this mean displacement. For some floats, the relatively long time delay (more than 10 days) before their first acoustic positioning was due to the failing of moorings M3 and M4 ALSs, in central Fram Strait. They were to keep track of the floats during this initial phase and later on. Consequently, this positioning was only possible after the subsequent deployment of the northern moorings M5 and M6. **Float 56**, the last float to be launched (September 7, 1988) at the southernmost position 76°45N, 12°E, over the 1500 m isobath, was located 12 days later at about 79°15N, above the 1000 m isobath. During this initial phase, it was the fastest (average speed, over the first 12 days: 28 cm.s^{-1}). The initial "black out" phase also lasted about 12 days for float 50 (Sept. 6-Sept. 17,1988), float 51 (August 22-Sept. 3,1988) and float 52 (Sept. 3-Sept. 15, 1988). Except for the slower "1000 m" deep floats, and the faster float 56, the "300 m" shallow floats average speeds, during the initial phase, were about 10 cm.s^{-1}.

The Hovgaard Fracture Zone eddy. The Return Atlantic Current (RAC) at 78°N.

Float 50 (Figure 19.a) detached from the main WSC core, at about 78°N, 12 days after its deployment on September 6, 1988, 100 km farther south-east. It then changed direction from Northwest to West in following the Hovgaard Fracture Zone. In mid-November 1988, it entered into the EGC area. This was a *confirmation of the existence of a main path, right above the Hovgaard Fracture Zone (HFZ), for East-West WSC recirculation* (branch C) at an average speed of 5 cm.s^{-1}. This recirculation (the so-called

TABLE 1. Launch data, f/H, mean speed between launch and first position.

			launch			f 10^{-4}	f/H 10^{-9}	mean speed			
	date	time	lat.	longi.	water depth			WE	SN	cape	speed
	1988		N	E	m	s^{-1}	m^{-1}.s^{-1}	cm.s^{-1}			cm.s^{-1}
46	29.8	12:55	79.463	5.495	2480	1.4353	57.87	- 5.35	3.59	303	6.45
47	6.9	11:28	78.005	2.853	2900	1.42812	49.24	-37.80	32.36	310	49.77
48	4.9	21:27	78.753	4.945	2367	1.43196	60.49	- 4.53	- 1.47	252	4.77
49	5.9	5:46	78.745	1.463	2490	1.43192	57.50	8.12	- 0.81	95	8.16
50	6.9	23:50	77.25	10.483	1496	1.42400	95.18	- 5.39	12.29	336	13.42
51	22.8	2:06	78.03	8.761	1472	1.42825	97.02	- 6.02	7.85	322	9.89
52	3.9	11:42	79.511	8.088	562	1.43560	255.44	- 4.14	7.77	331	8.81
53	29.8	18:59	79.483	6.543	1412	1.43547	101.66	- 1.38	-1.69	219	2.18
54	28.8	18:16	79.000	6.483	1432	1.43317	100.08	- 3.73	8.36	335	9.15
55	28.8	14:40	79.006	7.401	1265	1.43320	113.29	- 8.96	0.21	271	8.96
56	7.9	5:19	76.748	12.015	1472	1.42112	96.54	- 9.81	26.71	339	28.45
57	21.8	13:37	78.486	8.325	1484	1.43061	96.40	- 4.05	6.67	328	7.80

RAC) was also substantiated (Figure 19.a) by the trajectory of a "1000 m" deep float (49), with half the mean speed of the "300 m" float 50. On each side of the HFZ, 2 "1000 m" deep floats (47 and 48) (Figure 20), launched respectively in the Boreas basin, South of the HFZ, and in the Molloy Deep area, North of the HFZ, indicated more variability in current direction and lower speed. In the HFZ area, Figure 22.a shows 6 oscillations of the time variation (mid-September to mid-November 1988) of the distance (350-400 km) separating float 50 from the southern mooring M1 (ALS7). Simultaneously, 6 oscillations in exact opposite phase are present in the time variation of the distance (200-250 km) between float 50 and ALS11 on northern mooring M5. This is clearly *identifying an eddy* (rather than a steady current) crossing the Fram Strait from East to West above the HFZ. However, the first three oscillations were different from the last three ones. The first 3 corresponding rotations were *anticyclonic*, with a *period of 6 days*, a *mean drift at 2-3 cm.s^{-1} in the South-West direction* (Figures 19.a and 22.a), while the last 3 rotations were *cyclonic* with a *12-day period*, and an *East-West mean drift at 5 cm.s^{-1}*. Thus, the mean drift accelerated between the third and the fourth oscillations, when rotation changed

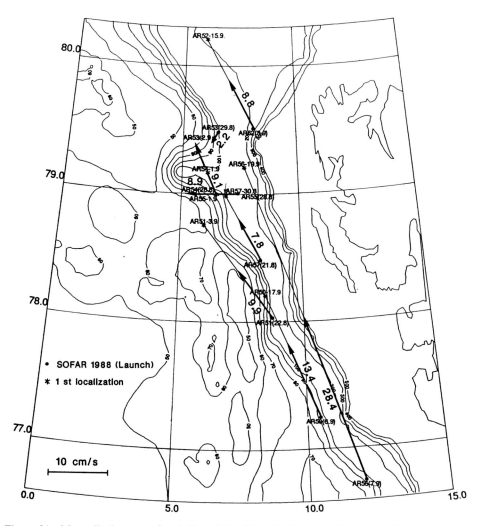

Figure 21. Mean displacement for shallow Sofar floats during the initial phase after deployment from August 22, 1988 until September 7, 1988. Float number, launching time (*dot*), time of first position (*star*) and mean speed (cm.s^{-1}) in between, are indicated. Bottom Topography in meters

direction. The tangential speed (calculated as the upper value of the time derivatives (Figure 22.b) of the radial distance between float and ALS (Figure 22.a), minus the mean speed projected along the float - ALS line) associated with the first rotation and the last 2, were comparable (order of 5 cm.s^{-1}) but doubled to 10 cm.s^{-1} for the 2nd, 3rd and 4th rotations. The eddy size changed from 10 km diameter at the beginning and at the end, up to 40 km in the middle. This big cyclonic eddy seems to correspond to the "gigantic" cyclonic eddy reported previously on section F (Figure 9), but we should note that almost two months passed between hydrology on section F (late August 88) and the float 50 drift

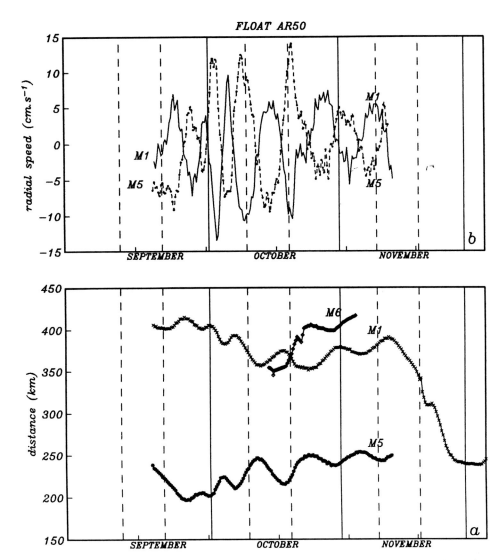

Figure 22. (a) Distance between float 50 and moorings M1, M5 and M6, as a function of time for the 3 month-period (September-November 1988). (b) Time derivative of distance (a): continuous line: from Mooring M1 - dashed line: from Mooring M5.

in this area (around mid-October). Our best interpretation, based on float 50 trajectory (Figure 19.a), is that there were *2 distinct eddies*. The transfer from one (anticyclonic) to the other (cyclonic) occurred at 78°N and 5°E before mid-October at the time the float 50 was crossing the HFZ. Initially the *anticyclonic* eddy (5 cm.s^{-1} tangential speed, 6-day period and 10 km diameter) drifted South-West at 3 cm.s^{-1}. The size of the eddy increased from 10 km diameter during the first loop to 20 km during the second and 40 km during the third loop. At the same time, the speed along the trajectory followed by the float increased from 5 cm.s^{-1} to 10 cm.s^{-1} and consequently the period of rotation went from 6 days to 12 days between the first and the third rotation. The fourth rotation had exactly the same characteristics except that it was *cyclonic* , suggesting that the *float was trapped in an other eddy* probably coming from the South-East, following the HFZ. In early November, this 12-day cyclonic eddy rapidly decayed between the fourth and the sixth loop (5 cm.s^{-1} speed, 20 km diameter) before joining the EGC. It appears that (Figure 23.a) float 50 was still indicating the presence of a 12-day eddy, after having been dragged along by the EGC from mid-November 1988 until early January 1989 at a possibly underestimated mean speed of 15 cm.s^{-1} (still calculated from the time derivative of the radial distance between float 50 and ALS7 on mooring M1). It could be (this is an hypothesis) the same eddy, advected downstream more than 500 km away in the southern Greenland Sea.

The Molloy Fracture Zone eddy at 79°05N, 7°E

Floats 54 and 55 launched at 79°N on August 28, 1988 (Figure 21), were located 4 days later on September 1st. Their mean drifts over this period indicated the presence of a WSC meander or an anticyclonic eddy, approximately 20 km in diameter and a tangential speed of 5 cm.s^{-1} in the Molloy Fracture Zone, North of 79°N. **Float 57**, launched at 78°30N on August 21, 1988, entered also in this eddy (or WSC meander) on August 30, 1988 (Figure 19.a) coming from the South at the time float 54 escaped from it to the North. According to the speed and the size of this eddy, the rotational period should be approximately 12 days. This was confirmed by the trajectories of floats 55 and 57 during the first week of September 1988 (Figure 24.a). On September 8, float 55 was joined by float 51 until September 11, 1988.

The WSC North of 79°30N (WSC branch B, West Yermak)

During the second and third weeks of September 1988 (Figure 24.a and .b), the northernmost floats 53 and 52, deployed at 79°30N, float 53 on the western side and float 52 on the eastern side, 30 km apart across the continental slope West of Spitsbergen, progressed northwards at a similar speed of nearly 10 cm.s^{-1}. The water depth under float 52 was about 600-700 m and 1500 m under float 53. From an average depth of 1 km with a cross distance of 30 km and a mean speed of 10 cm.s^{-1}, one could estimate a transport of 3 Sverdrups associated with this current, the WSC branch B. Taking into account all of the float trajectories, we gauge that time and space variability in the circulation associated with this WSC branch B and the 3 Sv transport is our best estimated mean value for this period. In the center of the current, where the water depth ranges from 800 to 1000m, float 51 was boosted at about 30 cm.s^{-1}. on September 12-15, 1988 (Figure 24.b), reaching a maximum speed of 45 cm.s^{-1}. on September 13, 1988. Only float 55 was close to moving at such a high speed (30 to 35 cm.s^{-1}) for a short time (September 17, 1988), 5 days later and one week after leaving the Molloy FZ eddy, in which it had been trapped for 12 days, after its launch at 79°N on August 28, 1988 (see § "The Molloy Fracture Zone eddy at 79°05N, 7°E").

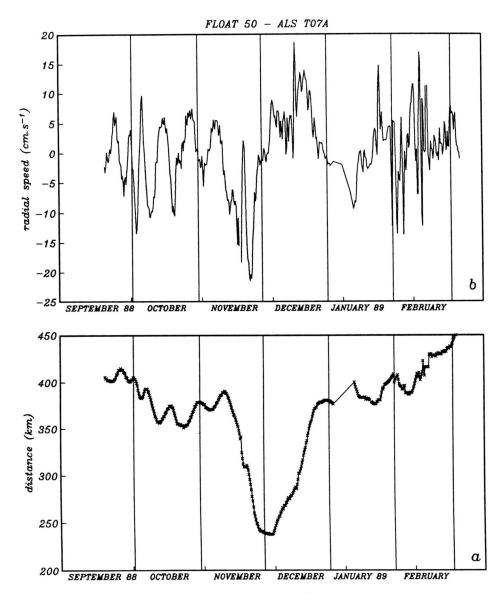

Figure 23. (a) Distance between Sofar float 50 and mooring M1 as a function of time for the period September 88 - March 89. (b) Time derivative of (a)

The U-turn to the East. The Yermak Shallows and the Yermak Pass

Between mid-September and mid-October, 1988 (Figures 24.c, 24.d and 34), 5 floats (51 to 56, except float 54) out of the 7 trapped in the WSC branch B, took a sharp "U-turn" to the East before reaching 81°N. This turn was obviously strongly influenced by the

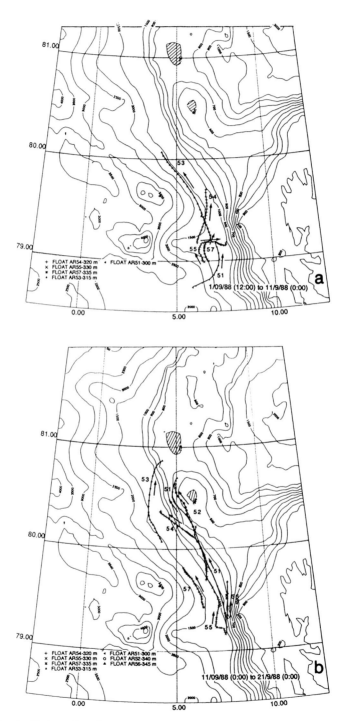

Figure 24. a, b. Trajectories of shallow floats by 10-day sequences from 1.09.88 until 21.09.88.

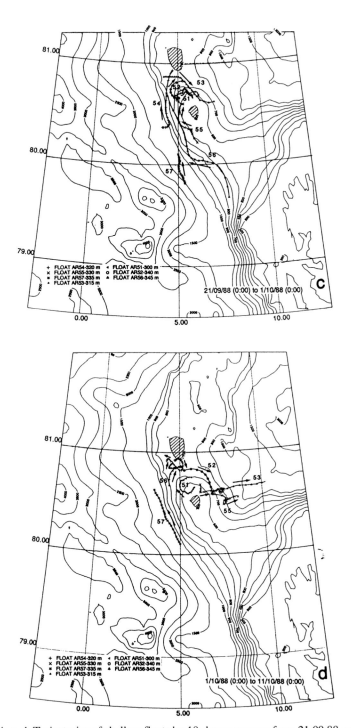

Figure 24. c, d Trajectories of shallow floats by 10-day sequences from 21.09.88 until 1.10.88.

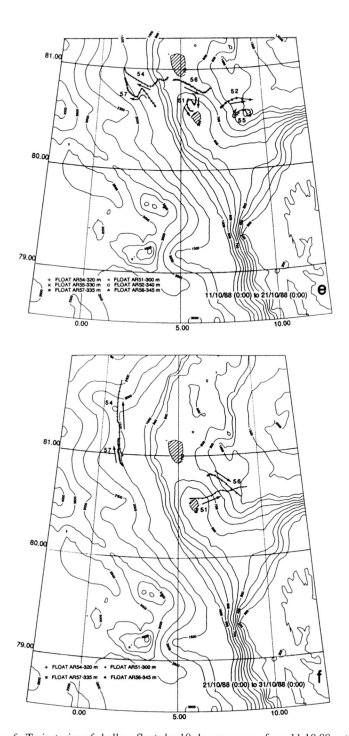

Figure 24.e, f. Trajectories of shallow floats by 10-day sequences from 11.10.88 until 31.10.88.

topography of the Yermak Plateau and more precisely related to the presence of shallows (less than 500 m depth) at about 80°30N and 6°E. Another shallow, less than 700 m deep, is located near 81°N, 5°E. In between is what we called the "Yermak Pass". Finally 2 floats only (57 and 54) out of the 7, did not get through the Pass, although one of them, float 54, went perilously close to it and the other one (float 57) was also influenced by the strong eastward drift associated with the U-turn. Late September 1988, this float drifted across isobaths from 1500 m up to 1100 m, but stabilised its course at this point and never crossed the 1000 m isobath like the other floats. This sharp turn to the right, contouring the Yermak Shallows to the North between 80°30 and 81°N, has never been reported in the literature before, despite all the detailed investigations in this area since Mizex 84. We firmly believe it is a quasi-permanent feature, characteristic of the WSC circulation North of Spitsbergen (branch B'), as we will explain in the discussion section.

Anticyclonic Vortices

Between the 2 Yermak shallows, some float trajectories show anticyclonic vortices (Figure 25). **Float 51**, for example, completed several anticyclonic rotations during one month (20 September-20 October, 1988), each rotation in about 12 days, on a circle 18 kilometers in diameter, centered at about 80°40N, 5°30E, on the northwest slope of the Yermak Shallow, approximately 18 kilometers away from the apex at 500 m depth. Further in the same direction at 80°50N, 5°E, floats **54 and 56** exhibited similar anticyclonic rotations (Float 54: 1-10 October 1988 and Float 56: 5-15 October 1988). **Float 52** was stalled for 6 days (September 21-27, 1988) just on the western side of the Float 51 anticyclonic eddy. We do not know exactly the origin of these vortices. They could be locally generated or advected and they probably influence the general circulation in the area. We will come back to this point in the discussion.

The splitting

On October 10, 1988, a split occurred between the eastward drift documented by 5 floats (51, 52, 53, 55, 56) and a westward drift documented by 2 floats (54 and 57). The split is best illustrated (Figure 24.e) by float 56 turning East, and float 54 turning West, at the same time and nearly at the same place, 80°50N, 4°E. Both floats drifted the same distance (about 55 kilometers) during the next 10 days, at about 5 cm.s^{-1}, in opposite directions. The transports associated respectively with the eastward and westward drifts on October 10 at 80°50N, 4°E, look similar as far as we can estimate them from these data. This means that the transport associated with WSC branch B continuing its northward progression along the continental slope West of the Yermak Plateau and North of 81°N, is divided by 2 (at least) compared with the transport calculated across 80°N. The remaining part deviates towards the East through the Yermak Pass (branch B').

The northern and eastern routes

On October 20, 1988, **floats 54 and 57** converged at 81°N above the 1500 m isobath, ending their westward drift before changing course and drifting North again for another

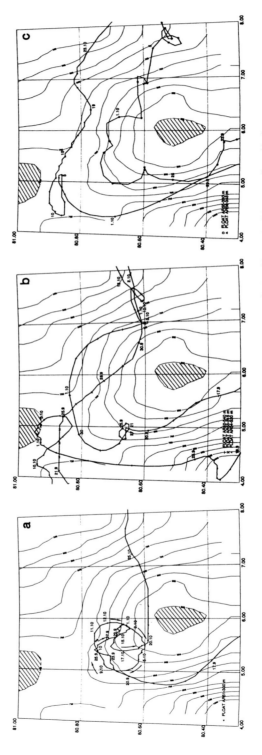

Figure 25. Trajectories of shallow Sofar floats on top of the Yermak Plateau. Detailed representation: (a) float 51. (b) floats 52, 53, 54. (c) floats 55 and 56.

month until we lost contact with them (Figure 24.f). During this period, they followed the continental slope (isobath 1500 m) West of the Yermak Plateau at an average speed of 5 to 10 cm.s^{-1}, *up to 82°N, confirming by the way, the presence of warm and salty WSC waters in this area*, as indicated by Perkin and Lewis [1984] and Muench et al. [1992].

Float 53 was the first to cross the Yermak Pass heading East on September 24, 1988, (Figure 24.c) shortly followed by floats **52 and 55** about one week later. Float 52 would have crossed the Yermak Pass at the same time as float 53, should it not have been delayed by stalling for a week at 80°40N and 5°E. **Float 51** was the last one to cross the Pass (Figure 24.f) on October 20, 1988, heading East. Even though it had been the first to approach the Pass from the western side one month earlier, it got trapped in an anticyclonic eddy Northwest of the Yermak shallow for one month. On the eastern side of the Pass, all the floats started to head southeast following isobath contours between 600 and 1000 m as before, until they reached 80°30N and turned sharply at 90° to the left remaining afterwards at constant latitude. The sharp 90° turn to the left was more likely due to the presence of the ice edge, oriented East-West at this latitude, and generating ice edge jet and eddies at the front. We will come back to this point in the discussion.

2 year-period : August 1988 - August 1990

The total duration and the period during which float signals were received by at least 2 ALSs and the locations could be computed are, for each float:

    Float 57: 21/8/88-1/12/88, 8/4/89-7/6/89, 22/9/89-19/11/89: 15 months, tracking: 7
    Float 50 : 6/9/88-13/11/88, 24/9/89-25/11/89:   15 months, tracking: 4
    Float 48 : 4/9/88-29/10/88, 26/10/89-1/11/89:   14 months, tracking: 2
    Float 86 : 14/4/89-7/6/89, 27/9/89-4/8/90:   16 months, tracking: 8
    Float 83 : 14/4/89-7/6/89, 9/5/90-3/6/90 :   14 months, tracking: 2

The 2-year results are shown in Figure 26. Float positions during the second year were computed by our student D.M. Mc Carren at Monterey (Naval Postgraduate School) in his thesis [McCarren, 1991], and part of his results were presented in Bourke et al. [1992].

Float 50 was not located after November 13, 1988. During winter, tracking floats acoustically under sea-ice became much more difficult. Nevertheless, some floats signals were still received by one ALS at least, from which it was possible to extract changes over time of the distance between float and ALS even if one cannot locate the floats. We were able to extract some information about float 50 large scale drift, in particular during the period when it was trapped in the EGC. In Figure 23.a, we plotted the distance between float 50 and ALS7 on M1 as a function of time for the entire period the acoustic signal from float 50 was received by ALS7 (6 months: September 1988 to March 1989). We can distinguish 3 phases. The *first phase* we already described from mid-September 88 until mid-November 88 (Figure 22.a), corresponds to the westward recirculation, from the WSC to the EGC along the Hovgaard FZ, of the Atlantic Waters carried on by the WSC. The *second phase,* from mid-November 88 until end of December 88, corresponds to the float drifting with the EGC. The drift speed increased up to 20 cm.s^{-1} (Figure 23.b); this is only an underestimation of the EGC velocity since it represents the projection of the true current velocity on the vector defined by float 50 and ALS7. During this phase, float 50 approached ALS7 on M1. The distance decreased from about 400 km

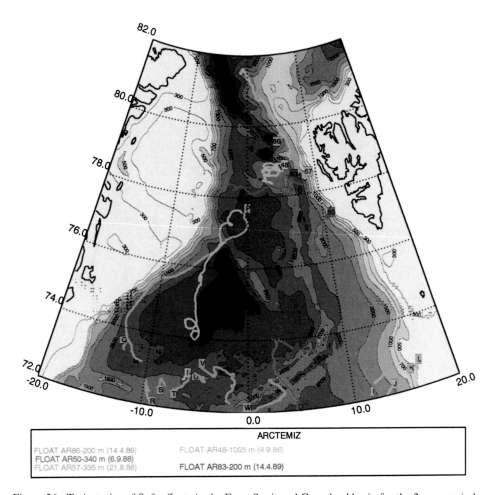

Figure 26. Trajectories of Sofar floats in the Fram Strait and Greenland basin for the 2-year period (August 88-August 90). Launch positions for the 5 floats are indicated by the float number. Segments of trajectories are labelled with letters in alphabetical order.

to about 230 km by the time the float was crossing the Greenland Fracture Zone and then it increased from 230 km up to 400 km, with about the same speed along the continental slope east of Greenland. The *third phase* is similar to the first one and can only be interpreted as a slow motion, including eddy-like motion. The float then probably drifted in the southwestern Greenland Sea, North of Jan Mayen, as indicated by other floats in the area. The contact with the float was lost in March 89, it was then 450 km away from ALS7 (M1). On September 24, 1989, acoustic signals from float 50 were received anew by the second ALS array and the float was located for another 2 months in the southern Greenland Sea (Figure 26). It seems the behavior of float 50 was quite comparable to float 86, which was circling around the Greenland gyre.

Float 57 : After being lost (acoustically) on early December 1988 at about 82°N, it was received 4 months later, on April 8, 1989, 400 km further South. It was then located

(Figure 26) during the following 2 months until June 7, 1989 when the ALSs stopped recording. It drifted *across* the Greenland gyre rather than around it. This float was received again as soon as the second array of ALSs was deployed in September 1989. It was then in the Lofoten basin, 200 km southwest of Bear Island, and it was tracked until mid-November 1989. This float accomplished a fantastic survey in about 1.5 years: launched in the West Spitsbergen Current in Fram strait, and cruising successively in the Arctic Ocean, up to 82°N, then in the East Greenland Current, crossing the Greenland gyre and finally returning to the Norwegian Current between northern Norway and Bear Island.

Float 86 : Ballasted initially for 200m, it was launched in April 89 in northern Fram Strait by Manley and Owens, at nearly the same place where floats 53 and 46 had been launched in August 1988 (Figure 26). Float 86 was located again, thanks to the deployment of the second array of ALSs, on September 27, 1989, 5 months after launch, at about 77°30N and 0°, in the vicinity of the Greenland Fracture Zone, close to the place where float 57 was located in April 1989 after the "black out" of the 4 winter months. According to its drift speed, it was trapped in the EGC from late September 89 until late November 89 when freezing started. In February 1990, float 86 was located again in the southern Greenland Sea, until August 1990, drifting eastwards at a much lower speed and more erratically .

Float 48 and Float 83 were located for a much shorter time than the 3 previous floats (Figure 26). Their positions in the South Greenland Sea, one year after their deployment in Fram Strait, confirmed the indications given by the 3 other floats. Float 48 was a deep float (1000 m), and it is interesting to note that it followed the same drift as the shallow ones, at almost the same mean speed.

Discussion

In the following discussion, we will closely reconsider topographic effects on the WSC multi-path current system, the Return Atlantic Current (RAC) and the interaction between WSC and the ice edge in Fram Strait.

Topographic effect . Yermak Plateau

The topography plays evidently a major role in the WSC circulation and recirculation schemes we discussed. We would like to understand why some floats moved northwards crossing over isobaths from 1000 m up to 700 m between 79°30N and 80°30N in the direction of the Yermak Shallow, instead of following isobaths and continuing to drift eventually in the North-West direction, and/or recirculating westwards and southwards along topographic fracture zones. Any dynamical reason explaining this behavior would represent the preconditioning stage for allowing, later on, the WSC (or part of it) to turn around the Yermak Shallow and to get across the Yermak Plateau through the Yermak Pass at 80°45N. Here we will reconsider two kinds of problems: large scale circulation along the continental slope West of the Yermak Plateau, and the mesoscale features around the Yermak Shallow.

Large Scale Circulation

To a first approximation, from September 88 until December 88, all the "300m" floats except float 50, more or less followed f/H isolines contours between 100 and 200.10^{-9} $m^{-1} s^{-1}$ corresponding approximately and respectively to isobaths H=1500 m and H=700 m (Figure 27.a and .b). Float 57, in particular, remained constantly in the vicinity of the 1500 m isobath during three months and float 52 close to 700 m isobath. For floats 51, 53, and 55, f/H doubled from 100 to 200.10^{-9} $m^{-1}s^{-1}$ in about 10 days between September 10 and 20, 1988 when floats moved from isobath 1000 m up to 700 m (Figure 28). It looks like floats were acquiring negative vorticity relative to a flow following f/H isocontours at the same time they were moving upslope, and this is compatible with a fluid conserving potential vorticity.

Within the quasi-geostrophy approximation, the conservation of potential vorticity in an homogeneous, barotropic fluid :

$$\frac{d}{dt}\left(\frac{\zeta+f}{H}\right) = 0 \quad \text{implies} \quad \frac{d\zeta}{dt} + \beta v = \left(\frac{\zeta+f}{H}\right)\left(\frac{dH}{dt}\right)$$

$$\text{with } f = f_0 + \beta y \quad \text{and} \quad v = \frac{dy}{dt}$$

Considering the approximations:
a) Quasi-geostrophy is valid since:

$$\frac{U}{fL} \# \frac{0.4 \, m.s^{-1}}{1.44 \times 10^{-4} \, s^{-1} \times 20.10^3 m} = 0.13$$

b) ζ is 10 times smaller than f. So, within 10% approximation, ζ is negligible compared to f.

c) βv (order of 10^{-11} s^{-2}) is 10 times smaller than $(d\zeta/dt)$ and $(f/H).(dH/dt)$ (order of 10^{-10} s^{-2}). So, within 10% approximation, βv, *advection of planetary vorticity*, is negligible compared to $(d\zeta/dt)$ *(vorticity growth rate)* and $(f/H).(dH/dt)$ *(divergence term)*.
The conservation of potential vorticity is then reduced to 2 terms: the vorticity growth rate, induced by shear (fluid acceleration) and the divergence term induced by topographic changes.

$$\frac{d\zeta}{dt} \approx \frac{f \, dH}{H \, dt} = f\frac{d}{dt}\left(Log \, H\right)$$

$$\frac{1}{f}\frac{d\zeta}{dt} \approx \frac{d}{dt}\left(Log \, H\right)$$

Is the rate of change of relative vorticity comparable to the rate of change of H and what would be the cause of this vorticity change inducing a variation in water depth following

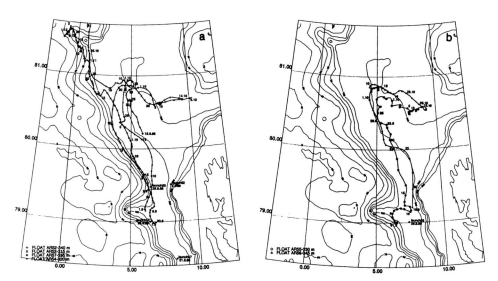

Figure 27. f/H isocontours (m^{-1}.s^{-1}.10^{-9}) and floats trajectories for the period September-December 1988: (a) floats 52, 53, 54 and 57. (b) floats 55 and 56.

floats? Changes in the circulation are evidenced, between September 10 and 20, 1988, by the floats 51 and 55 velocities computed along their trajectories (Figure 29). On September 11, float 51 started to accelerate from about 10 cm.s^{-1} up to 30 cm.s^{-1} and reached jet-like speed (40 cm.s^{-1}) on September 12 and 13, 1988, before slowing down to 10 cm.s^{-1} until September 20 at a smaller rate. Speeds of neighbouring floats are clearly lower, and the jet-like feature, evidenced by float 51, injects shear vorticity in the WSC circulation. A rough estimate of the vorticity increase is given by comparing the acceleration of floats 51 and 55. Prior to the jet event, they were located very close to each other (Figure 24.a). Float 55 had just completed an anticyclonic loop 10 miles in diameter in about 10 days and was joined by float 51 on the loop track, 2 days before the end. On September 11, both floats changed course, turning 90° to the left and started to head North (Figure 24.b). The jet came 12 hours later and boosted float 51, 15 miles away North of float 55 in about 24 hours, that is at a speed of 30 cm.s^{-1} relative to float 55. At the same time, float 55 speed was approximately 5 to 10 cm.s^{-1} North (Figure 29). A mooring at 79°26 N and 6°48 E, laid in 1270 m of water and equipped with current meters at 100, 200, 500 and 1000 m, was installed by the R-TEAM group (WHOI) on September 1, 1988 (Figure 18). Current meters were located 10 miles West of the main axis of the WSC current. Their data indicated (Figure 30) bursts of current early and mid-September 1988 in particular at 100 m level (up to 30 cm.s^{-1}) and less (20 cm.s^{-1}) at greater depths (200-500). At 1000 m depth, near the bottom (1270 m), speed rarely exceeded 10 cm.s^{-1}. This seems to indicate a *strong barotropic component*, at least down to 1000 m depth. Float 46, at 1000 m depth, launched on August 28, 1988, about 10 km west of R-TEAM mooring (Figure 20), indicated a slow motion in the northwest direction, more or less parallel to isobaths 2000-2500 m, until early October 1988. Its average speed was less than 3 cm.s^{-1}.

From these observations, one can estimate the mean shear vorticity across half of the WSC at about: 0.3 m.s^{-1}/20 km = 1.5.10^{-5} s^{-1} (absolute value). The current width was

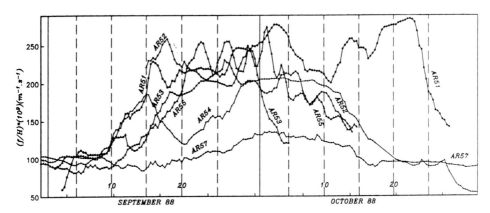

Figure 28. f/H (m^{-1}. s^{-1}.10^{-9}) calculated along floats 51, 52, 53, 54, 55 and 57 total trajectories as a function of time for the period September-October 1988.

approx. 20 miles and the speed difference between the center and the eastern edge of the current was also 30 cm.s^{-1} approx. Our estimation for the shear vorticity is based on the assumption that *only the inner part of the current* is under acceleration. Before the fluid acceleration, shear vorticity was nil across the current. Under acceleration, shear vorticity was positive (cyclonic) on the left side and negative on the right side of the main axis of the current looking downstream.

The rate of change of vorticity with time is important to estimate the balance of vorticity. On September 11, float 51 speed increased, in 8 hours, by 10 cm.s^{-1} (Figure 29). This gives an absolute value of $1.7.10^{-10}s^{-2}$ for the *vorticity growth rate* relative to the edge of the current, 20 km away, where no acceleration occurred at the same time. This value is positive to the left and negative to the right of the current looking downstream. Scaled by f, that one can approximate as constant in this case (f is not the dominant term, but H is),

$$1/f. \, d\zeta/dt \text{ is approximately} = -1.2 \, . \, 10^{-6} \, s^{-1}$$

We can estimate now the other dominant terms of the potential vorticity equation for comparison. Considering our approximations, the only remaining term should be the *divergence term generated by topography* (depth change), assuming the fluid is barotropic. This term (1/H).(dH/dt) can be expressed as the Lagrangian time derivative d_t(Log H), H being the water depth at the vertical of the float at time t. In Figure 31, we plotted H versus time for float 51 trajectory and we calculated the best fit:

$$H = K. \, 1/t$$

with $K = 10^9$ m.s.   (H in meters and t in seconds)

we can deduce :

$$(1/H).(dH/dt) = -1/t = -H(m)/10^9(m.s)$$

From the best fitted curve, we calculate at time :

t=200 h, H = 1350m and (1/H).(dH/dt) = - 1350/10^9 = - 1.35 . 10^{-6} s^{-1}
t=225 h, H = 1200m and (1/H).(dH/dt) = - 1200/10^9 = - 1.20 . 10^{-6} s^{-1}

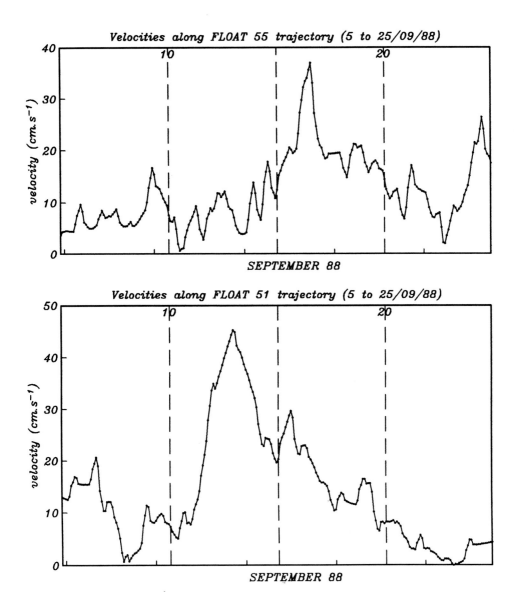

Figure 29. Velocities (cm.s⁻¹) calculated along the floats 51 and 55 trajectories as a function of time, for the period September 5-25, 1988

which corresponds to the *estimated vorticity growth rate* calculated previously. It indicates that both terms in the vorticity equation balanced each other at the time the WSC was accelerating from September 10 to 15 (t=216 to 336) and moving across 1350 m isobaths up to 900 m isobaths and corresponding to a change in f/H from 100 to $200.10^{-9} m^{-1}.s^{-1}$. This calculation applies also for other floats like floats 55 and 54. Consequently, at about 79°30N, it appears that *negative vorticity generated by a strong acceleration of the inner part of the WSC*, located above the 800-1000 m isobaths, relative to the outer part of the current extending over the 700 m isobath, to the East, is in balance with the divergence term of the vorticity equation assuming barotropic condition. In other words, an acceleration creating a jet-like stream in the core of the WSC such as observed in September 1988, is more likely to be *responsible for the initial slight eastward deviation of part of the WSC forcing the flow to cross isobaths in direction of the Yermak Plateau*. In consequence, this part of the WSC, **"the 700 m WSC branch"**, is able to cross the Plateau at about 80°40N before heading East. This important result is also supported (a contrario) by the northwestward and continuous drift of float 57, at a constant and low speed for the entire 3-month period. It constantly followed the 1500 m isobath, except for a slight deviation towards the East by the time it came close to the area where the other part of the WSC was deviated towards the East.

From our estimation, the *transport* related to **"the 700 m WSC branch"** drifting eastwards across the Yermak Plateau, was about 1 Sverdrup based on a width of 15 km, a mean depth of 650 m and a mean speed of 10 cm.s^{-1}. The transport relative to the other part, **"the 1500 m WSC branch"** progressing northwestwards and circulating cyclonically around the western and northern slope of the Yermak Plateau following the 1500 m isobath preferentially, was comparable.

Mesoscale Features

During Mizex 84, we had already observed the strong influence of the "500m" Yermak Shallow on the ocean circulation in this area. One float, launched right on the top of the Shallow on June 11, 1984, stayed over it until July 17, 1984 (Figure 32). Ballasted for 220 m, it drifted at mid-depth just over the bump. Other floats, launched over the Yermak Pass during summer 84, showed a slow motion with no clear trend (neither East or West). High frequency motion was mainly at the diurnal frequency as suggested by some floats trajectories. For example, float 5001 (Figure 32) indicated diurnal motions across the slope clearly modulated by spring-neap tides. The spring tides occurred on July 1-2, 1984 (tidal coefficients 88 to 86). Tidal coefficients were > 70 from June 28 to July 5. The previously mentioned role of these tidal currents, by rectification, might be to generate a circulation [Muench et al., 1992]. As shown in Figure 32, the mean northward drift, at about 45° to the bottom slope, occurred at a speed of 5 cm.s^{-1}, around the spring tide period. In 1988, float 53 cruised in the same area but 100 m deeper than float 5001. The mean current was much higher in 1988 but the direction relative to the bottom slope was the same. Float 5001 started also to experience East-West drift when approaching 81°N. In 1988, floats were located only 3 times per day and we could not resolve the high frequency associated with the diurnal motions as in 1984, when floats were located every hour.

In 1988, floats 51 and 55 showed interesting mesoscale features on the Yermak Plateau. **Float 51**, after being boosted by the jet from about 79°10N up to 80°30N in less than 10 days, entered in an anticyclonic loop, we already described Figure 25.a (cf. § "Anticyclonic vortices"). The float completed the loop in about 12 days above 600 - 650 m isobaths at a mean speed of 5 cm.s^{-1}. This feature can be compared with what we termed the "Molloy FZ Eddy" (cf § "The Molloy Fracture Zone eddy at 79°05N, 7°E") at 79°05N and 7°E from August 28 to September 10, 1988 (12 days). It could be the same eddy, assuming it had been advected downstream. Two other floats (54 and 56) were also

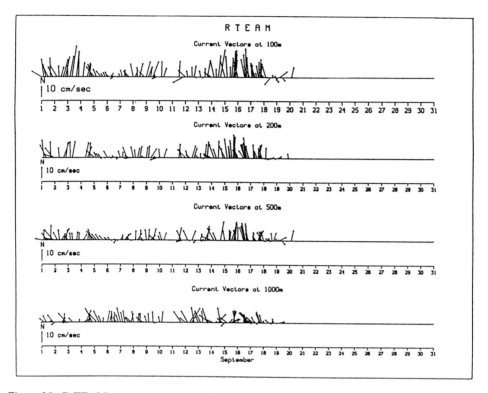

Figure 30. R-TEAM current meter vector diagram at 100 m, 200 m, 500 m and 1000 m depth for the period September 1-20, 1988. (water depth 1270 m).

trapped in anticyclonic loops on the Yermak Plateau, having more or less similar characteristics than the Molloy eddy : mesoscale, slow currents, negative vorticity. **Float 55** completed a 3/4 turn around the Yermak Shallow from September 23 to October 4, 1988 (12 days) describing, at the same time, 4 anticyclonic loops (Figure 25.c). This motion was obviously a composite between a mean drift current (10 cm.s[-1]) following the 600 m isobath and a rotational component with 3-day periodicity at a tangential speed of about 10 cm.s[-1] and a radius of curvature of 4 km. Evidently, this kind of "3-day" eddy (2nd kind) is quite different from the first kind, the "12-day" eddy which we described previously, even if both are anticyclonic. Our best guess is that *the first kind eddies are topographically generated* in Fram Strait and are *advected* by the boundary current. The *second kind could more likely be generated by current shear vorticity initially and by curvature vorticity subsequently* in the vicinity of the Yermak Shallows. It looks like thefirst kind of anticyclonic eddies might play a crucial role on the general circulation in defining the "*splitting*" conditions we already mentioned above the Yermak Plateau at about 81°N (see § "The splitting"). Upstream, the part of the current which has not been deviated upslope towards the Yermak Shallow would not be captured, later on, by the U-turn around the Shallows topography, which greatly accentuates the anticyclonic rotation of the flow and ultimately the WSC eastward drift at 80°30N. Rather, the anticyclonic vorticity generated locally or advected by the main flow in this area would contribute, in

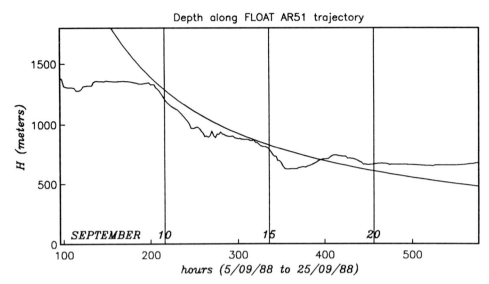

Figure 31. Water depth H along float 51 trajectory for September 5-25, 1988. The smooth curve is representing the best fit for H as a function of K/t (t is time in seconds and $K = 10^9$m.s).

absence of a strong local topographic effect, to the ultimate direction of the flow as it happened for floats 56 and 54 on October 10, 1988 (Figure 24.e).

Return Atlantic Current

Topographic Fracture Zones

Among the eight "300 m" shallow floats launched in Fram Strait in August 1988, two were trapped in the Return Atlantic Current (RAC). One of these floats (Float 50) just after being launched in the WSC, followed the Hovgaard Fracture Zone from East to West. It then merged into the East Greenland Current on the other side of the strait and circulated around the Greenland Sea during the following year (November 88-November 89). **Float 50** shows clearly the role played by WSC eddies in the RAC dynamics, as suggested by Gascard et al. [1988]. **Float 57**, after drifting at a low speed (5 cm.s^{-1}) for three months along the continental slope West of Spitsbergen and West of the Yermak Plateau, reached approx. 81°30N early December 1988. Then it started to drift away from the continental slope above the last topographic fracture zone before the Yermak continental slope changes orientation by 90° to the right. Floats 57 and 50, like two out of the four "200 m" floats launched in Fram Strait in spring 89 by Manley and Owens, recirculated in the Greenland Sea one year later (Figure 26). **Float 48**, one of the four "1000 m" deep floats launched in the Molloy Deep in August 1988, recirculated also one year later in the southern Greenland Sea indicating that *deep flow was also concerned by the recirculation*. In 1988, between 78°N and 81°30N, no more RAC branches in Fram Strait were evident, except the eddy observed in the Molloy Fracture Zone at 79°N early

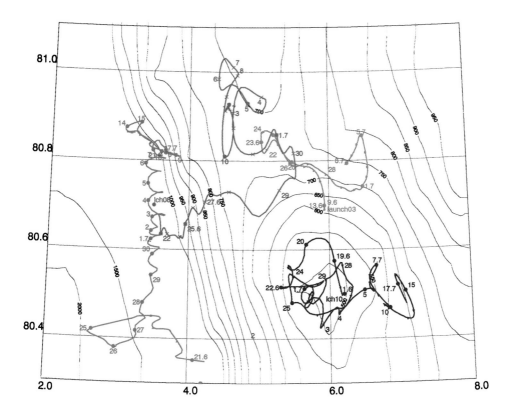

FRAM STRAIT TOPOGRAPHY

MIZEX 84 EXPERIMENT

FLOATS 5001, 5003, 5008, and 5010 trajectories

```
►  FLOAT 5003 -126m
×  FLOAT 5008 -110m
▼  FLOAT 5010-220m
▲  FLOAT 5001-225m
```

Figure 32. Trajectories of Mizex 84 floats on top of the Yermak Plateau (5003, 5008, 5010) and along the continental slope (5001).

September 1988. It is interesting to compare these results with Mizex 84 [Manley et al., 1987], where westward recirculation was observed in summer 84 at a latitude of 80°30N and between 0 and 4°E in the Spitsbergen Fracture Zone.

It seems clear that the RAC is basically composed of *eddies* which are generated by the WSC and which interact with the bottom topography, in particular along topographic fracture zones at 78°, 79°, 80°30 and 81°30N. These eddies, mainly composed of Atlantic warm and salty waters, interact with the East Greenland boundary current and are advected by it. They represent the main source of heat and salt for the Greenland Sea. In winter, they bring in the necessary potential energy and vorticity to favorably precondition the deep convection in the Greenland gyre.

The *total transport* associated with all these WSC recirculating branches is hard to estimate. The basic information coming out of our observations concerns more the space and time scales associated with the large scale recirculation than the total transport. A proper estimation of the transport would require more floats over a longer period, to overcome the high frequency variability of this complex circulation pattern. As a first order of approximation, one could suggest 1 Sverdrup for each of the 3 types of WSC branches : B (northwards), C (westwards) and the newly discovered eastward branch B' crossing the Yermak Pass at about 80°30N.

Basin Scale Circulation

The most striking result coming out of the 2-year float tracking was undoubtedly the *large scale recirculation* of some of the floats launched in Fram Strait in August 1988 and April 1989 and still recirculating in the Greenland Sea one year or more later. Five floats were drifting in the Greenland Sea 14 to 16 months after launch. There were numerous interruptions in the localization, due to winter ice conditions in surface, in particular from December to March (both in 1988 and 1989). An other interruption of nearly four months occurred from June to September 1989, when we had to recover the first ALSs array and redeploy the second one. In all, the effective number of months for tracking floats during the 14-16-month period was reduced to 8 months in the best case (thus 50% efficiency). The results are surprising. Floats 50, 57 and 86 give, for the first time, *the proper time and space scales characterizing, from a lagrangian point of view, the basin scale circulation in the Greenland Sea - Fram Strait area.* They also yield some information on the types of interaction between the circulation in the interior and the boundary currents, namely the EGC, the Jan Mayen current, the WSC-Norwegian Current and the Greenland gyre.

Ice Edge North of Spitsbergen. Ocean - Ice Interaction

North of Spitsbergen, the ice edge changes direction quite abruptly. Instead of a nearly North-South orientation in Fram Strait and West of Spitsbergen, it is almost zonal in an area located between 80°N and 81°N and East of 5°E. It is clear that *the WSC*, flowing North and East at this latitude, carries heat and salt, and *has a strong impact on the marginal ice zone*, also oriented East-West North of Spitsbergen. AVHRR images, collected in September 88, during MIZEX 83, in July 89 and many others, are also particularly illuminating (Figures 17 and 33 for example).

We attempted to *estimate the ocean heat flux* associated with the WSC between 79°30N and 80°30N, West of the Yermak Plateau. This estimation is based on our 1988 floats data and on MIZEX 84 hydrological data which provided a good coverage over the Yermak Plateau and surroundings, down to 500 m depth [Quadfasel and Ungewiβ, 1986]. In the layer 100 - 500 m, there is a general 200 m upslope trend of the 2°C-3°C

Figure 33. AVHRR infrared image on July 27, 1989. A light grey mask is applied to clouds. In red, sea surface temperature > 4°C. *Superimposed:* Trajectories of ARCTEMIZ 88 floats 52 (purple-pink), 53 (green), 54 (white) and 57 (red).

isotherms over a distance of 100 km downstream. The mean temperature in each 100 m layer, between 100 and 500 m, is decreasing by 0.5°C over this 100 km distance. With a mean speed of 10 cm.s^{-1} in each layer, as confirmed by the R-TEAM measurements (Figure 30) and by our 300 m floats drift, and assuming a heat loss occurring primarily through the surface, we estimated at about 200 W/m^2 the upward heat flux for each layer. This assumption is doubtful, since the total heat losses from the 100 to 500 m layer alone would generate a 800 W/m^2 surface heat flux . This value is high (considering that we have also to add the heat loss for the surface layer) but similar to what Aagaard et al. [1987] calculated for the 100-200 m layer in the WS coastal current.

For the surface layer alone (0-100m), Quadfasel et al. [1987] estimated the heat loss at about 12 kcal/cm^2 by the time the WSC had gone from an ice-free area (77°30N, 9°E) to the ice melting zone (80°20N, 4°E), based on the following arguments: (a) the ice edge is remarkably stationary in the area north of Spitsbergen, (b) the ice pack advances at 5 km per day normal to the ice edge, (c) the ice is 1.5 m thick. 12 kcal/cm^2 (equivalent to 5000 Watts/m^2) is not very realistic. The highest melting rate ever reported was 50 cm per day [Josberger, 1987], equivalent to 1670 Watts/m^2, measured in the marginal ice zone (Mizex 83). According to the floats and current meters data, one can estimate *at about 10 days* (rather than 1), the time interval over which heat losses really took place. The heat flux estimated by Quadfasel et al. [1987] should then be reduced to 500 Watts/m^2 for the surface layer alone. It is wrong to say that there would be no contribution to the total surface heat flux from the deeper layer below 100 m depth [Quadfasel et al., 1987]. Consequently, the total heat flux including all layers, associated with the WSC flowing North, West of the Yermak Plateau and between 79°30N and 80°30N, would be over 1 kW/m^2.

This very large heat flux explains why the ice cover completely disappeared over an area located between 80°N and 80°30N and 5°E to 8°E, 30 km wide and 60 km long, as shown by the AVHRR images. Looking at the September 4, 1988, AVHRR image (Figure 17), two facts are quite surprising : *(1)* the 2000 km^2 polynia was located Northeast of the warm core, upstream of the ice drift, normally coming from the Northeast. *(2)* temperatures of the surface water inside the polynia were quite high (4°C) and much higher than one would expect from waters stemming from melting ice (<0°C). This could be due to a special event that we witnessed: a Polar Low at 990 mbar north of Greenland (84°N and 20°E) on September 2, 1988. Moving swiftly northeastwards at 15 knots and contrasting with a high pressure (1010 mbar) over Svalbard, this Low created strong southwesterlies (40 knots) North of Svalbard on September 2-3, 1988, blowing away ice, melting waters towards the northeast and creating a strong divergence in the ocean. On September 3, a cold front associated with the low pressure system passed over the Yermak Plateau area. Air temperatures dropped from +9°C to -2°C. On September 4, 1988, the AVHHR image indicated that the ice edge on the northeastern side of the polynia was quite compact compared to the more diffuse western side, in agreement with what southwesterlies would causer. Another AVHRR image taken on July 27, 1989 (Figure 33), represents a more normal situation with no wind effect. *The polynia is then on the downstream side of the ice drift*, southwest relative to the trajectories of the floats materializing the path of the warm WSC. This image shows surface water inside the polynia which is *colder* than in 1988.

Ice-ocean interactions over this area are quite complex. Simple analytical models like proposed by Untersteiner [1988] should be reconsidered since the general ocean circulation of water under the ice looks significantly different from what had been commonly accepted. The ice wedge length used in Untersteiner's model, should be modified. From 5 km suggested by Quadfasel et al. [1987] up to 150 km proposed by Untersteiner [1988], we think *50 km* would be a more appropriate value corresponding to a proper time scale of about *10 days* at a mean drift rate of *5 km per day*. Anyway, oceanic heat fluxes estimations appear to be abnormally high, due to the assumption concerning vertical heat transfer. In an area full of eddies like the Fram Strait, the contribution of eddies for enhancing ice edge melting should not be forgotten. It was estimated by Johannessen et al. [1987] of the order of 1-2 km per day.

Conclusion

The EGC observations made by Gascard et al. [1988] (their figures 15.a and 15.b, p. 3631) and our Arctemiz 88 data related to WSC show that both currents are described as jets, 20 miles wide, with high speeds in the center (up to 40 cm.s^{-1}) and lower speeds on the edges (10 cm.s^{-1}), positive shear vorticity on the left-hand side looking downstream (order of 10^{-5}.s^{-1}) and negative to the right, strong acceleration between 79°N and 80°N (10 cm.s^{-1} up to 40 cm.s^{-1} in 1 or 2 days over 50 to 100 km distance) and equal mean transport (3 Sv). Despite such similarities, there are strong differences. EGC is a boundary current with a strong baroclinic component and vertical shear between Polar waters, Atlantic Waters and Arctic Intermediate Waters. WSC is a boundary current with a strong barotropic component and horizontal shear between Atlantic Waters and Arctic Intermediate Waters. As a direct consequence from barotropy, WSC strongly interacts with bottom topography. A great difference between EGC and WSC concerns stability versus variability. EGC looks like a steady boundary current compared with the much more variable WSC. *Unsteadiness, barotropy and bottom topography explain why WSC has a strong propensity for "branching" along topographic fracture zones and eddying almost everywhere in Fram Strait.*

In launching SOFAR floats in the WSC in summer 1988, our main intention was to verify one of the conclusions presented by Gascard et al. [1988] concerning the recirculation of the WSC warm and salty waters across Fram Strait. According to these authors, the Recirculation Atlantic Current (RAC) would occur as eddies generated by instabilities of the WSC. These eddies would follow preferentially topographic fractures, quite abundant in Fram Strait (Hovgaard FZ, Molloy FZ, Spitsbergen FZ; Greenland FZ) and crossing the strait in the southeast-northwest direction. The RAC had been recognised by many authors in the past [Perkin and Lewis, 1984; Quadfasel et al., 1987; Bourke et al. 1988], but it had been usually described in terms of elongated features like "branches" rather than eddies as suggested by Gascard et al. [1988]. *The RAC is indeed a very important process associated with the WSC and intrusions of Atlantic warm and salty waters in the Arctic Ocean and in the Greenland Sea. The pattern of the RAC circulation corresponding to a multi-path current system strongly influenced by topographic features, is also confirmed.* The main recirculation areas are located between the Hovgaard FZ and the Molloy FZ, both FZ separating deep basins: the Boreas basin to the South and the Molloy Deep to the North. We discovered that some recirculation path still occurred 300 km farther North as far as 82°.

One of the eight floats did follow the continental slope West of the Yermak Plateau and supports the observations of Perkin and Lewis (1984) and Muench et al. [1992] on a *branch of warm and salty water circulating clockwise around the Plateau.* It is impossible at this point to make any estimate of transport associated with it and/or of the persistence of this branch structure over time.

Our most surprising finding is the discovery of a *major path of circulation for WSC water masses on top of the Yermak Plateau* (Figure 34), which has never been referred to before in the literature. It is quite clear that this pattern of circulation directly and consistently influences the ice edge's location, as illustrated by AVHRR images. Ocean heat fluxes which melt ice are more important than previous estimates.

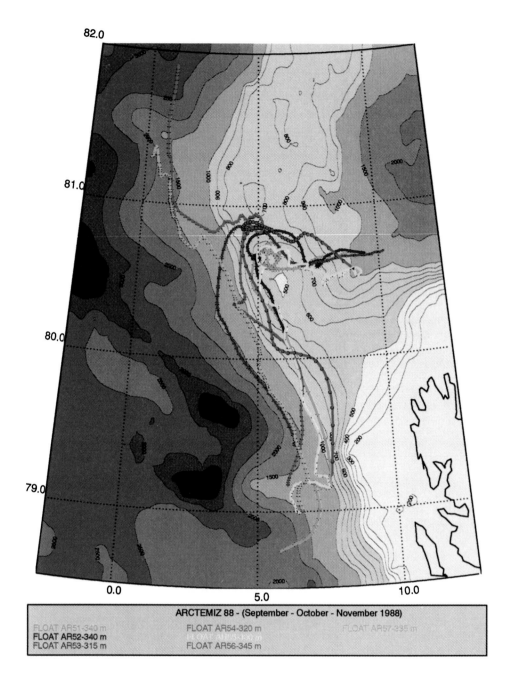

Figure 34. Lagrangian circulation revealed by floats drifting at about 330m depth on top of Yermak Plateau and along the continental slope West of the Plateau in September-November 1988.

Was such a detailed study about ocean circulation in one of the most complex region of the world ocean useful? As mentioned by Untersteiner [1988], the typical large-scale model calculations described Fram Strait area by no more than a couple of grid points. Fram Strait obviously deserves more attention, due to *the phenomenal impact water mass circulation and transformation occurring there, has over the entire Arctic Basin to the North and to the North Atlantic Gyres to the South,* from both an oceanographical and climatological point of view.

Acknowledgments. We would like to thank Commandant Paugam, the Crew of the French Research Vessel "Cryos", and the scientific party from LODYC, who operated successfully in sea ice, up to 81°21' N, during ARCTEMIZ 88. We are very grateful to Professor Michel Durand, Scientific Head at the French Embassy at Oslo, who encouraged and supported us enthusiastically during the 1988-90 Arctemiz Experiments, and unfortunately died prematurely one year later. Claude Kergomard, from the University of Lille, took care of the AVHRR images in real time at the Tromso Satellite Station. Henri Berteaux, from the Woods Hole Oceanographic Institution, provided us with the R-TEAM current meter results. Bert Rudels, from the North Polar Institute, provided two deep SOFAR floats, thanks to the French-Norwegian Foundation. Many thanks to the Director of the Tromso Satellite Station and to M. Schreuder, French Consul at Tromso (Norway), who helped us to solve a lot of logistical and technical problems. We will thank also Ursula Schauer, on board the R/V Valdivia (IFM Hamburg), Bert Rudels on board the Norwegian Coast Guard Andenes, and Robert Bourke, on board the US Navy Ship Bartlett, who helped for the recovery and redeployment of the ALS in the Greenland Basin and Fram Strait. Breck Owens and Chris Wooding were very helpful in allowing us to use the WHOI Sofar Float software package and also the WHOI facilities for ballasting the floats. This work was funded by: DRET 87/178, CEE/EV4C 1/00100, INSU 8850N176, IFREMER, 87/1/430006, DRCI FD/MCL/88/1446.

References

Aagaard, K. and E.C. Carmack, The role of sea ice and other fresh water in the Arctic circulation, *J. Geophys. Res., 94,* C10, 14485-14498, 1989.

Aagaard, K., A. Foldvik, and S.R. Hillman, The West Spitsbergen Current: Disposition and water mass transformation, *J. Geophys. Res., 92,* C4, 3778-3784, 1987.

Bourke, R.H., R.G. Paquette, and R.F. Blythe, The Jan Mayen Current of the Greenland Sea, *Deep-Sea Res., 97,* C5, 7241-7250, 1992.

Bourke, R.H., A.M. Weigel, and R.G. Paquette, The Westward turning Branch of the West Spitsbergen Current, *J. Geophys. Res., 93,* C11, 14065-14078, 1988.

Dickson, R.R., J. Meincke, S.-A. Malmberg, and A.J. Lee, The "Great Salinity Anomaly" in the Northern North Atlantic 1968-1982, *Prog. Oceanogr., 20,* 103-151, 1988.

Foldvik, A., K. Aagaard, and T. Tørresen, On the velocity field of the East Greenland Current, *Deep-Sea Res., 35,* 8, 1335-1354, 1988.

Gascard, J.-C., C. Kergomard, P.-F. Jeannin, and M. Fily, Diagnostic study of the Fram Strait marginal ice zone during summer from 1983 and 1984 Marginal Ice Zone Experiment. Lagrangian observations, *J. Geophys. Res., 93,* C4, 3613-3641, 1988.

Häkkinen, S., An Arctic source for the Great Salinity Anomaly: A simulation of the Arctic ice-ocean system for 1955-1975, *J. Geophys. Res., 98*, C9, 16397-16410,1993.

Hanzlick, D.J., The West Spitsbergen Current: Transport, forcing and variability, Ph.D. Thesis, University of Washington, Seattle, 1983.

Johannessen, J.A., O.M. Johannessen, E. Svendsen, R. Shuchman, T. Manley, W.J. Campbell, E.G. Josberger, S. Sandven, J.C. Gascard, T. Olaussen, K. Davidson, and J. Van Leer, Mesoscale eddies in the Fram Strait marginal ice zone during the 1983 and 1984 Marginal Ice Zone EXperiments, *J. Geophys. Res., 92*, C7, 6754-6772, 1987.

Josberger, E.G., Bottom ablation and heat transfer coefficients from the marginal ice zone of the Greenland Sea, *J. Geophys. Res., 92*, C7, 7012-7016, 1987.

Manley, T.O., R.H. Bourke, and K.L. Hunkins, Near-Surface circulation over the Yermak Plateau in northern Fram Strait, *J. Mar. Syst., 3*, 107-125, 1992.

Manley, T.O., J.Z. Villanueva, J.-C. Gascard, P.-F. Jeannin, K.L. Hunkins, and J. Van Leer, Mesoscale oceanographic processes beneath the ice of Fram Strait, *Science, 236*, 432-434, 1987.

McCarren, D.H., Analysis of drifting SOFAR buoys in the Greenland Sea, Dissertation Thesis, Naval Postgraduate School, Monterey, California, 1991.

Muench, R.D., M.G. McPhee, C.A. Paulson and J.H. Morison, Winter oceanographic conditions in the Fram Strait-Yermak Plateau region. *J. Geophys. Res.,* 97, C3, 3469-3484 , 1992.

Perkin, R.G., and E.L. Lewis, Mixing in the West Spitsbergen Current, *J. Phys. Oceanogr., 14*, 8, 1315-1325, 1984.

Quadfasel, D., J.-C. Gascard, and K.-P. Koltermann, Large-Scale oceanography in Fram Strait during the 1984 Marginal Ice Zone Experiment. *J. Geophys. Res., 92*, C7, 6719-6728, 1987.

Quadfasel, D., and M. Ungewiß, Large-scale hydrographic structure of the upper layers in Fram Strait during the Marginal Ice Zone Experiment, 1984, *Institut für Meereskunde der Universität Hamburg Tech. Rept* 2-86, 41 pp., 1986.

Rudels, B., On the mass balance of the Polar Ocean, with special emphasis on the Fram Strait, *Norsk Polarinstitutt Skrifter, 188*, 1-53, 1987.

Swift, J.H. and K. Aagaard, Seasonal transitions and water mass formation in the Iceland and Greenland Seas, *Deep-Sea Res., 28*, 1107-1129, 1981.

Untersteiner, N., On the ice and heat balance in Fram Strait. *J. Geophys. Res., 93*, C1, 527-531, 1988.

Vinje, T., and O. Finnekasa, The ice transport through the Fram Strait, *Norsk Polar Institute, Oslo, Norway , 186*, 36 pp., 1986.

Wadhams, P., Sea-Ice thickness distribution in Fram Strait, *Nature*, 305, 5930, 1983.

5

Chemical Oceanography of the Arctic and its Shelf Seas

Leif G Anderson

Abstract

The chemical signature of most water masses within the Arctic Ocean is the result of processes occurring in the vast shelf seas. More than 0.1 Sv of river-runoff is added, containing large amounts of organic and inorganic carbon as well as nutrients. The inorganic carbon added to the surface layer of the Siberian shelf seas is both incorporated into the Beaufort gyre and the Siberian branch of the transpolar drift. The large runoff input to the Kara Sea seems to first flow to the east into the Laptev Sea, before entering the interior of the Arctic Ocean. The fresh water supply also stimulates primary productivity, both by the supply of nutrients and by increasing the stratification. This primary productivity primarily takes place in the shelf seas, where the ice cover is at its minimum during the summer. The primary production and the input of particulate organic matter by the runoff results in high organic content in the shelf sediments. The decay of organic matter in the sediment causes a recycling of nutrients and total carbonate back into the overlying bottom water. In the winter, brine is rejected during the formation of ice, which in places form a high salinity bottom water. When such a water is overlying the organic rich sediments, the decay products will not be mixed back into the surface water, but will be concentrated in bottom water. This high salinity, nutrient rich bottom water has been found in the Chukchi Sea, in the eastern Barents Sea and in the Storfjorden, southern Svalbard. With time the shelf bottom water flows out into the deep interior of the Arctic Ocean, where its signature is striking in the halocline of the Canadian Basin. That the shelf bottom water largely ends up in the halocline is obvious as it covers a large fraction of the density range. However, if the density of the shelf bottom water is high enough it will, while entraining surrounding waters, penetrate down into the deepest layers. This mechanism is important for the ventilation of both the Eurasian and the Canadian basins. The ventilation of the latter is a combination of direct shelf plumes and spill over from the Eurasian Basin.

Arctic Oceanography: Marginal Ice Zones and Continental Shelves
Coastal and Estuarine Studies, Volume 49, Pages 183–202
Copyright 1995 by the American Geophysical Union

Introduction

Water enters the Arctic Ocean, from the rivers, through Bering Strait and as Atlantic Water both over the Barents Sea and through Fram Strait. Except for the latter all of these waters flow over enormous shelf areas before entering the deep central regions (Figure 1) and consequently processes occurring over the continental shelves are of outmost importance in modifying the chemical signatures of most water masses encountered within the Arctic Ocean. Also the formation or melting of sea ice is important in modifying the chemical signature, either directly or indirectly. The direct effect of sea ice formation is the release of brine that is added to the surrounding water, while melting adds relatively fresh water with its chemical signature. One indirect effect of ice formation is the production of high salinity bottom waters of freezing temperature whose chemical signature could be changed by sediment interaction. Another effect of sea ice melting is the physical stabilization of a surface water, governing the biological primary production and therefore resulting in a change of the chemical signature. It is the objective of this chapter to present and discuss in detail various biogeochemical processes and show how they affect the chemical signature of the water masses of the central Arctic Ocean. The chemical signatures are then used to trace the water masses, resulting in a plausible circulation pattern of the different water masses.

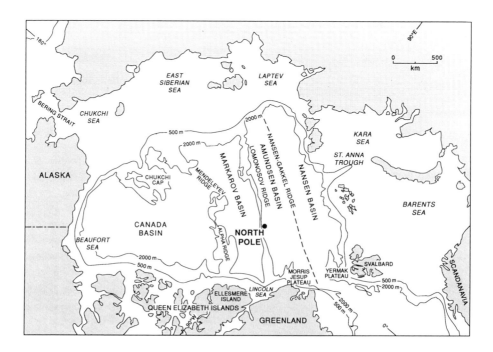

Figure 1. Map of the Arctic Ocean

River runoff

The Arctic Ocean is second only to the Atlantic Ocean in the amount of fresh water it receives through river runoff. The largest volume is from the Siberian rivers, but also the Mackenzie River and the Yukon River (the latter via the Bering Strait) are significant sources. Furthermore, it has been estimated that about half of the fresh water runoff to the Arctic Ocean from Canada is through creeks and streams, thus more or less doubling the volume from large rivers draining the American continent. There might be a similar situation on the Siberian side.

Chemical Signature

The chemical signature of the river water is dependent on the environment in the drainage basin, for example, the type of minerals and extent of biological production and decay. The tundra is, for instance, thought to have a vast capacity for long-term storage of carbon [Oechel and Strain, 1985, and references therein]. The large amount of organic matter will during the decay process release carbon dioxide, which in turn will react with the different mineral types. The summary reaction can be simplified by the following.

$$MeCO_3(s) + CH_2O(org) + O_2 \rightarrow Me^{2+} + 2HCO_3^{-} \qquad (1)$$

From reaction (1) it is seen that the river water adds total alkalinity or total carbonate (sum of dissolved inorganic carbon) and metals to the surface water over the shelves.

In Table 1 the concentrations of total alkalinity are summarized for different rivers. As the weighted mean concentration is high, the mixing with seawater will give elevated values which makes alkalinity and total carbonate excellent tracers of runoff within the Arctic Ocean. The southern boarder of the Siberian Branch of the Transpolar Drift, for instance, is clearly seen by a strong front in total alkalinity and total carbonate within the Eurasian Basin (Figure 2) [Anderson et al., 1989; Anderson and Jones, 1992; Anderson et al., 1994a]. To the south of the front, the fresh water source is dominated by sea ice melt water, indicating that the northern Barents Sea is very little affected by river runoff. Hence, the chemical signature of the surface waters in this region is mainly a result of the inflowing water of Atlantic origin.

Time Estimates

The front between fresh water from river runoff and fresh water from sea ice melt is also evident in the oxygen isotope signal, with the lighter isotopic composition in the river runoff as this has a meteoric origin. The meteoric water was contaminated by tritium during the bomb tests in the beginning of the sixties and an estimate of the time since the water left the drainage basin can thus be made from tritium measurements. The resulting "age" is about 10 years for the fresh water in the surface and halocline waters to reach the Fram Strait [Östlund, 1982; Östlund and Hut, 1984]. In a recent investigation Schlosser et al.

[1994] have used the combination of the oxygen isotopic composition and both the tritium and helium-3 concentration to derive "age" estimates. The tritium clock starts when the precipitation reach the earth surface (close in time to when the runoff leaves the river mouth), while the helium clock starts when a water parcel last was in contact with the atmosphere. In Arctic shelf seas the latter occurs when the surface water is covered by sea ice. Generally the shelf seas are ice free in the summers and largely ice covered in the winters, thus the helium clock starts when the surface water reach thecontinental slope.Hence, the difference in age between the tritium and helium "ages" will reflect the mean residence time over the shelf. The estimates by Schlosser et al. [1994] yield a mean residence time of the river runoff on the shelves of the Barents and Kara Seas of about 3.5±2 years.

TABLE 1. Annual averaged discharge and total alkalinity concentrations of the major rivers entering the Arctic Seas. [from Anderson et al., 1983, and references therein]

River	Discharge ($m^3 s^{-1}$)	Total Alkalinity (mmol/l)
Mackenzie	12,000	1.74
Yukon	5,100	2.4±0.4
Kolyma	3,800	0.37
Indigirka	1,800	0.72
Yana	1,000	0.45±0.07
Lena	15,800	1.08
Pyasina	2,500	0.56
Yenisey	18,600	0.95±0.25
Ob	12,500	1.2±0.2
Pechora	4,100	0.5±0.1
Mezen	800	0.68
N. Dvina	3,500	2.1±0.1
Total	105,600	1.13[a]

[a]Expressed as the weighted mean concentration.

Shelf processes

General

One result of the large river input to the shelf seas is a high supply of nutrients. Added to the supply from the rivers is that from the entering seawater, Atlantic water with its moderate nutrient content to the Barents and Kara Seas and Pacific water with its high nutrient content to the Chukchi and East Siberian Seas. Consequently there is a significant primary production when the light conditions are favorable. For the shallow Siberian Seas (maximum depth around 50 m) light limitation is normally not a problem as the stratification is rather strong. The main cause of shading is either turbid water from the rivers or the plankton themselves. Hence, primary production in these seas can occur over wide areas of the shelf and is mainly limited by the nutrient supply in the summer. On the

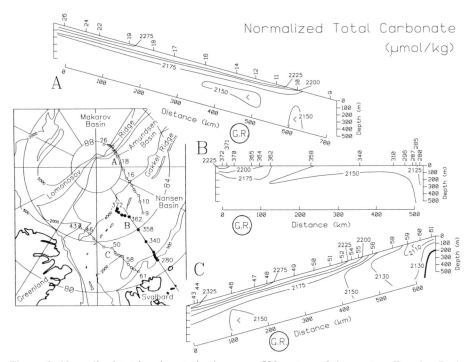

Figure 2. Normalized total carbonate in the upper 500 meters of the western Eurasian Basin. Adopted from Anderson and Jones [1992] and Anderson et al., [1994]. The high normalized total carbonate concentration in the surface water to the north (left part of the sections) reflect the river runoff signal. The front is seen where the isolines for 2175 and 2225 μmol/kg outcrops. In the three sections the position of the Gakkel Ridge is indicated by G.R.

other hand the depth of the Barents and Kara seas exceeds 200 m in large areas and vertical stratification is often quite weak. Light conditions are therefore not optimal, but primary production is concentrated to the marginal ice zone where a stable surface layer is formed by the addition of ice melt water. In general, primary production is quite high in all Arctic shelf seas, resulting in extensive biological production also at higher trophic levels.

Production is mostly confined to the summer period with its more or less constant light supply. In the winter, darkness is amplified in the sea by the growing ice cover. One result of this seasonal production is a significant sedimentation of particles that largely ends up at the sediment surface in the shallow shelf seas. During ice production, the brine generated forms a high salinity water that is trapped along the bottom if the topographic conditions are right. The chemical constituents that are formed during the decay of the organic matter in the upper sediment will, to a large degree, be released into the overlaying water. If this overlaying water is the high salinity bottom water, the chemical constituents will not be mixed up into the shallower waters over the shelves, at least not directly. Instead, they will stay with the high density water as it flows along the bottom towards the deep interior. This chain of processes, summarized in Figure 3, is very important in the formation and modification of most water masses within the deep central Arctic Ocean.

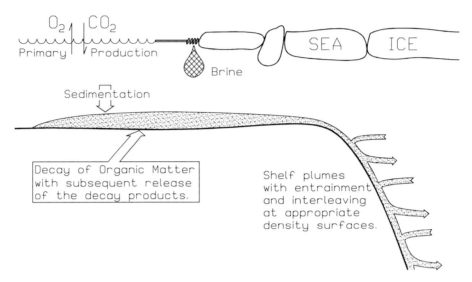

Figure 3. Schematic illustration of different processes taking place over the year in the shallow shelf seas. During the summer season primary production fixate nutrients and CO_2 and release oxygen. At the end of the productive season, organic matter sinks to the bottom where it decays. Brine is released from sea ice when it forms. In shallow seas the brine form a cold, high salinity bottom water which slowly flows out into the deep interior.

Barents Sea

The ecology of the Barents Sea has been extensively studied within the Norwegian Research Program for Marine Arctic Ecology (Pro Mare) and the results have been summarized in volume 10(1) of Polar Research [1991]. The general picture is that primary production follows the retreating ice zone, starting in the south and moving to the north during the summer season. Added to this ice zone production are patchy regions of high production within the pack ice. One result of biological production is formation of halogenated hydrocarbons. The exact biochemical processes for the formation of these halogenated hydrocarbons are not yet established, but different investigations show that the concentrations are very high in polar waters, especially for bromoform as is seen in a depth plot for three stations in the north eastern Barents Sea (Figure 4.)

The main source of bromoform is pelagic and/or ice algae, resulting in a maximum close to the surface. Hence, there is a significant release of halides to the atmosphere, which for bromoform has been estimated to be in the order of 10^6 kg Br year^{-1} from the Arctic Ocean [Krysell, 1991].

These halogenated hydrocarbons can also be used to trace water masses that have experienced extensive biological production. This could be especially valuable in ice covered regions, where the ice hampers the escape of these volatile compounds to the atmosphere.

Primary production changes the chemical signal of the surface water by consuming nutrients and producing oxygen. Often the surface water becomes super-saturated with oxygen

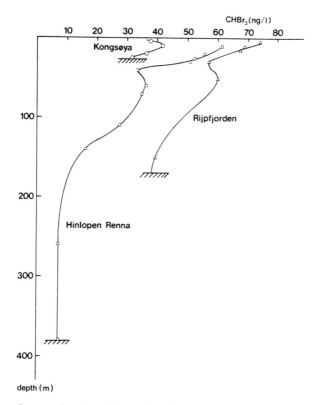

Figure 4. Bromoform profiles from three stations in north western Barents Sea [from Dyrssen and Fogelqvist, 1981].

causing degassing to the atmosphere. The loss of oxygen from the surface water will result in a lowering of the NO content. (NO is a conservative relationship, defined by Broecker [1974] as $O_2 + 9NO_3$.) NO will not be affected in the deeper layers where the organic matter decays as the decrease in oxygen will be compensated by the release of nitrate, as long as the water is well oxygenated.

Part of the water that enters the Arctic Ocean through Barents Sea will hence be modified in its NO content. The water that is found at the surface in the central Barents Sea during the productive season will be cooled during the following fall, before entering the Kara Sea for further flow north into the deep Arctic interior. During the transit through the Kara Sea it will meet lower density water and be forced down below the surface. The Atlantic water that enters the central Arctic Ocean through the Kara Sea turns east and follows the continental slope before heading back towards Fram Strait in several branches following the ridges [Quadfasel et al., 1993; Rudels et al., 1994].

Centered at a salinity of about 34.2 in the north eastern Eurasian Basin is a minimum in NO, the Lower Halocline Water (Figure 5), which is thought to have a Barents Sea origin [Jones and Anderson, 1986; Anderson and Jones, 1992; Anderson et al., 1994b]. This NO distribution is one of the indicators to the postulated flow pattern of the intermediate deep

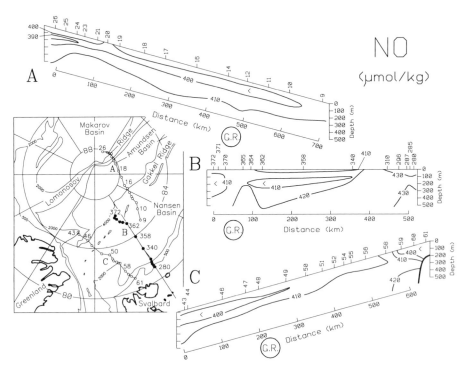

Figure 5. NO in the upper 500 m of the western Eurasian Basin. From Anderson and Jones [1992] and Anderson et al. [1994b]. The minimum at 100 to 200 m depth is a signature of Lower Halocline Water (salinity ≈ 34.2). In the three sections the position of the Gakkel Ridge is indicated by G.R.

water masses of Atlantic origin. It is apparent that NO also will be lowered in the surface waters of the other shelf seas when oxygen is lost to the atmosphere during primary production [Wilson and Wallace, 1990]. However, in the more eastern seas (the Chukchi, the East Siberian and the Laptev seas) the salinity of the surface water is too low for the water to penetrate below the Surface Mixed Layer (SML) of the central Arctic Ocean. Hence, this low NO water will be incorporated in the SML and thereby making it possible to have a continuos exchange of oxygen with the atmosphere, with the result that NO is not necessarily conserved.

Shelf waters play a part in the formation of water at all depth in the Arctic Ocean through dens water produced during ice formation. The existence of high salinity bottom water as a result of brine release from sea ice production have been shown in the eastern Barents Sea [Midttun, 1985] and in the Storfjorden, southern Svalbard [Anderson et al., 1988]. In the Storfjorden the release of nutrients into the bottom water and the consumption of oxygen from the bottom water was also shown. The drainage of the high salinity bottom water found in the eastern Barents Sea is thought to take place through the St. Anna Trough, northern Kara Sea. The existence of these high salinity shelf plumes was shown by Quadfasel et al. [1988] when the outflow from the Storfjorden was traced down the Norwegian Sea continental slope and further north into the Fram Strait.

Chukchi and East Siberian Seas

It is believed that the Chukchi and East Siberian seas are the most productive of the Arctic shelf seas. The nutrient supply is large, both from the rivers and through the Bering Strait. Reported concentrations are highest around the Wrangel Island, with bottom water levels above 50 μM, 15 μM and 2.5 μM for silicate, nitrate and phosphate, respectively [Codispoti and Richards, 1968]. In that study, only the bottom water around Wrangel Island had salinities above 33. More recent investigations document primary production of over 10 g C m^{-2} day^{-1} in some areas of the Chukchi Sea (Figure 6a.) [Springer and McRoy, 1993]. These high production rates are reflected by the content of organic matter in the sediment, between 1 and 2 % with a C/N ratio of about 6 (Figure 6b and c) [Grebmeier, 1993]. In the same study the sediment oxygen uptake rate was estimated to be on the order of 20 to 40 mmol m^{-2} day^{-1} (Figure 6d.). These investigations are consistent with those of several other scientists [e.g., Rusanov, 1980 and references therein] who have suggested that regeneration during winter enriches the waters of the Chukchi Sea with silicate (and other nutrients). The nutrient enriched water formed on the shelf interleave on appropriate density surfaces and can be traced throughout the Canadian Basin in the Upper Halocline Water (UHW) centered around a salinity of about 33.1 [Jones and Anderson, 1986]. All these investigations indicate the importance of the Chukchi and East Siberian seas in the transformation of chemical constituents through the chain of processes summarized in Figure 3.

The penetration of the high nutrient signal from the Chukchi Sea into the UHW of the Canadian Basin is presented in a section adopted from Treshnikov [1985] (Figure 7.). From this investigation, along with the ones from ice camps, T3 [Kinney et al., 1970], LOREX [Moore et al., 1983], CESAR [Jones and Anderson, 1986] and the sections sampled from surface ships, Polarstern 1987 [Anderson et al., 1989] and Oden 1991 [Anderson et al., 1994b], it could be concluded that the nutrient rich Upper Halocline Water is more or less confined to the Canadian Basin. However, the outflow of UHW has been identified, by its pronounced silicate signal, as following the continental slope north of Greenland [Rudels et al., 1994] and further south through the eastern Fram Strait [Anderson and Dyrssen, 1981; Jones et al., 1991].

During the decay of organic matter, total carbonate is also released with the nutrients as part of the decay process, and oxygen is consumed. The relative shift of these constituents during the decay process is quite constant in the marine environment as the P/N/C ratio of organic matter does not vary much between different species. The differences between the measured and preformed values in the UHW at ice station CESAR are very similar to the theoretical shifts expected during decay of organic matter [Jones and Anderson, 1986]. This is a further strong indication that decay processes are responsible for the nutrient signal in the UHW. Using the profiles at ice station CESAR, Anderson et al. [1990] calculated the excess total carbonate of shelf sea origin and converted this excess to new production by using the whole deep Arctic Ocean as the extent of the UHW and literature data of the water mass residens times. If the new production is evenly distributed over the Laptev, Chukchi and East Siberian seas, it would be about 45 g C m^{-2} a^{-1}, if no burial of organic

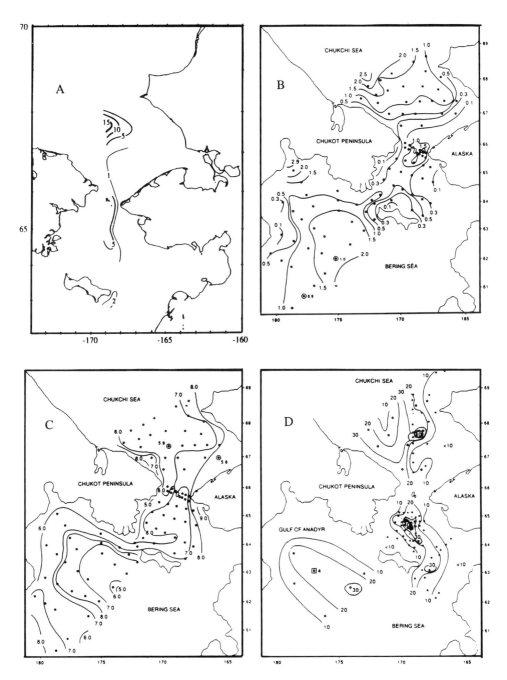

Figure 6. Distribution in the Bering Strait region of (A) depth integrated primary productivity (g C m^{-2} d^{-1}) [Springer and McRoy, 1993], (B) organic carbon (%), (C) C/N ratio (w/w) in the surface sediments and (D) sediment oxygen uptake rates (mmol O$_2$ m^{-2} day^{-1}) [Grebmaier, 1993]

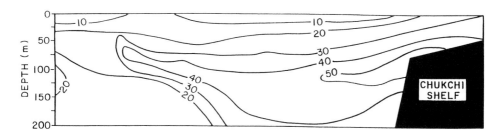

Figure 7. Section of silicate across the Canadian Basin from Alaska to Greenland in late winter 1991. Adopted from Treshnikov [1985].

matter in the shelf sediment is assumed. Recent investigations [Anderson et al., 1994a] have shown that the UHW is more or less confined to the Canadian Basin, thereby decreasing the calculated mean new production to about 25 g C m^{-2} a^{-1}, using the same production area. These values are not unrealistic, since Hansell et al., [1993] estimated the new production in the Anadyr water, both north and south of the Bering Strait, to be 288 g C m^{-2} a^{-1}.

Beaufort Sea and the Shelves North of the Canadian Archipelago

The shelves in the Arctic Ocean north of the American continent are very narrow compared to those north of Siberia. They are also to a large extent ice covered the year around. This results in relatively low total biological production, although high production occurs in some regions outside the Mackenzie River where the new production has been estimated to 20 g C m^{-2} a^{-1} [Macdonald et al., 1987]. The UHW as defined by the nutrient maximum is seen also at the shelf break outside the Mackenzie estuary but is thought to be advected to this region even if high salinity bottom water is formed on the shelf [Macdonald et al., 1987; Macdonald et al., 1989]. Hence, the same chain of processes that occur in the other shelf seas, also occurs in the Beaufort Sea, but the quantitative importance to the Arctic Ocean as a whole is much less. Further to the northeast, north of Ellesmere Island, both UHW and LHW have been identified over the shelf at about 300 m bottom depth [Jones and Anderson, 1990]. Direct measurements of nutrient regeneration at that site indicated little supply of organic matter, consistent with the limited primary production in this mostly ice covered region.

It is remarkable to find the uniform distribution of the nutrient maximum, as a signal of UHW, and the NO minimum, as a signal of the LHW, along the rims of the Canadian Basin north of the American continent, without any likely production areas for these waters. This point to a circulation pattern that distribute these water masses all around the Canadian Basin without much mixing with over and under-lying waters.

Marginal ice zone

The Marginal Ice Zone (MIZ) are important for the transformation of chemical constituents in the shallow shelf seas, as has been mentioned in the *Shelf Process* section. In Fram Strait we find the only MIZ that is over the deep ocean. This zone is allied with the front between the north flowing West Spitsbergen Current and the south flowing East Greenland Current. In the summer season extensive primary production is often associated with the MIZ, hence affecting the chemical composition of the surface water. For instance, carbon dioxide is consumed and oxygen released. This is illustrated in Figure 8 with the lowest pCO_2 values very close to the MIZ. The oxygen levels do not have the same large variation but generally the high values are found in the open water to the east. To the west, close to the northeast water polynya, oxygen is super-saturated and carbon dioxide is under-saturated, a signal of the high production in the polynya. A few stations with high pCO_2 values are found in the northwest, reflecting the outflow of some of the nutrient rich water from the Arctic Ocean.

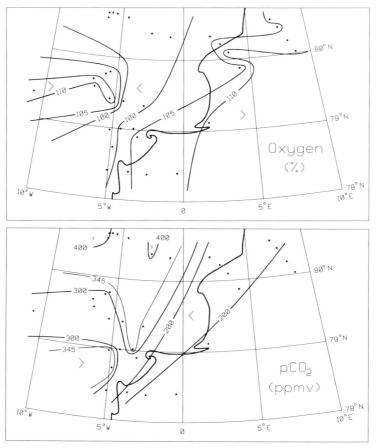

Figure 8. pCO_2 and oxygen saturation at 5 m depth along the marginal ice zone in Fram Strait. Data from MIZEX 84 (Anderson and Jones, unpublished data). The ice edge as of June 30 is illustrated by the thick line.

Interaction with the deep central basin

General

The interactions of the shelf waters with the upper layers of the central Arctic Ocean have been discussed above. The waters produced over the shelves, however, will not only interleave into the halocline, but will also be incorporated into deeper layers.

For instance, some of the Atlantic water that enters the Arctic Ocean over the Barents Sea will even increase its salinity as a result of input of brine from sea ice production. This water forms the Barents Sea branch in the Arctic Ocean and spans a great density interval and, thus, also span a large depth interval [e.g. Rudels et al., 1994]. Some of the water coming from the Barents Sea will even be dense enough to penetrate through the Atlantic Layer, traditionally defined by the potential temperature (Θ) > 0 ^{o}C. This water of shelf origin will carry its chemical signature, which may be especially pronounced for the transient tracers. High density shelf water most probably penetrates to the intermediate and deep layers of all the central Arctic Ocean basins.

Atlantic Layer

The Atlantic Layer can be traced all around the central Arctic Ocean by its high temperature. In general the temperature decreases cyclonically from Fram Strait around the deep Arctic Ocean, a feature earlier deduced as being a result of the circulation pattern, with the water entering through Fram Strait and progressing eastward as a boundary flow along the continental slope [Coachman and Barnes, 1963]. More recent investigations have shown the existence of another inflow through the Barents Sea [Rudels, 1987; Blindheim, 1989]. The two sources of the Atlantic Layer, the Fram Strait branch and the Barents Sea branch, have slightly different characteristics and clear interleaving of the two branches occur within the central basin [Quadfasel et al., 1993; Rudels et al., 1994]. After the two branches meet north of the Kara Sea, they follow the continental slope to the east into the Laptev Sea where a fraction of the water splits off and returns westward along the Gakkel Ridge. That which is left continuos along the continental margin until it approaches the Lomonosov Ridge where another fraction splits off and returns along this ridge, while the rest continues into the Canadian Basin. As the two branches are not fully mixed, the return flows will not contain a homogenous mixture, but mostly water that has the "inner lane" when the separation occur [Rudels et al., 1994].

Transient tracers (e.g. the CFCs) and nonconservative chemical constituents (e.g. silicate) can give a time perspective on the different return flows as well as an idea of the processes the waters have undergone during their transit within the Arctic Ocean. Rudels et al. [1994] assigned a general circulation pattern of the Atlantic Layer like the one in Figure 9. The two loops in the Canadian Basin were deduced by the existence of two cores of outflowing Atlantic Layer water, each with different chemical signatures, along the continental slope of the Morris Jesup Plateau. One of the cores had a lower concentration of CFCs (older

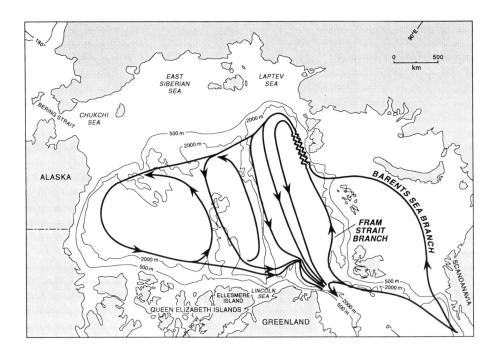

Figure 9. The inferred circulation of the Atlantic Layer of the Arctic Ocean [from Rudels et al., 1994]. For a more detailed labeling, see Figure 1.

water) and higher concentration of silicate compared to the other core. However, both cores had lower CFC and higher silicate concentrations than the Atlantic Layer water that recirculate within the Eurasian Basin. The elevated silicate concentrations are most likely a result of the interleaving of some shelf-derived water, which is high in silicate (same reason as the UHW is high in nutrients). The most plausible way to achieve the observed chemical signature of the two cores is to have two loops within the Canadian Basin as presented in Figure 9.

The presence of several loops in the Atlantic Layer is also supported by the age distribution for the Atlantic Layer water as determined by transient tracers at different sites within the Arctic Ocean (Table 2) [Wallace et al., 1992]. The only value reported from the Canadian Basin (Makarov + Canada basins) is from ice station CESAR, which has a much greater age than those from the Eurasian Basin, confirming at least one loop within this area. The age in the water above the continental slope, northeast of Greenland (stn. 330, ARK II/3) has the greatest age of the Eurasian Basin samples, likely a result of water flowing out from the Canadian Basin, mixing with Atlantic Layer water in the Eurasian Basin. The increased apparent age to the north within the ARK IV/3 stations illustrate the flow pattern in the northern Nansen Basin area (Figure 9).

TABLE 2. Tracer ratio ages in the core of the Atlantic Layer [modified from Wallace et al., 1992]. The ARK IV/3 stations are located across the Nansen Basin along the 30°E longitude, the CESAR on top of the Alpha Ridge, the FRAM3 at the northern tip of the Yermak Plateau and the ARK II/3, stn 330 at the shelf slope northeast of Greenland.

Station	Θ_{max}	CCl_4/F11	^{3}H/^{3}He	F11/F12
ARK IV/3[a]				
358	1.89	12	8	
362	1.82	13	10	
365	1.44	15	12	
370	1.64		15	
371	1.19		13	
CESAR[b]	0.44			28
FRAM3[c]	1.72		12	
ARK II/3[d] 330	0.64			18

[a] Wallace et al. [1992]
[b] Wallace and Moore [1985]
[c] Östlund et al. [1982]
[d] Smethie et al. [1988]

Intermediate and Deep Waters

Historically it was believed that the Arctic Ocean deep waters were renewed advectively through Fram Strait. More recent investigations have shown that neither the Θ-S, nor the chemical signature of the Arctic Ocean deep waters could be explained by advective renewal alone, so a high salinity shelf source was suggested [e.g. Swift et al., 1983; Aagaard et al., 1985]. Increased concentrations of CFCs close to the bottom (followed by increased salinities) in the Eurasian Basin, northeast of Greenland, support the convective shelf source [Smethie et al., 1988]. The Eurasian Basin deep and bottom water have a higher temperature than the inflowing Norwegian Sea Deep Water (NSDW) and the high salinity shelf water, which starts at feezing temperature on the shelf as it is formed by sea ice production, must therefore be modified before entering the deep layers. Aagaard et al. [1985] showed that mixing with Atlantic Layer water at the shelf break is the source of the heat to the descending plume.

Estimates of the turnover time of Eurasian Basin Bottom Water (EBBW) at the southern tip of the basin are about 20 years, based on CFC11/CFC12 ratio [Smethie et al., 1988], and about 50 years, from box models using several transient tracers [Heinze et al., 1990]. The content of transient tracers are much lower in the Eurasian Basin further away from Fram Strait, where the current pattern is known to be complicated, Carbon tetrachloride is for instance the only CFC that is detectable in the bottom water of the central Nansen Basin [Krysell and Wallace, 1988]. Calculation of the mean age from the ^{14}C concentration with the application of a continuous mixing concept results in 200 to 250 years [Schlosser et al., 1990]. The distribution of the transient tracers in a section across the Nansen Basin, here represented by CFC11 (Figure 10) clearly indicates the existence of boundary currents. This was further confirmed during Arctic 91, when measurements over four sections in the Nansen and Amundsen Basins of the Eurasian Basin and into the Makarov Basin of the Canadian Basin were carried out [Anderson et al., 1994b].

Calculations of the mean age of the deep and bottom waters of the Canadian Basin result in 500 - 800 years, using ^{14}C and with the application of a similar continuous mixing concept as the one for the Eurasian Basin [Östlund et al., 1987]. The problem is to explain the excess salt and heat in the deep Canadian Basin, considering the long residence time and the limited shelf contribution that the [18]O and ^{3}H content imply. Comparing the nutrients and oxygen concentration profiles in the deep Amundsen and Makarov Basins, phosphate and nitrate are much the same, while the silicate concentration is higher and oxygen is lower in the Makarov Basin [Oden-91 data].

This result could not have been obtained if in situ remineralization was responsible, as have been suggested for the Canada Basin [Macdonald and Carmack, 1991]. In order to fit both the transient tracer distribution and the salt, temperature, oxygen and nutrient distribution, Jones et al. [1994] used data from the Oden-91 cruise and adopted a plume model with a significant entrainment (doubling of the volume every 150 m) and two plumes, one originating on the shelf and the other as an overflow from the Eurasian Basin. The total flow of water that starts at 200 meters depth of the continental break and enters the layers below 1700 m, is 0.011 Sv. The salinity at 200 m depth needed to penetrated to the deepest layer (3800-3950 m) is 36.113 and the maximum silicate concentration needed to fit the observed profile is 80 to 90 μmol/kg. It is the entrainment of water, and especially Atlantic Layer water, that causes the relative high temperature in the Canadian Basin. That the bottom waters in the Eurasian Basin are colder is explained by the fact that the shelf plumes here to a large degree follow deep canyons where they partly bypass the relatively warm Atlantic Layer [Jones et al., 1994].

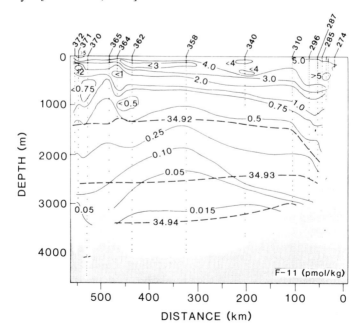

Figure 10. CFC11 distribution across the Nansen Basin at about 30°E. [from Wallace et al., 1992].

Input of atmospheric carbon dioxide

The large continental shelves of the Arctic Ocean are key to the transport of atmospheric CO_2 to deeper regions. As the atmospheric pCO_2 is rather constant over a short time, the processes in the surface water are the most important for determining the flux. Of these, the biological primary production is the most significant, followed by the cooling of surface water. (CO_2 has higher solubility in cold water than in warm.) Decay of the biogenic matter produced over the shelves and incorporation of river runoff (containing dissolved carbonate) into dense water formed on the shelves provide the mechanisms for the transport of atmospheric CO_2 into the central Arctic Ocean.

In addition to the natural variation of the atmospheric pCO_2, there is a continuous increase from burning of fossil fuel and deforestation. This anthropogenic increase will cause a net flux from the atmosphere to the ocean, proportional to the increased partial pressure difference. This flux is denoted anthropogenic input, as opposed to the natural steady state.

Using the same technique as when estimating the new primary production on the Arctic shelves, Anderson et al. [1990] calculated the steady state CO_2 input to different water masses of the Arctic Ocean. By combining the ratios of total carbonate, total alkalinity and calcium they resolved the source regions; river runoff and shelves (Table 3).

The input of anthropogenic carbon dioxide into the Arctic Ocean is mainly through convective deep water formation. A water parcel, at around freezing temperature, in equilibrium with the atmosphere will today have a total carbonate concentration that is about 35 μmol/kg higher than in preindustrial times as the pCO_2 has increased from about 280 ppmv to 360. With this information the anthropogenic input is simply calculated by multiplying the excess total carbonate and the volume involved in the deep convection. The main question is to get this volume, especially as it includes entrainment of water at different depth layers.

TABLE 3. Steady state flux of carbon into the different surface layers of the Arctic Ocean [from Anderson et al., 1990].

Water mass	Net flux (10^{12} g C a^{-1})		
	River runoff	Shelf	Total
Surface mixed layer	57	0	57
Halocline	24	96	120
Atlantic Layer	0	33	33
Total			210

Conclusions

Even if we during the last five years have gained much information about the Arctic Ocean there are still a lot to be learned. We only have a crude picture of the general circulation of different water masses. The knowledge is even less when it comes to the formation and modification of these water masses. We know that the shelves play an significant role and the interaction between the shelves and the deep interior is hence of outmost importance. A very powerful tool in achieving the necessary information is the chemical signature of the different water masses. If we learn more about the biogeochemical processes in the shelf seas different models can be constructed that will shed more light on the whole Arctic Ocean system. It is obvious that the future investigations thus have to have access to all regions within the Arctic Ocean, including the shelf seas.

Acknowledgment. The author thank all colleagues I have cooperated with during my polar research. A special thank is to D. A. Hansell and E. P. Jones for helpful reviews during the manuscript preparation. I am also grateful for support from the Swedish Natural Research Council, the Bank of Sweden Tercentenary Foundation, the Knut and Alice Wallenberg Foundation and the Marianne and Marcus Wallenberg Foundation.

References

Aagaard, K., J. H. Swift, and E. C. Carmack, Thermohaline circulation in the Arctic Mediterranean Sea, *J. Geophys. Res., 90,* 4833-4846, 1985.

Anderson, L., and D. Dyrssen, Chemical constituents in the Arctic Ocean in the Svalbard area, *Oceanol. Acta, 4,* 305-311, 1981.

Anderson, L. G., and E. P. Jones, Tracing upper waters of the Nansen Basin in the Arctic Ocean, *Deep-Sea Res., 39,* S425-S434, 1992.

Anderson, L. G., D. Dyrssen, and E. P. Jones, An assessment of the Transport of Atmospheric CO_2 into the Arctic Ocean, *J. Geophys. Res., 95,* 1703-1711, 1990.

Anderson, L. G., K. Olsson, and A. Skoog, Distribution of dissolved inorganic and organic carbon in the Eurasian Basin of the Arctic Ocean, in *The Role of the Polar Oceans in Shaping the Global Environment,* edited by R. Muench and O. M. Johannessen, pp. 255-262, AGU, Washington, D. C., 1994a.

Anderson, L. G., D. Dyrssen, E. P. Jones, and M. G. Lowings, Inputs and outputs of salt, fresh water, alkalinity and silica in the Arctic Ocean, *Deep-Sea. Res., 30,* 87-94, 1983.

Anderson, L. G., E. P. Jones, R. Lindegren, B. Rudels, and P-I. Sehlstedt, Nutrient regeneration in cold, high salinity bottom water of the Arctic shelves, *Continental Shelf Res., 8,* 1345-1355, 1988.

Anderson, L. G., E. P. Jones, K. P. Koltermann, P. Schlosser, and D. W. R. Wallace, An oceanographic section across the Nansen Basin in the Arctic Ocean, *Deep-Sea Res., 36,* 475-482, 1989.

Anderson, L. G., G. Björk, O. Holby, G. Kattner, P. K. Koltermann, E. P. Jones, B. Liljeblad, R. Lindegren, B. Rudels, and J. H. Swift, Water Masses and Circulation in the Eurasian Basin: Results from the Oden 91 Expedition, *J. Geophys. Res., 99,* 3273-3283, 1994b.

Blindheim, J., Cascading of Barents Sea bottom water into the Norwegian Sea, *Rapp. P.-v. Reun. Cons. int. Explor. Mer., 188,* 49-58, 1989.

Broecker, W. S., "NO", A conservative-mass tracer, *Earth Planet. Sci. Lett., 23,* 100-107, 1974.

Coachman, L. K., and C. A. Barnes, The movement of Atlantic water in the Arctic Ocean, *Arctic, 16,* 8-16, 1963.

Codispoti, L. A., and F. A. Richards, Micronutrient distribution in the East Siberian and Laptev seas during summer 1963, *Arctic, 21,* 67-83, 1968.

Dyrssen, D., and E. Fogelqvist, Bromoform concentrations of the Arctic Ocean in the Svalbard area, *Oceanol. Acta, 4,* 313-317, 1981.

Grebmeier, J. M., Studies of pelagic-benthic coupling extended onto the Soviet continental shelf in the northern Bering and Chukchi seas, *Continental Shelf Res., 13,* 653-668, 1993.

Hansell, D. A., T. E. Whitledge, and J. J. Goering, Patterns of nitrate utilization and new production over the Bering-Chukchi shelf, *Continental Shelf Res., 13,* 601-627, 1993.

Heinze, Ch., P. Schlosser, K. P. Koltermann, and J. Meinke, A tracer study of deep water renewal in the European polar seas, *Deep-Sea Res., 9,* 1425-1453, 1990.

Jones, E. P., and L. G. Anderson, On the origin of the chemical properties of the Arctic Ocean halocline, *J. Geophys. Res., 91,* 10,759-10,767, 1986.

Jones, E. P., and L. G. Anderson, On the origins of the properties of the Arctic Ocean halocline North of Ellesmere Island: Results from the Canadian Ice Island, *Continental Shelf Res., 10,* 485-498, 1990.

Jones, E. P., L. G. Anderson, and D. W. R. Wallace, Chemical tracers of near-surface, halocline and deep waters in the Arctic Ocean, *J. Marine Systems, 2,* 241-255, 1991.

Jones, E. P., B. Rudels, and L. G. Anderson, Deep waters of the Arctic Ocean: Origins and circulation, *Deep-Sea Res.,* in print, 1994.

Kinney, P., M. E. Arhelger, and D. C. Burrell, Chemical characteristics of water masses in the Amerasian Basin of the Arctic Ocean, *J. Geophys. Res., 75,* 4097-4104, 1970.

Krysell, M., Bromoform in the Nansen Basin in the Arctic Ocean, *Mar. Chem., 33,* 187-197, 1991.

Krysell, M., and D. W. R. Wallace, Arctic Ocean ventilation studied with a suite of anthropogenic halocarbon tracers, *Science, 242,* 746-748, 1988.

Macdonald, R. W., E. C. Carmack, Age of Canada Basin Deep Water: A way to estimate primary production for the Arctic Ocean, *Science, 254,* 1348-1350, 1991.

Macdonald, R. W., C. S. Wong, and P. E. Erickson, The distribution of nutrients in the southeastern Beaufort Sea: Implication for water circulation and primary production, *J. Geophys. Res., 92,* 2939-2952, 1987.

Macdonald, R. W., E. C. Carmack, F. A. McLaughlin, K. Iskei, D. M. Macdonald, and M. C. O'Brien, Composition and modification of water masses in the Mackenzie shelf estuary, *J. Geophys. Res., 94,* 18,057-18,070, 1989.

Midttun, L., Formation of dense bottom water in the Barents Sea, *Deep-Sea Res., 32,* 1233-1241, 1985.

Moore, R. M., M. G. Lowings, and F. C. Tan, Geochemical profiles in the central Arctic Ocean: Their relation to freezing and shallow circulation, *J. Geophys. Res., 88,* 2667-2674, 1983.

Oechel, W. C. and B. R. Strain, Native species responses to increased atmospheric carbon dioxide concentration, in *Direct Effects on Increasing Carbon Dioxide on Vegetation,* edited by B. R. Strain and J. D. Cure, pp. 117-154, Office of Energy Research, United States Department of Energy, Washington, D. C., 1985.

Östlund, H. G., The residence time of the fresh water component in the Arctic Ocean, *J. Geophys. Res., 87,* 2035-2043, 1982.

Östlund, H. G., and G. Hut, Arctic Ocean water mass balance from isotope data, *J. Geophys. Res., 89,* 6373-6381, 1984.

Östlund, H. G., Z. Top, and V. E. Lee, Isotope dating of waters at Fram III, *Geophys. Res. Lett., 9,* 1117-1119, 1982.

Östlund, H. G., G. Possnert, and J. H. Swift, Ventilation rate of the deep Arctic Ocean from carbon 14 data, *J. Geophys. Res., 92,* 3769-3777, 1987.

Quadfasel, D., B. Rudels, and K. Kurz, Outflow of dense water from a Svalbard fjord into the Fram Strait, *Deep-Sea Res., 35,* 1143-1150, 1988.

Quadfasel, D., A. Sy, and B. Rudels, A ship of opportunity section to the North Pole: Upper ocean temperature observations, *Deep-Sea Res., 40,* 777-789, 1993.

Rudels, B., On the mass balance of the Polar Ocean with special emphasis on the Fram Strait, *Norsk Polar. Skr., 188,* 1-53, 1987.

Rudels, B., E. P. Jones, L. G. Anderson, and G. Kattner, On the Origin and Circulation of the Intermediate Depth Waters of the Arctic Ocean, in *The Role of the Polar Oceans in Shaping the Global Environment,* edited by R. Muench and O. M. Johannessen, pp. 33-46, AGU, Washington, D. C., 1994.

Rusanov, V. P., Hydrochemical characteristics of surface water of the Arctic Basin, translated to English by Canadian Translation of Fisheries and Aquatic Sciences from, *Biology of the central Arctic Basin,* pp 15-33, "Nauka" Publishing House, Moscow, USSR, 1980.

Schlosser, P., G. Bönisch, B. Kromer, and K. O. Münnich, Ventilation rates of the waters in the Nansen Basin of the Arctic Ocean derived from a multitracer approach, *J. Geophys. Res. 95,* 3265-3272, 1990.

Schlosser, P., D. Grabitz, R. Fairbanks, and G. Bönisch, Arctic river-runoff: mean residence time on the shelves and in the halocline, *Deep-Sea Res., 41,* 1053-1068, 1994.

Smethie, W. M. Jr, D. W. Chipman, J. H. Swift, and K. P. Koltermann, Chlorofluoromethanes in the Arctic Mediterranean seas: evidence for formation of bottom water in the Eurasian Basin and deep-water exchange through Fram Strait, *Deep-Sea Res., 35,* 347-369, 1988.

Springer, A. M., and C. P. McRoy, The paradox of pelagic food webs in the northern Bering Sea-III. Patterns of primary production, *Continental Shelf Res., 13,* 575-600, 1993.

Swift, J. H., T. Takahashi, and H. D. Livingston, The contribution of the Greenland and the Barents Sea on the deep water of the Arctic Ocean, *J. Geophys. Res., 88,* 5981-5986, 1983.

Treshnikov, A.F., Arctic Atlas (in Russian), 204 pp. Arkt. Antarkt. Nauchno Issled. Inst., Moscow, 1985.

Wallace D. W. R., and R. M. Moore, Vertical profiles of CCl_2CF_2 (F-12) and CCl_3F (F-11) in the central Arctic Ocean basin, *J. Geophys. Res., 90,* 1155-1166, 1985.

Wallace, D. W. R., P. Schlosser, M. Krysell, and G. Bönisch, Halocarbon ratio and tritium/[3]He dating of water masses in the Nansen Basin, Arctic Ocean, *Deep-Sea Res., 39,* S435-S458, 1992.

Wilson, C., and D. W. R. Wallace, Using the nutrient ratio NO/PO as a tracer of continental shelf waters in the central Arctic Ocean, *J. Geophys. Res., 95,* 22,193-22,208, 1990.

6

DOC Storage in Arctic Seas: The Role of Continental Shelves

John J. Walsh

Abstract

An analysis of extensive data sets from the Bering/Chukchi Seas during 1979-1983, 1985-1988, and 1990-93 suggests that perhaps half of the nitrate, fueling the respective spring and summer blooms within the Bering and Chukchi Seas, might be derived from shelf-break exchange, if *in situ* nitrification during winter periods of the shelf ecosystems is not ignored. The new particulate nitrogen of these blooms may then be transferred to a labile pool of dissolved organic nitrogen, rather than to metazoan tissue or phytodetritus, with incomplete bacterial utilization as a function of the cold temperatures. Optical properties of the water column and a few DOC observations suggest that an excess amount of atmospheric CO_2 may be stored under the ice cap as spring and summer pulses of dissolved organic matter in larger than Redfield DOC/DON ratios.

Introduction

The continental shelves may host either heterotrophic ecosystems [Smith and Mackenzie, 1987], or only recent autotrophic sinks of atmospheric CO_2 since the onset of the Industrial Revolution [Wollast and MacKenzie, 1989]. Consideration of oceanic pCO_2 observations during January-April and July-October and atmospheric pCO_2 gradients [Tans et al., 1990], i.e. ignoring late spring and fall blooms in the deep sea [Taylor et al., 1991] and the entire production cycle on the shelves, suggests, for example, that an unknown terrestrial sink of CO_2 may occur in the northern hemisphere. Perhaps an unknown marine sink prevails there as well.

Arctic Oceanography: Marginal Ice Zones and Continental Shelves
Coastal and Estuarine Studies, Volume 49, Pages 203–230
Copyright 1995 by the American Geophysical Union

Such calculations further ignore interhemispheric exchange of both CO_2 [Broecker and Peng, 1992] and dissolved organic carbon, DOC [Walsh et al., 1992], within sinking North Atlantic Deep Water. They also do not consider the pre-industrial outgassing of CO_2 from the ocean to balance 1) weathering losses of terrestrial carbonate and silicate cycles and 2) riverine delivery of organic carbon to coastal seas [Sarmiento and Sundquist, 1992].

If such an outgassing were restricted to shelves, the partial pressure of carbon dioxide, pCO_2, of surface waters might be 20-74 μatm greater than that of the overlying atmosphere [Sarmiento and Sundquist, 1992]. The pCO_2 of the Columbia River and the Strait of Juan de Fuca are indeed ≥ 900 μatm [Park et al., 1969; Kelley and Hood, 1971a], but few pCO_2 data exist for most continental shelves to test this hypothesis. Exceptions are time series of pCO_2 observations obtained in the Bay of Calvi [Frankignoulle, 1988], off the Oregon coast [Gordon, 1973], around the Antarctic peninsula [Karl et al., 1991], and in the Bering Sea [Kelley and Hood, 1971a,b; Kelley et al., 1971; Gordon et al., 1973; Codispoti et al., 1986; Chen, 1985, 1991; Takahashi et al., 1993], which has been the subject of intense investigations within the PROBES [McRoy et al., 1986], ISHTAR [Walsh et al., 1989a], and BERPAC [Nagle, 1992] projects.

A recent analysis of both nitrogen and carbon uptake by phytoplankton in these polar time series [Codispoti et al., 1986; Karl et al., 1991] suggests luxury consumption of CO_2 in weight ratios of 8.7-10.3 [Sambrotto et al., 1993]. Before an algal cell divides with an idealized Redfield et al. [1963] POC/PON ratio of "5.7", it presumably must lose the excess carbon as respired CO_2 or excreted DOC. A mean C/N uptake rate of 9.5 and no respiratory loss, for example, coupled with respective POC/PON and DOC/DON ratios of 6 and 15, suggest that as much as 39% of the total photosynthate (dissolved + particulate) may be released as DOC.

Through this process, a net storage of atmospheric CO_2 would ensue, at least until this form of organic carbon was consumed by bacteria. Most phytoplankton cells have DOC excretion rates of $\sim$4% of their photosynthesis [Reid et al., 1990], however, such that a 40% loss of DOC would usually imply other processes at the population level of biotic interactions, i.e. cell lysis and/or inefficient feeding of herbivores [Jumars et al., 1989]. An exception is the colonial prymnesiophyte, *Phaeocystis pouchetii*, whose formation of a gelatinous cell matrix leads to DOC secretion rates of $\sim$40% [Reid et al., 1990]; at times, these phytoplankton are very abundant in polar seas [Smith et al., 1991].

In an attempt to explore this DOC sequestration hypothesis of complex biotic interactions on Arctic shelves, the PROBES 1979-1981 data on nutrients, CO_2, light penetration, phytoplankton species composition and productivity, and zooplankton abundance on the middle shelf of the southeastern Bering Sea [Area (0) of Fig. 1] are first reanalyzed with respect to additional ISHTAR and BERPAC 1985-1988 data on bacterial activities in the northern Bering/Chukchi Seas. After discussion of the combined data from PROBES, ISHTAR, and BERPAC, bio-optical and nitrogen budgets then lead to predictions of time-dependent storage of atmospheric CO_2 within labile pools of DOC, produced by plankton blooms of these shelf regions.

The results of these budgets are compared to both the output of a numerical model [Walsh and Dieterle, 1994] and sparse DOC validation data [Hansell, 1993]. None of the early DOC measurements [Table 1] were made with the recent platinum technique [Suzuki et al., 1992], but no attempt is made here to update these prior estimates, since Suzuki [1993] suggested that they may yield similar values. Finally, some recent 1992-93 surveys of DOC stocks within the Bering/Chukchi Seas are used to infer the time scales of northward transfer of DOC to the Arctic basins.

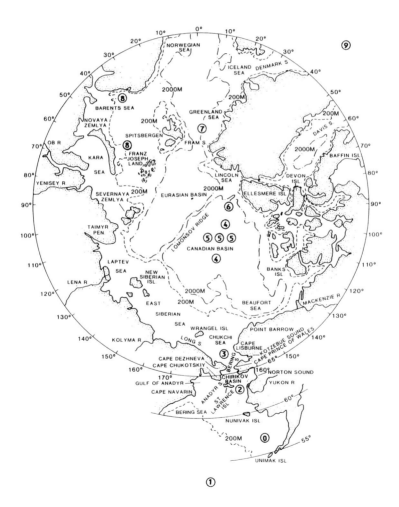

Figure 1. The location of the southeastern Bering Sea (SBS) time series above a nominal depth of 74 m [Area (0)] in relation to prior observations of DOC within Arctic Seas [Areas (1)-(9)].

TABLE 1. The distribution of dissolved organic carbon (mg DOC l^{-1}) within surface waters of the Arctic Ocean and the adjacent Bering, Chukchi, Lincoln, Greenland, Barents, and Iceland Seas.

Area	Date	Amount	Source
0 - 57°N, 165°W	4-5, 6-7/81	0.6, 1.3^{+}	This analysis
1 - 53°N, 177°W	10/73	0.7	Gordon et al. (1974)
2 - 63°N, 170°W	6/90	0.9-1.1	Hansell (1993)
3 - 66-69°N, 165-172°W	7/68	0.7-2.0	Loder (1971)
4 - 82°N, 157°W	4/68	1.3	Kinney et al. (1971)
5 - 82-84°N, 155-168°W	5-10/75	1.0-3.0	Melnikov/Pavlov (1978)
4 - 85°N, 125°W	5/69	1.5	Kinney et al. (1971)
6 - 86°N, 111°W	5/83	0.8-1.2	Gordon/Cranford (1985)
7 - 79°N, 5°W	7/88	1.1^{++}	Kattner/Becker (1991)
8 - 76-79°N, 47-58°E	8-9/76	1.9	Belyaeva et al. (1989)
8 - 70-73°N, 35-38°E	8-9/76	4.0	Belyaeva et al. (1989)
9 - 58-60°N, 44°W	9/58	0.6-0.8	Duursma (1965)

$^{+}$Mean of bloom and post-bloom calculations - see Figure 5e.
$^{++}$DON x 15 = DOC

Observations

The March-October data of PROBES capture the seasonal transition of plankton dynamics, through the autotrophic stage of the phytoplankton spring bloom to the heterotrophic stage of respiration by the bacterioplankton, zooplankton, and benthos. The July-August data of ISHTAR and BERPAC on the western and eastern sides of Bering Strait instead respectively represent arrested stages of bloom and oligotrophic conditions, as a spatial consequence of continuous upwelling along the Siberian coast.

1. Phytoplankton

With a seasonal decline of wind forcing (Fig. 2a), initiation of diatomaceous spring blooms in the southeastern Bering Sea (SBS) begins during April periods of both relative calm and greater incident radiation [Sambrotto et al., 1986]. Prolonged weak winds of <3 m sec^{-1} during May-June, however, both prevent resuspension of sinking diatoms and curtail vertical resupply of shelf-break stocks of nitrate to the euphotic zone. Nitrate stocks of the euphotic zone are thus negligible by the end of May (Fig. 2c).

Within bloom conditions of >8 μg l^{-1} of chlorophyll and diatom abundance of >1 x 10^{6} cells l^{-1} during 15 April-15 May (Figs. 3-4), the C/chl ratio (Figure 2f) was a mean of 46 in 1979-1980, similar to ratios of 45-47.5 measured within diatom-dominated spring populations of the Middle Atlantic Bight [Walsh et al., 1978; Malone et al., 1983]. In contrast, the phytoplankton populations of pre- and post-bloom conditions of the SBS in 1979-1981, when diatoms constituted <10% of the algal abundance (Figs. 3b and 4b),

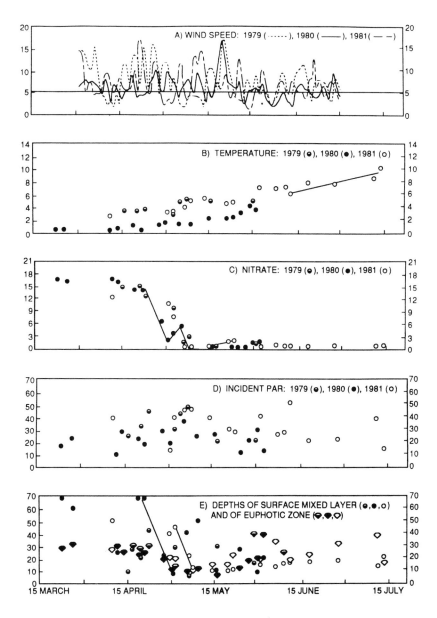

Figure 2. Time series of a) geostrophic wind speed (m sec^{-1}), b) temperature (°C), c) nitrate (μM),
d) incident (PAR) photosynthetically active radiation (Einsteins m^{-2} day^{-1}), e) depths (m) of the
surface mixed layer (0.02 σ_t criterion) and euphotic zone (1% isolume), f) carbon/chlorophyll ratio
(μg μg^{-1}), g) phytoplankton carbon-specific growth rate (day^{-1}), h) primary production (g C m^{-2}
day^{-1}), i) herbivore biomass (g C m^{-2}), and j) partial pressure of carbon dioxide (μatm) during
March-July 1979-1981 at the 74-m isobath of the SBS.

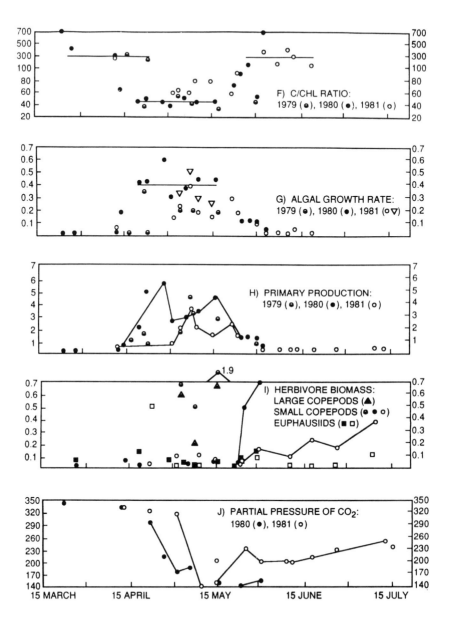

Figure 2. Time series of a) geostrophic wind speed (m sec^{-1}), b) temperature (°C), c) nitrate (μM), d) incident (PAR) photosynthetically active radiation (Einsteins m^{-2} day^{-1}), e) depths (m) of the surface mixed layer (0.02 σ_t criterion) and euphotic zone (1% isolume), f) carbon/chlorophyll ratio (μg μg^{-1}), g) phytoplankton carbon-specific growth rate (day^{-1}), h) primary production (g C m^{-2} day^{-1}), i) herbivore biomass (g C m^{-2}), and j) partial pressure of carbon dioxide (μatm) during March-July 1979-1981 at the 74-m isobath of the SBS.

had respective C/chl mean ratios of 309 and 320. These ratios are typical of both post-bloom situations in the Barents Sea [Martinez, 1991] and of cultures of *Phaeocystis pouchetii*, under both temperature-regulated and nutrient-limited situations [Verity et al., 1988; 1991].

During 1980, the populations of *P. pouchetii* constituted about 1% of the SBS spring bloom at the 70-m isobath (Fig. 3d), however, compared to ~10% in 1981 on the middle shelf of the SBS [Barnard et al., 1984], and to almost 100% in oceanic regions of the Greenland [Smith et al., 1991] and Barents [Wassmann et al., 1990] Seas. With respective C/chl ratios of 45 and 320 for diatoms and prymnesiophytes, and spring abundances of 1% and 10% for *P. pouchetii*, the combined C/chl ratios are computed to be 47.8 and

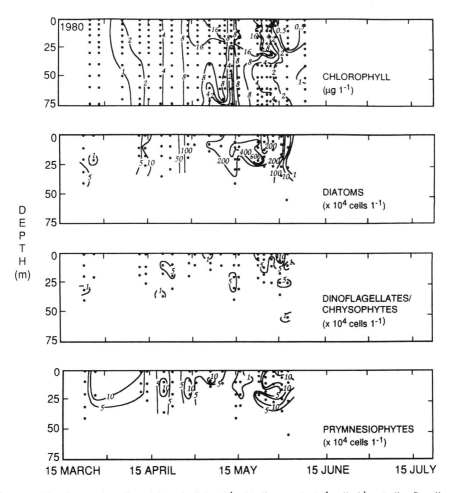

Figure 3. The time series of a) chlorophyll (μg l^{-1}), b) diatoms (x 10^4 cells l^{-1}), c) dinoflagellates and chrysophytes (x 10^4 cells l^{-1}), and d) prymnesiophytes (x 10^4 cells l^{-1}) during March-June 1980 at the 74-m isobath of the SBS.

72.5 during April-May 1980 and 1981 in the SBS, compared to measured means of 45.9
and 69.8 (Fig. 2f). I therefore conclude that phytoplankton excretion, at a diatom rate of
4% of photosynthesis, in the SBS may be a small source of DOC, unlike perhaps the
P. pouchetii-dominated blooms of the Barents and Greenland Seas.

2. Zooplankton

Warmer temperatures of the euphotic zone in 1981, compared to 1980 (as much as 4°C -
see Fig. 2b), may have led to larger growth rates of *P. pouchetii* [Verity et al., 1991],
greater grazing losses of the diatoms [Walsh and McRoy, 1986], and more production
of DOC through "sloppy" feeding of zooplankton [Banse, 1992]. Based on the abundance of the naupliar, copepodite, and adult stages of the larger copepods (*Neocalanus*

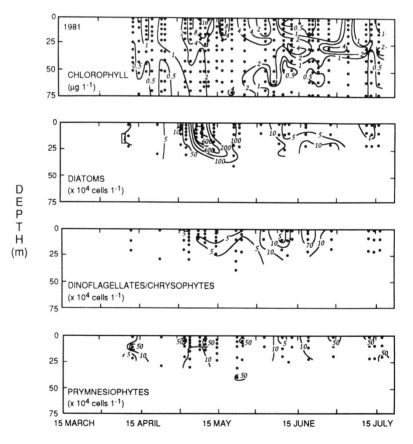

Figure 4. The time series of a) chlorophyll (μg l^{-1}), b) diatoms (x 10^4 cells l^{-1}), c) dinoflagellates
and chrysophytes (x 10^4 cells l^{-1}), and d) prymnesiophytes (x 10^4 cells l^{-1}) during April-July 1981
at the 74-m isobath of the SBS.

plumchrus, N. cristatus, and *Metridia pacifica*), the smaller copepods (*Calanus marshallae, Acartia* spp., *Pseudocalanus* spp., and *Oithona* spp.) and the euphausiids (*Thysanoessa raschi* and *T. inermis*) above the 70-80 m isobaths, estimates were made of the daily biomass (Fig. 2i), grazing stress, respiration, and DOC supply rates of the dominant herbivores of the SBS in 1979-1981.

Because of differences in size of these herbivores, the individual dry weight (dw) biomass of adult females of *Thysanoessa, Neocalanus, Calanus, Pseudocalanus (Acartia),* and *Oithona* are taken to be approximately 10000, 1000, 600, 12, and 1 μg [Vidal and Smith, 1986; Smith, 1991]. Assuming a constant carbon content of 0.45 μg C (μg dw)$^{-1}$ for each crustacean then leads to perhaps a two-fold interannual variation of April-May total copepod biomass, 90 mg C m^{-2} in 1981 and 48 mg C m^{-2} in 1980 (Fig 2i) above a depth of 74 m. Over the same time periods, this depth integral of mean euphausiid biomass was 189 mg C m^{-2} in 1981 and 73 mg C m^{-2} in 1980 (Fig. 2i).

Assuming a daily growth rate of 3% body weight day^{-1} for the euphausiids [Vidal and Smith, 1986; Smith, 1991], an assimilation efficiency of 70%, and that 50% of the assimilated material is respired with the rest added as body tissue, these herbivores might have consumed 16.2 mg POC m^{-2} day^{-1} of the spring bloom in 1981 and 6.2 mg POC m^{-2} day^{-1} in 1980. Similarly, with a 13% day^{-1} growth rate for the copepods [Vidal and Smith, 1986] and the same assimilation efficiency and respiration cost, the copepods may have ingested 33.4 mg C m^{-2} day^{-1} during 15 April-15 May 1981 and 17.7 mg C m^{-2} day^{-1} in the same time period of 1980.

However, the combined POC ingestion by the copepod and euphausiid herbivores in either year may have only removed 1-2% of the mean 2.14 g C m^{-2} day^{-1} fixed by phyto-plankton during April-May 1979-1981 (Fig. 2h). If we assume that only 50% of the grazing flux is actually ingested, with the other 50% lost as DOC during inefficient feeding [Walsh and Dieterle, 1994], an equivalent amount of DOC might be released from crushed, but not ingested diatoms, i.e. 0.3-0.7 mg DOC m^{-3} day^{-1} in 1980 and 1981.

These herbivore-mediated release rates of DOC are similar to a diatomaceous excretion rate of 4% of the net primary production of the spring bloom, i.e. 1.2 mg DOC m^{-3} day^{-1}. However, at the combined plankton release rate of 1.5-1.9 mg DOC m^{-3} day^{-1}, it would take 368-467 days to build-up a labile, unutilized DOC stock of 0.7 g m^{-3} (mg l^{-1}). Our bio-optical calculations will suggest that this may actually occur within 30 days in the SBS (Fig. 5).

At other times, of course, a larger fraction of the particulate production was consumed by metazoans of the SBS. After the spring bloom, cohorts of *Calanus marshallae* matured in June-July 1980-1981 (the solid lines of Fig. 2i), constituting 60-80% of the maximum copepod biomass of 380-790 mg C m^{-2} during the respective summers. Euphausiid biomass then averaged 67 mg C m^{-2}. During these periods, the same growth, assimilation, and respiration rates suggest that the two groups of crustaceans might ingest as much as 224 mg C m^{-2} day^{-1}. This amounts to 77% of the mean photosynthesis of

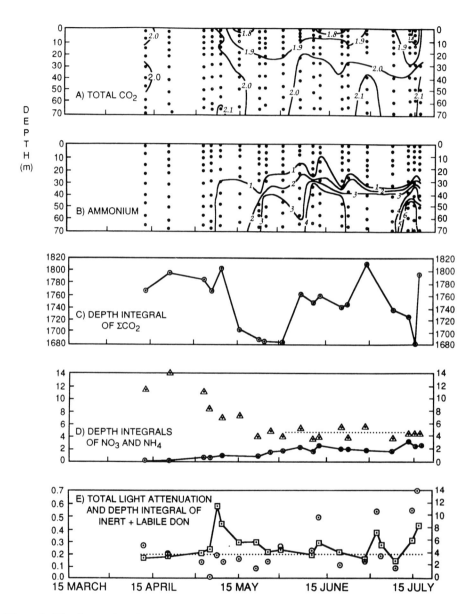

Figure 5. The time series of a) total carbon dioxide (μM), b) ammonium (μM), c) depth-integrated total carbon dioxide (g C m^{-2}), d) depth-integrated nitrate and ammonium (g N m^{-2}), and e) total diffuse light attenuation (m^{-1}) and depth integral of inert + labile dissolved organic nitrogen (g N m^{-2}) during April-July 1981 at the 74-m isobath of the SBS.

the 0.29 g C m^{-2} day^{-1} in June-July 1980-81, implying a ten-fold larger DOC release rate of 3.1 mg DOC m^{-3} day^{-1}.

As a result of the early spring bloom during April 1979 [Walsh and McRoy, 1986], the mid-shelf copepod biomass was a 20-fold higher 1500 mg C m^{-2} in May 1979 [Dagg et al., 1982], compared to May 1981 (Fig. 2i). Ship-board grazing experiments [Dagg et al., 1982] then suggested a combined crustacean grazing stress of 721 mg C m^{-2} day^{-1}, or 34% of the primary production during April-May 1979-1981. Such a herbivore-mediated release rate of 9.7 mg DOC m^{-3} day^{-1} would instead yield a labile, unutilized DOC stock of 0.7 mg l^{-1} within 72 days.

Despite the 20-fold increments of zooplankton biomass and grazing stress in May 1979, however, 2-fold larger nitrate uptake occurred, compared to May 1981 [Sambrotto et al., 1986]. By the end of May 1979, > 30 μg chl l^{-1} was still left behind to fall out of the spring water column [Walsh and McRoy, 1986], initiating summer remineralization of particulate and dissolved matter by both the benthos and the bacterioplankton. In fact, the near-bottom ammonium concentrations were 3-4 fold greater in mid-June 1979, than at the same time in 1981 [Whitledge et al., 1986], suggesting that perhaps the benthos is also a major source of DOC each year.

3. Benthos

During the post-bloom period of 1-15 June 1981, as much as 138 mg POC m^{-2} day^{-1}, 47% of the daily photosynthesis during June-July 1980-81, fell out of the top 12 m of the mid-shelf water column [Rowe and Phoel, 1992]. At depths of 72-78 m, 6-12 m above bottom, however, 640-1474 mg POC m^{-2} day^{-1} (221-508% of the overlying carbon fixation) were instead caught by sediment traps, reflecting resuspension of unutilized debris of the spring bloom. Indeed, bottom respiration at 3 mid-shelf stations, clustered around the 75-m isobath, was a mean of only 57 mg CO$_2$ m^{-2} day^{-1} at this time [Rowe and Phoel, 1992].

An assimilation efficiency of 50% suggests a benthic ingestion of 228 mg POC m^{-2} day^{-1} during June 1981, half of which could be met by just the summer influx of POC. Since the burial rate of siliceous diatoms in the SBS sediments may be 105-210 mg POC m^{-2} day^{-1} [Banahan and Goering, 1986], i.e. 5-10% of the bloom's primary production, most of the spring influx of POC to the sediments is apparently neither eaten nor buried. Cell lysis to DOM is an alternate fate, with release of DOC to the overlying water column [Burdige et al., 1992].

At the 70-m isobath off Plymouth, England, for example, near-bottom pulses of DOC are found in May and November, after the spring and fall influxes of POC [Banoub and Williams, 1973]. Farther offshore at the 2000-m isobath of the Porcupine Bight, sediment release of DOC is 10-fold that of CO$_2$ during June [Lampitt et al., 1994]. We assess both the metabolic demands of the bacterioplankton and bio-optical estimates of seasonal DOC accumulation in the water column to determine whether a similar 10-fold sediment efflux of 570 mg DOC m^{-2} day^{-1} (7.7 mg DOC m^{-3} day^{-1} over the 74-m water column) might occur in the SBS.

4. Bacterioplankton

No estimates of bacterial biomass or metabolic activities were made during the 1978-1981 PROBES studies, but such measurements were obtained during the 1985-1988 ISHTAR and BERPAC analyses of post-bloom situations within the oligotrophic Alaska Coastal Water of the NBS [Walsh et al., 1989a; Nagel, 1992]. Under the same surface temperatures in the eastern NBS and SBS of 8-10°C (Fig. 2a), at which bacterial growth might be maximal [Pomeroy and Deibel, 1986; Billen and Becquevort, 1991], direct measurements of thymidine incorporation, frequency of dividing cells [Hanson and Robertson, 1991], and concurrent dark uptake of ^{14}C-carbonate [Kudryatsev et al., 1991] were made at two stations within Alaska Coastal Water during August 1988.

These data yield estimates of microbial secondary production of respectively 6.4, 17.8, and 7.8 mg C m^{-3} day^{-1}, or a mean bacterial growth of 10.7 mg C m^{-3} day^{-1}. Such a carbon-specific growth rate of ~ 0.01 hr^{-1} is ten-fold less than that employed in a model of bacterial dynamics of subtropical waters around Bermuda [Fasham et al., 1990]. Above the 74-m isobath of the SBS, this microbial growth is equivalent to the same amount of depth-integrated respiration of 792 mg CO_2 m^{-2} day^{-1} (Table 2), if their secondary production is 50% of a substrate uptake of 21.4 mg DOC m^{-3} day^{-1}.

TABLE 2. Components of the bulk water column respiration (mg CO_2 m^{-2} day^{-1}) and depth-averaged DOC (mg DOC m^{-3} day^{-1}) production(+) and consumption(-) during possible transient and steady-state conditions of bacterial dynamics within a post-bloom ecosystem above the 74-m isobath of the southeastern Bering Sea.

	Transient Stage		Steady-state Stage	
Components:	CO_2	DOC	CO_2	DOC
Bacterioplankton	792	- 21.4	37	-1.0
Zooplankton	78	+ 9.7	78	+0.3
Benthos	57	+ 7.7	57	+0.8
Phytoplankton	29	+ 1.2	29	+0.2
Total community respiration	956		201	
Bulk:				
Net community respiration	567[1]		-100[2]	
Photosynthesis	290		290	
Total community respiration	857		190	

[1]Mean over June 1981 (Figure 15 of [Codispoti et al., 1986]).
[2]Mean over 3 June - 18 July 1981 (Table 4 of [Codispoti et al., 1986]).

Such a summer microbial DOC demand can almost be met by the larger May 1979 contribution of DOC by zooplankton, by a diatom DOC release rate of 4% during April-May, and by a 10-fold larger June sediment release of DOC, compared to the CO_2 efflux (Table 2). A prior ISHTAR estimate [Walsh et al., 1989a] of 0.5 mg C m^{-3} day^{-1} for bacterial growth during July-August 1987 in the NBS, however, instead assumed a steady state between the bacterioplankton and their predators, e.g. the choanoflagellate, *Diaphanoeca sphaerica*, and the ciliate, *Lohmaniella oviformis*, over a longer time period than daily BERPAC incubations. These different views of transient and steady-state dynamics of microbial heterotrophy are resolved by analysis [Codispoti et al., 1986] of sequential inventories of ΣCO_2 in the water column, e.g. Figure 5a.

The larger estimate of microbial respiration suggests that bacterioplankton dominate the community respiration of the first stage of the post-bloom state of the SBS ecosystem, constituting 83% of the total of 956 mg CO_2 m^{-2} day^{-1}. Perhaps only 29 mg CO_2 m^{-2} day^{-1} of the community respiration is then effected by the phytoplankton (10% of particulate production in June-July), 57 mg CO_2 m^{-2} day^{-1} by the benthos, and 78 mg CO_2 m^{-2} day^{-1} by the zooplankton (Table 2).

Allowing for invasion of atmospheric CO_2 and changes of carbonate alkalinity, the net community respiration (total respiration - photosynthesis) was estimated by Codispoti et al. [1986] to be a mean of 567 mg CO_2 m^{-2} day^{-1} (Table 2) above the 64-77 m isobaths during June 1981. A daily photosynthesis of 290 mg C m^{-2} day^{-1} then yields a total community respiration of 857 mg CO_2 m^{-2} day^{-1} (Table 2), similar to our estimate of 956 mg CO_2 m^{-2} day^{-1}.

In contrast, the net community respiration between the longer period of 3 June and 18 July 1981 was instead -100 mg CO_2 m^{-2} day^{-1} , i.e. a withdrawal of carbon dioxide, yielding a chemical estimate of total respiration of 190 mg CO_2 m^{-2} day^{-1} (Table 2). The smaller ISHTAR respiration loss of 37 mg CO_2 m^{-2} day^{-1} by the microbiota instead constitutes 18% of the biological estimate of net community respiration of 201 mg CO_2 m^{-2} day^{-1} (Table 2).

The microbial DOC demand of 1.0 mg DOC m^{-3} day^{-1} in the steady-state case can then be met by the smaller April-May 1980 contribution of DOC by zooplankton, by a diatom DOC release rate of 4% during June-July, and by a 10-fold smaller June sediment release of DOC (Table 2). If the larger release rate of sediment DOC prevailed in this second scenario, we would instead expect seasonal accumulation of DOC within the water column. Heterotrophic microbiota indeed lag the phytoplankton bloom by 15-30 days in the Weddell, Davis [Billen and Becquevort, 1991] and Barents [Thingstad and Martinussen, 1991] Seas, reflecting a need for sufficient build-up of small monomeric substrates [Pomeroy et al., 1990].

5. *Dissolved Organic Matter*

After passage through the western, productive (100-350 g C m^{-2} yr^{-1}) parts of the Bering and Chukchi shelves, May surface waters of the ice-covered Canadian Basin at 82-86°N

[Areas (4)-(6) of Fig. 1] contain 0.8-1.5 mg DOC l[-1] (Table 1), as measured by both wet combustion and ultraviolet oxidation [Kinney et al., 1971; Gordon and Cranford, 1985]. Larger observations of 1.0-3.0 mg DOC l[-1] (Table 1) have been reported for seasonal studies of the Canadian Basin [Melnikov and Pavlov, 1978]. These values are to be compared to only ~0.7 mg DOC l[-1] found by the first technique within surface waters of the Bering Sea Basin [Area (1) of Fig. 1] at 53°N [Gordon et al., 1974]. The same amount of deep-sea 0.7 mg DOC l[-1] is upwelled onto the northwestern outer shelf of the Bering Sea during summer [Handa and Tanoue, 1981].

Such a poleward increment implies both greater release and/or inefficient utilization of DOC by shelf plankton and benthic communities within the fast moving (15-25 cm sec[-1]), cold (2-3°C in August), and eutrophic (> 10 μM NO_3, > 900 mg chl m[-2], ~1-10 g C m[-2] day[-1]) Anadyr Water along the Siberian coast, compared to less release and/or more efficient consumption within slow moving (1-5 cm sec[-1]), warm (up to 22.5°C in August), and oligotrophic (< 1 μM NO_3, < 50 mg chl m[-2], ~0.25 g C m[-2] day[-1]) Alaska Coastal Water to the southeast of St. Lawrence Island. At a mean flow of 7.5 cm sec[-1] on the outer shelves and of 15 cm sec[-1] between Anadyr Strait and Wrangel Island, it might take a water parcel about 1 year to travel the distance of ~3000 km from Unimak Pass at the Bering shelf-break, through the Gulf of Anadyr, and past Wrangel Island to the Chukchi shelf-break (Fig. 6); transit through the quiescent (~1 cm sec[-1] currents) inner shelves along 2000 km of the Alaskan coast might take 2000 days instead.

With phytoplankton sinking rates of 1-10 m day[-1] [Walsh and Dieterle, 1994], shelf export from the middle and outer parts of the SBS may thus be mainly in the form of DOC [Walsh et al., 1992], rather than POC [Rowe and Phoel, 1992], towards Bering Strait and the Chukchi Sea. Early reconnaissance [Area (3) of Fig. 1], in fact, found as much as 2.0 mg DOC l[-1] within a July plankton bloom in western Bering Strait and as little as 0.7 mg DOC l[-1] in unproductive regions of the eastern Chukchi Sea [Loder, 1971].

Similarly, within shelf waters of the Barents Sea [Area (8) of Fig. 1], DOC stocks of 4.0-4.1 mg DOC l[-1] were found in August-September [Belyaeva et al., 1989], after the usual diatom and prymnesiophyte blooms in May [Slagstad and Stole-Hansen, 1991], compared to 1.5-2.0 mg DOC l[-1] within polar waters near Franz Josef Land (Table 1). Prymnesiophyte and/or diatom blooms are thus linked with high DOC concentrations within the Chukchi and Barents Seas, but no data exist for the SBS. We thus compute DOM from bio-optical budgets of the water column during April-July 1981 at the 74-m isobath of the SBS (Fig. 5e), for comparison with both model output from the same region (Fig. 7) and June 1990 observations at the 35-m isobath near St. Lawrence Island [Area (2) of Fig. 1].

Bio-optical Budget

The estimate of labile and inert DOM stocks on each day is obtained from the deconvolution of materials contributing to the total attenuation of light, measured within the 1981

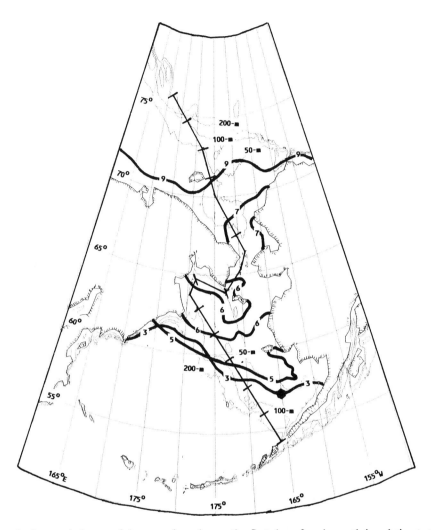

Figure 6. Seasonal change of the mean ice-edge on the first day of each month in relation to the PROBES time series (●) and hypothetical transit of water parcels from Unimak Pass at the shelf-break of the Bering Sea, through the Gulf of Anadyr, and past Wrangel Island to the shelf-break of the Chukchi Sea.

euphotic zone (depth of the 1% isolume - Table 3). The subsurface light field, L(z), at depth z can be described as the usual function of the incident PAR, I_o, by

$$L(z) = I_o e^{-kz} \text{ with} \tag{1}$$

$$kz = [k_w z + k_m \int_0^z P(z)\ dz + k_d \int_0^z .45\ DOC(z)\ dz] \tag{2}$$

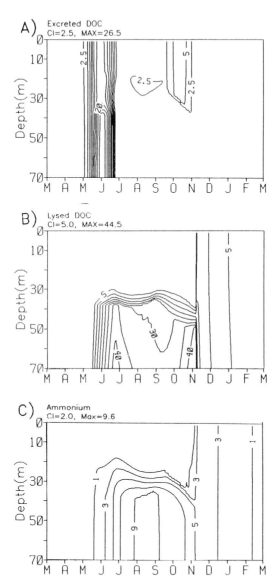

Figure 7. The simulated annual cycles of A) excreted DOC (μM), B) lysed DOC (μM), and C) ammonium (μM) above the 70-m isobath of the southeastern Bering Sea.

where the total diffuse attenuation coefficient, k, is partitioned into k_w, k_m, and k_d, which represent respectively: 1) the attenuation coefficient of water and associated detritus, 2) the specific attenuation coefficient of different size classes of phytoplankton pigments, P, and 3) the specific attenuation coefficient of labile colored dissolved organic carbon (CDOC), taken to be 0.45 DOC.

I chose k_w to be 0.130 m^{-1} for turbid shelf waters [Walsh et al., 1988], based on the residuals of the observed total attenuation and pigment concentrations during a March 1979 bloom in the MAB [Walsh et al., 1987], when labile DOC stocks may have been negligible; a k_w of 0.100 m^{-1} has been derived for other coastal waters [Taylor et al., 1986]. Because k_{w1} for pure water is ~ 0.030 m^{-1} over 400-600 nm [Morel, 1988], the attenuation of light by both suspended particulate detritus, k_{w2}, and inert CDOC, k_{w3}, within neritic regions (taken to be 0.065 m^{-1} - [Walsh, 1988]) is thus included within the larger value of 0.130 m^{-1} for k_w in this budget, where $k_w = k_{w1} + k_{w2} + k_{w3}$.

A k_m of 0.023 l μg^{-1} chl m^{-1} [Bannister, 1974; Jamart et al., 1977] represents a netplankton bloom community, while a k_m of 0.055 l μg^{-1} chl m^{-1} [Walsh et al., 1989b] is employed for the post-bloom conditions of nanoplankton chlorophytes (Table 3), reflecting the inverse relationship of pigment absorption coefficients and cell size [Morel and Bricaud, 1981].   Similarly, different C/chl ratios of 45 and 150 [Malone et al.,

Table 3. The role of dissolved organic nitrogen, estimated from optical properties of the water column, in a nitrogen budget of plankton dynamics during the 1981 spring bloom above the 74-m isobath of the southeastern Bering Sea.

Parameter	Early-bloom 4/23-24/81	Mid-bloom 5/8/81	Post-bloom 7/15/81
Euphotic zone(EZ) depth(m)	27	8	15
EZ chlorophyll mean (μg l^{-1})	1.75	19.43	1.11
Σ chlorophyll (mg m^{-2})	65.1	1077.0	45.8
C/chl ratio (mg mg^{-1})	45	45	150
POC/PON ratio (mg mg^{-1})	6	6	6
DOC/DON ratio (mg mg^{-1})	15	15	15
Σ NO$_3$ (g N m^{-2})	14.19	6.68	4.36
Σ NH$_4$ (g N m^{-2})	0.09	0.65	2.24
Σ NO$_2$ (g N m^{-2})	0.34	0.24	0.21
Σ urea (g N m^{-2})	-	-	0.64
Σ PON (g N m^{-2})	0.49	8.08	1.15
Σ inert DON (g N m^{-2})	3.90	3.84	3.90
Σ labile DON (g N m^{-2}	0.06	0.00	6.95
Σ Total (g N m^{-2})	19.07	19.49	19.45
Chl specific attenuation coefficient (l μg^{-1} m^{-1}	0.023	0.023	0.055
CDOC specific attenuation coefficient (l mg^{-1} m^{-1})	0.183	0.183	0.183
Total attenuation (m^{-1})	0.171	0.576	0.307
Chlorophyll attenuation(m^{-1})	0.040	0.447	0.061
Water attenuation (m^{-1})	0.030	0.030	0.030
Detrital attenuation (m^{-1})	0.100	0.100	0.100
Labile CDOC attenuation(m^{-1})	0.001	-0.001	0.116
Labile CDOC (mg l^{-1})	0.006	-0.006	0.634
Labile DOC (mg l^{-1})	0.012	-0.012	1.409
Σ labile DOC (g m^{-2})	0.88	-0.88	104.27

1983] are applied to the respective spring diatom and summer chlorophyte blooms of the SBS (Table 3).

A specific CDOC absorption, a_d, of 0.15 l mg^{-1} CDOC m^{-1} was also assumed yielding a k_d of 0.183 l mg^{-1} CDOC m^{-1} (Table 3), since $a_d = 0.82 k_d$ [Walsh et al., 1992]. We further assume that both labile and inert CDOC are 45% of DOC. An inert CDOC attenuation of 0.063 m^{-1} and these assumptions yield an inert DOC stock of 0.77 mg DOC l^{-1}, for example, similar to that measured in unproductive regions of the Bering-Chukchi Seas [Loder, 1971; Gordon et al., 1974; Handa and Tanoue, 1981]. If the computed attenuation of labile CDOC, k_d, became negative after subtraction of k_{w1}, k_{w2}, k_{w3}, and k_m from the observed k (Fig. 5e), the assumed background estimate of inert DOC/DON (dotted line of Fig. 5e) was then reduced by that amount (Table 3).

From this budget, the computed sum of the inert and labile stocks on 15 July 1981 (Table 3) is a depth-averaged DOC concentration of 2.2 mg DOC l^{-1}, similar to that of 2.0 mg DOC l^{-1} found [Loder, 1971] within a July bloom in the Chukchi Sea (Table 1). Over the SBS bloom period of April-May 1981, the mean total DOC stock of the SBS is estimated to be 0.6 mg DOC l^{-1}, compared to a mean of 1.3 mg DOC l^{-1} during post-bloom conditions of June-July (Table 1). During June 1990, the DOC stocks at the 35-m isobath off St. Lawrence Island, measured with the HTCO technique [Hansell, 1993], were 0.9-1.1 mg DOC l^{-1} (Table 1).

Summer increments of DOC have been observed in the Irish [Mantoura and Woodward, 1983], North [Banoub and Williams, 1973], and Wadden [Laane, 1982] Seas as well. Such a labile DOC increment of 0.7 mg DOC l^{-1} in the SBS, or 51.8 g DOC m^{-2} over 30 days, i.e. 1.73 g DOC m^{-2} day^{-1}, requires almost full solubilization of the April-May particulate production of 2.14 g POC m^{-2} day^{-2}, allowing bacteria to utilize perhaps 20% of the plankton release of DOC. A simulation model sheds light on the time-dependent nature of these processes.

A quasi-two dimensional model of seasonal carbon/nitrogen cycling in the SBS at the 70-m isobath assumed an initial inert DOC stock of 0.7 mg DOC l^{-1}, a 4% excretion rate of diatoms, a 50% DOC excretion rate of the POC grazed by zooplankton, a cell lysis of 90% of the phytodetritus and fecal pellets entering the sediments, and DOC uptake by a dynamic bacterial population, subjected to substrate limitation and predation [Walsh and Dieterle, 1994]. The model replicated the observed monthly cycles of nitrate, ammonium, carbon dioxide, and chlorophyll displayed in Figures 2-5.

As a consequence of the biotic interactions within this complex model, the predicted vertical patterns of seasonal DOC stocks of both lysed sediment origin and excreted plankton origin are shown in Figures 7a,b. The combined sum of 1.1 mg DOC l^{-1} for the depth-averaged inert DOC (0.7 mg l^{-1}), the lysed DOC_2 (0.2 mg l^{-1}), and the excreted DOC_1 (0.2 mg l^{-1}) pools of the model during the month of June matches both the total DOC of 0.9-1.1 mg l^{-1} measured by HTCO in the SBS during June 1990 [Hansell, 1993] and the June-July mean of 1.3 mg DOC l^{-1} from the bio-optical budget.(Fig. 5). Note that the seasonal maximum of simulated ammonium of 9.6 μM NH_4 (Fig. 7c), recycled by the bacteria, lags the influxes of DOC both from excretion by phytoplankton and zooplankton and from detrital lysis within the sediments; during the summers of 1978-1981, near-bottom values of 6-15 μM NH_4 were indeed observed at mid-shelf of the SBS [Whitledge et al., 1986]. The annual results of the baseline case of this model suggest that 50% of the net invasion of atmospheric CO_2 to the SBS is exported seawards as labile DOC, with incomplete bacterial utilization as a function of the cold temperatures [Pomeroy and Deibel, 1986].

With a DOC/DON ratio of 15 [Benner et al., 1992], the estimated stocks of inert DON in the bio-optical budget vary from zero to 3.90 g N m^{-2} between April and July (below the dotted line of Fig. 5e). The post-bloom stocks of labile DON increase from 0.00-0.06 g N m^{-2} on 23-24 April and 8 May to 6.95 g N m^{-2} on 15 July (Table 3; Fig. 5e). A nitrogen budget provides another check on the validity of these results.

Nitrogen Budget

The 1981 time series of biochemical and optical properties of the water column at mid-shelf of the SBS appear to yield a reasonable mass balance in a region of weak residual currents. During late April, most of the nitrogen is in the form of unutilized nitrate and inert DON (Table 3). About 57% of the nitrate stock of 14.19 g N-NO_3 m^{-2} on 23-24 April is then converted to a phytoplankton particulate nitrogen (chlorophyll x C/chl ratio x POC/PON ratio) of 8.08 g PON m^{-2} by 8 May. This PON is subsequently converted to both stocks of a labile pool of 6.95 g DON m^{-2} and of recycled ammonium of 2.24 g N-NH_4 m^{-2} by 15 July 1981, with some additional uptake of nitrate.

If nitrifying bacteria were as metabolically active as the ammonifying microbes during summer in the SBS, we might expect an *in situ* increment of the NO_3 stocks and a further decrement of the CO_2 stocks [Billen, 1976] by mid-July. Once surface waters are depleted of nitrate by the end of May (Fig. 2c), however, the depth-integral at the 74-m isobath remains static, with a mean of 4.4 g N-NO_3 m^{-2}, until at least 17 July 1981 (the dotted line of Fig. 5d). The depth-integrated ammonium stocks display a seasonal increase (Fig. 5d), from 0.2 g N-NH_4 m^{-2} on 12 April to 2.5 g N-NH_4 m^{-2}, with short-term reductions associated with pulses of algal biomass (Fig. 4a), in response to injections of the recycled nitrogen within the euphotic zone (Fig. 5b).

A June 1981 nitrate release of 497 μM NO_3 m^{-2} day^{-1} from the sediments [Rowe and Phoel, 1992] would supply 136.7 mM NO_3 m^{-2} to the 74 m water column over 275 days between June and March. This amounts to a depth-average of only 1.8 μM NO_3, when the surface mixed layer intersects the bottom in March (Fig. 2e). Since 15 μM NO_3 are uniformly distributed within the March water column (Fig. 2c) and similar summer nitrification rates of 300-450 μM NO_3 m^{-2} day^{-1} were measured in the sediments of the NBS [Walsh et al., 1989a], 90% of the mid-shelf nitrate stocks must be supplied by cross-isobath diffusion from the shelf-break [Coachman and Walsh, 1981], unless water column nitrification is significant in the SBS. Microbial nitrification rates of 0.016-0.068 μM day^{-1} within eutrophic waters off California [Ward et al., 1982], for example, could provide initial spring stocks of 4.4-18.7 μM NO_3, after 275 days within the aphotic zone. Our simulation model [Walsh and Dieterle, 1994] indeed suggests that as much as 50% of the resupply of nitrate to the SBS may actually result from *in situ* processes of nitrification by water and sediment microbes. The other half of the spring initial condition is instead provided by cross-shelf mixing of nitrate from the shelf-break, as a consequence of the strong tidal regime. If the SBS is partially restocked with deep-sea nitrate each fall-winter period, what is the fate of the previous year's supply of "new" nitrogen [Walsh, 1991]?

Discussion

The seasonal cycles and amounts of lysed and excreted labile DOC in the SBS (Fig. 7) are functions of the timing of the spring bloom and the initial stock of nitrate. Continuous upwelling of nutrients of Pacific origin within the stronger advective regime of the western part of the Chukchi Sea prolongs spring blooms of the emigre SBS phytoplankton over the entire summer [Walsh et al., 1989a]. Even larger phytoplankton removal of CO_2 would occur in the Chukchi Sea, if the ice cover were of the same duration there as in the SBS (Fig. 6).

[14]C estimates of carbon uptake suggest an annual production of at least 324 g POC m^{-2} yr^{-1} in the Chukchi Sea [Sambrotto et al., 1884], compared to a net synthesis of ~140 g POC m^{-2} yr^{-1} in the SBS [Walsh and Dieterle, 1994]. At similar relative rates of carbon burial/production, the surface sediments of the Chukchi Sea should also have larger reservoirs of organic carbon than in the SBS. They do - >1.5 % dw carbon contents of surficial sediments are found north of Bering Strait, compared to 0.1-0.5% dw values within the SBS (Fig. 8).

At steady-state, relative benthic effluxes of DOC should reflect influxes of POC to the bottom as well. Above the carbon-poor sediments to the east of St. Lawrence Island (Fig. 8), 0.9-1.1 mg DOC l^{-1} were found in June 1990 [Hansell, 1993]. In contrast, 1.9-5.8 mg DOC l^{-1} were measured, with a UV and persulfate oxidation method [Collins and Williams, 1977], at a grid of 25 stations above the carbon-rich sediments of the Gulf of Anadyr (Fig. 8) in June 1992 [Agatova et al., 1994]. The DOC/DON ratios in the latter samples [Agatova et al., 1994] ranged from 7 to 25, reflecting the storage of excess atmospheric CO_2 in this form of organic matter, compared to a mean POC/PON ratio of 6.4 in phytoplankton of the Bering/Chukchi Seas [Handa and Tanoue, 1981].

As suggested by the seasonal cycle of DOM in the SBS (Fig. 7), however, the high June 1992 values of 1.9-5.8 mg DOC l^{-1} within the Gulf of Anadyr were replaced by lower values of 1.1-1.7 mg DOC l^{-1} in the same area (Fig. 8) during September 1993 [D. Stockwell, personal communication]. Higher values of 3.0-3.7 mg DOC l^{-1} were instead measured with the HTCO technique [Benner et al., 1993] in September 1993 near Wrangel Island and at the shelf-break of the Chukchi Sea; values of 1.5-3.0 mg DOC l^{-1} were also found here and in the northern part of the Gulf of Anadyr (Fig. 8).

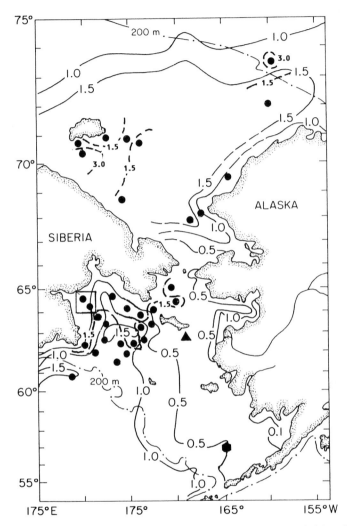

Figure 8.   The distribution of organic carbon (% dry weight) within surficial sediments of the
Bering/Chukchi Seas in relation to DOC observations (mg l⁻¹) during 1990 (▲), 1992 (□), 1993
(●), and the location of the PROBES time series at the 74-m isobath (◗).

At a mean current of 7.5 cm sec⁻¹ on the outer shelf (Fig. 6), the high June 1992 values
of DOC in the Gulf of Anadyr may have originated from the spring blooms on the outer
shelf of the SBS (Fig. 7). Alternatively, seasonal runoff of the Anadyr River may be the
source of the high 1992 DOC values, since values of 5.2-9.5 mg l⁻¹ of terrestrial DOC
are found in the Mackenzie, Lena, Yenisey, and Ob Rivers [Telang et al., 1990]. The
high September 1993 concentrations of DOC in the northern Chukchi Sea may thus
reflect a combination of the summer blooms within the Southern Chukchi Sea (SCS) and

advection of unutilized terrestrial and marine DOC stocks from the Gulf of Anadyr. Attempts to oxidize DOC within a plankton model of oxygen cycling in the Chukchi Sea failed [Penta and Walsh, 1994]. Bacterial utilization of DOC in cold waters of the western Chukchi shelf of this model did not allow replication of the observed oxygen supersaturations, implying that labile stocks of both the SBS and SCS may exit Bering Strait with some impunity to microbial attack.

If the labile DOM of plankton and benthos origin from the spring and summer blooms of the SBS and SCS survives fall and winter respiration demands of microbiota within Arctic slope waters as well, it may constitute a net sink of atmospheric CO_2, with storage in the polar basins [Walsh, 1989]. To avoid exchange of remineralized DOC with the atmosphere by evasion of biogenic CO_2, surface waters and their DOC stocks must either sink [Walsh et al., 1992], or otherwise be isolated from the sea surface. Formation of winter ice in the Chukchi Sea leads to brine rejection and intermittent export of cold, high-salinity shelf waters into the halocline of these ice-covered Arctic basins [Aagaard et al., 1985; Aagaard and Roach, 1990].

Entrainment of plankton debris from the adjacent polar shelves is required in steady state budgets of the AOU demands in the Canadian basin, since the *in situ* primary production is negligible [English, 1961; Subba Rao and Platt, 1984; Pomeroy et al., 1990]. Offshelf export of the particulate and dissolved residues of summer primary production within the Chukchi Sea [Walsh, 1989], during the following winter, may supply 50% of the respiration demands of 4.8-12.7 g CO_2 m^{-2} yr^{-1} within the 100-m halocline [Wallace et al., 1987; Anderson et al., 1990] of the adjacent heterotrophic ecosystems within the Canadian and Eurasian basins. Here in the deep Arctic Ocean, subsurface maxima of DOC coincide with oxygen minima within the halocline [Melnikov and Pavlov, 1978], and may also represent a large storage pool of organic carbon [Kinney et al., 1971].

Few particles survive oxidation losses in the Arctic basins, since 1) reactive radionuclides suggest unusually low particle scavenging rates [Bacon et al., 1989], 2) POC stocks are less than those of the central gyres of the Atlantic and Pacific Oceans [Gordon and Cranford, 1985], and 3) sedimentation rates are among the lowest found in the ocean [Hargrave et al., 1989]. Consequently, the organic carbon content of Holocene sediments in the Eurasian Basin is ~0.5% dw and may be of terrestrial origin [Stein et al., 1994], compared to 1.5-2.0% dw (Fig. 8) and a marine origin in the western Chukchi Sea [Walsh et al., 1989a]. Export of shelf-derived organic carbon is thus more likely to be in the form of DOC rather than POC, with longshore variation of Arctic shelf-break fluxes as a function of ice-cover mediated primary production [Penta and Walsh, 1994].

Although gas exchange may take place across the sea-ice cover [Fanning and Torres, 1991], CO_2 stocks build up, albeit slowly, at presumably the expense of DOC stocks, during the 4-5 year transit of the shelf export from Bering Strait to Fram Strait [Walsh et al., 1989a]. On the continental slope of the Chukchi Sea during January 1967, for example, seawater, at temperatures of -1.5 to -1.7°C, had supersaturated pCO_2 values, which were 13% greater than the atmospheric values of pCO_2 overlying the pack-ice [Kelley, 1968].

Smaller values of 0.8-1.2 mg DOC l[-1] are found (Table 1) at the surface towards the Lincoln Sea [Area (6) of Fig. 1], where the oxygen stocks were undersaturated at 89% within the upper 50 m of the slope water column during June 1988 [Pomeroy et al., 1990], similar to Sverdrup's [1929] observations in the Chukchi/East Siberian Seas. Labile DOC in the form of dissolved free amino acids were also undetectable, unlike Alaskan waters [Button and Robertson, 1988], implying prior removal during transit from Bering Strait.

South of Fram Strait in the Greenland Sea, the pCO_2 of ice-covered waters of the East Greenland Current is then 20% greater than atmospheric values [Anderson et al., 1990], unlike recent ice-covered, undersaturated waters of the Bering Sea [Chen, 1985, 1991]. The DOC stocks (Table 1) are then ~1.1 mg DOC l[-1] (15 x 70 μg DON l[-1]) under the ice [Area (7)] of Fig. 1) in the East Greenland Current [Kattner and Becker, 1991]. By the time this Current is south of Denmark Strait in the Iceland Sea [Area (9) of Fig. 1], however, the DOC stocks have been reduced to 0.6-0.8 mg DOC l[-1] [Duursma, 1965], i.e. the same as those found in the deep Bering Sea [Gordon et al., 1974].

Polar CO_2 stocks may thus now evade the ocean within the Greenland and Iceland Seas and invade within shallow portions of the Bering and Chukchi Seas, with transit between the two regions in the form of DOC. As atmospheric values of pCO_2 have increased above the SBS, in response to anthropogenic forcings, however, this shelf ecosystem may have only changed from a source to a sink of atmospheric CO_2 over the last 250 years [Walsh and Dieterle, 1994]. Solution of missing carbon puzzles [Tans et al., 1990; Walsh, 1991; Broecker and Peng, 1992; Sarmiento and Sundquist, 1992] may thus require more study of the couplings between shelves and basins, at the expense of isolated studies of either ecosystem.

Acknowledgments. This research was funded by the National Aeronautics and Space Administration under Grant NAGW-3459 and the National Science Foundation under Grant DPP-9215329.

References

Aagaard, K., and T. Roach, Arctic ocean-shelf exchange: measurements in Barrow Canyon, *J. Geophys. Res.*, *95*, 18,163-18,175, 1990.

Aagaard, K., J. H. Swift, and E. C. Carmack, Thermohaline circulation in the Arctic Mediterranean Seas, *J. Geophys. Res.*, *90*, 4833-4846, 1985.

Agatova, A. I., N. V. Arzhanova, E. V. Dafner, V. V. Sapozhnikov, and N. I. Torgunova, Dissolved organic forms of carbon, nitrogen, and phosphorus in the Bering Sea, *Cont. Shelf Res.*, in press, 1994.

Anderson, L. G., D. Dyrssen, and E. P. Jones, An assessment of the transport of atmospheric CO_2 into the Arctic Ocean, *J. Geophys. Res.*, *95*, 1703-1711, 1990.

Bacon, M. P., C. A. Huh, and R. M. Moore, Vertical profiles of some natural radionuclides over the Alpha Ridge, Arctic Ocean, *J. Geophys. Res.*, *95*, 15-22, 1989.

Banahan, S., and J. J. Goering, The production of biogenic silica and its accumulation on the southeastern Bering Sea shelf, *Cont. Shelf Res., 5*, 199-214, 1986.

Bannister, T. T., A general theory of steady state phytoplankton growth in a nutrient saturated mixed layer, *Limnol. Oceanogr., 19*, 13-30, 1974.

Banoub, M. W., and P. J. Williams, Seasonal changes in the organic forms of carbon, nitrogen, and phosphorous in sea water at E1 in the English Channel during 1968, *J. Mar. Biol. Ass. U.K., 53*, 695-703, 1973.

Banse, K., Grazing, temporal changes of phytoplankton, and the microbial loop in the open sea, in *Primary Productivity and Biogeochemical Cycles in the Sea*, edited by P. G. Falkowski and A. D. Woodhead, pp. 409-440, Plenum Press, New York, 1992.

Barnard, W. R., M. O. Andreae, and R. L. Iversen, Dimethylsulfide and *Phaeocystis pouchetii* in the southeastern Bering Sea, *Cont. Shelf Res., 3*, 103-113, 1984.

Belyaeva, A. N., A. I. Daniushevskaya, and E. A. Romankevich, Organic geochemistry of Barents Sea sediments, in *The Arctic Seas*, edited by Y. Herman, pp. 761-798, Van Nostrand Reinhold, New York, 1989.

Benner, R., J. D. Pakulski, M. McCarthy, J. I. Hedges, and P. G. Hatcher, Bulk chemical characteristics of dissolved organic matter in the ocean, *Nature, 255*, 1561-1564, 1992.

Benner, R., B. von Bodungen, J. Farrington, J. Hedges, C. Lee, F. Mantoura, Y. Suzuki, and P. M. Williams, Measurement of dissolved organic carbon and nitrogen in natural waters: Workshop report, *Mar. Chem., 41*, 5-10, 1993.

Billen, G., Evaluation of nitrifying activity in sediments by dark ^{14}C-bicarbonate incorporation, *Water Res., 10*, 51-57, 1976.

Billen, G., and S. Becquevort, Phytoplankton-bacteria relationship in the Antarctic marine ecosystem, *Polar Res., 10*, 245-254, 1991.

Bricaud, A., A. Morel, and L. Prieur, Absorption by dissolved organic matter of the sea (yellow substance) in the UV and visible domains, *Limnol. Oceanogr., 26*, 43-53, 1981.

Broecker, W. S., and T.-H. Peng, Interhemispheric transport of carbon dioxide by ocean circulation, *Nature, 356*, 587-589, 1992.

Burdige, D. J., M. J. Alperin, J. Homstead, and C. S. Martens, The role of benthic fluxes of dissolved organic carbon in oceanic and sedimentary carbon cycling, *Geopys. Res. Lett., 19*, 1851-1854, 1992.

Button, D. K., and B. R. Robertson, Distribution, specific affinities, and partial growth rates of aquatic bacteria by high resolution flow cytometry, *EOS, 69*, 1106, 1988.

Chen, C.-T. A., Preliminary observations of oxygen and carbon dioxide of the wintertime Bering Sea marginal ice zone, *Cont. Shelf Res., 4*, 465-483, 1985.

Chen, C.-T. A., Carbonate chemistry of the wintertime Bering Sea marginal ice zone, *Cont. Shelf Res., 13*, 67-88, 1993.

Coachman, L. K., and J. J. Walsh, A diffusion model of cross-shelf exchange of nutrients in the Bering Sea, *Deep-Sea Res., 28*, 819-837, 1981.

Codispoti, L. A., G. E. Friederich, and D. W. Hood, Variability of the inorganic carbon system over the SE Bering Sea shelf during spring 1980 and spring-summer 1981, *Cont. Shelf Res., 5*, 133-160, 1986.

Collins, K. G., and P. G. Williams, An automated photochemical method for the determination of dissolved organic carbon in sea and estuarine waters, *Mar. Chem., 5*, 123-141, 1977.

Dagg, M. J., J. Vidal, T. E. Whitledge, R. L. Iversen, and J. J. Goering, The feeding, respiration, and excretion of zooplankton in the Bering Sea during a spring bloom, *Deep-Sea Res., 29*, 45-63, 1982.

Duursma, E. K., The dissolved organic constituents of sea water, in *Chemical Oceanography, Vol. 1*, edited by J. P. Riley and G. Skirrow, pp. 433-477, Academic Press, New York, 1965.

English, T. S., Some biological oceanographic observations in the central north Polar Sea, Drift Station Alpha, 1957-58, *Arctic Inst. North Am. Res. Pap. 13*, 1-80, 1961.

Fanning, K. A., and L. M. Torres, ^{222}Rn and ^{226}Ra: indicators of sea-ice effects on air-gas exchange, *Polar Res., 10*, 51-58, 1991.

Fasham, M. J., H. W. Ducklow, and S. M. McKelvie, A nitrogen-based model of plankton dynamics in the oceanic mixed layer, *J. Mar. Res., 48*, 591-639, 1990.

Frankignoulle, M., Field measurements of air-sea CO_2 exchange, *Limnol. Oceanogr., 33*, 313-322, 1988.

Gordon, D. C., and P. J. Cranford, Detailed distribution of dissolved and particulate organic matter in the Arctic Ocean and comparison with other oceanic regions, *Deep-Sea Res., 32*, 1221-1232, 1985.

Gordon, L. I., A study of carbon dioxide partial pressures in surface waters of the Pacific Ocean, Ph.D. Thesis, pp. 1-216, Oregon State University, Corvallis, Oregon, 1973.

Gordon, L. I., P. K. Park, J. J. Kelley, and D.W. Hood, Carbon dioxide partial pressures in North Pacific surface waters. 2. General late summer distribution, *Mar. Chem., 1*, 191-198, 1973.

Gordon, L. I., E. A. Siefert, L. I. Barstow, and P. K. Park, Organic carbon in the Bering Sea, in *Bering Sea Oceanography: An Update*, edited by D. W. Hood and Y. Takenouti, pp. 239-244, Inst. Mar. Sci. 75-2, Fairbanks, Alaska, 1974.

Handa, N., and E. Tanoue, Organic matter in the Bering Sea and adjacent areas, in *The Eastern Bering Sea Shelf: Oceanography and Resources, Vol. 1*, edited by D. W. Hood and J. A. Calder, pp. 359-382, Univ. Washington Press, Seattle, 1981.

Hansell, D. A., Results and observations from the measurement of DOC and DON in seawater using a high-temperature catalytic oxidation technique, *Mar. Chem., 41*, 195-202, 1993.

Hanson, R. B., and C. Y. Robertson, Thymidine incorporation, frequency of dividing cells, and growth rates of bacteria, in *Results of the Third Joint US-USSR Bering & Chukchi Seas Expedition (BERPAC), Summer 1988*, edited by P. A. Nagel, pp. 60-74, USFWS, Washington, D.C., 1992.

Hargrave, B. T., B. v. Bodungen, R. S. Conover, S. G. Phillips, and W. P. Vass, Seasonal changes in sedimentation of particulate matter and lipid content of zooplankton collected by sediment trap in the Arctic Ocean off Axel Heiberg Island, *Polar Biol., 9*, 467-475, 1989.

Jamart, M., D. F. Winter, K. Banse, G. C. Anderson, and R. K. Lam, A theoretical study of phytoplankton growth and nutrient distribution in the Pacific Ocean off the northwestern U.S. coast, *Deep-Sea Res., 24*, 753-773, 1977.

Jumars, P. A., D. L. Penry, J. A. Baross, M. J. Perry, and B. W. Frost, Closing the microbial loop: dissolved carbon pathway to heterotrophic bacteria form incomplete ingestion, digestion, and absorption in animals, *Deep-Sea Res., 36*, 483-496, 1989.

Karl, D. M., B. D. Tilbrook, and G. Tien, Seasonal coupling of organic matter production and particle flux in the western Bransfield Strait, Antarctica, *Deep-Sea Res., 38*, 1097-1126, 1991.

Kattner, G., and H. Becker, Nutrients and organic nitrogenous compounds in the marginal ice zone of the Fram Strait, *J. Mar. Syst., 2*, 385-394, 1991.

Kelley, J. J., Carbon dioxide in seawater under the Arctic ice, *Nature, 218*, 862-864, 1968.

Kelley, J. J., and D. W. Hood, Carbon dioxide in the Pacific Ocean and Bering Sea: upwelling and mixing, *J. Geophys. Res., 76*, 745-752, 1971a.

Kelley, J. J., and D. W. Hood, Carbon dioxide in the surface water of the ice-covered Bering Sea, *Nature, 229*, 37-39, 1971b.

Kelley, J. J., L. L. Longerich, and D. W. Hood, Effect of upwelling, mixing, and high primary productivity on CO_2 concentrations in surface waters of the Bering Sea, *J. Geophys. Res., 76*, 8687-8693, 1971.

Kinney, D. J., T. C. Loder, and J. Groves, Particulate and dissolved matter in the Amerasian Basin of the Arctic Ocean, *Limnol. Oceanogr.*, *16*, 132-137, 1971.

Kudryatsev, V. M., V. O. Mamaev, and T. F. Strigunkova, Bacterial production and destruction of organic matter, in *Results of the Third Joint US-USSR Bering & Chukchi Seas Expedition (BERPAC), Summer 1988*, edited by P. A. Nagel, pp. 74-78, USFWS, Washington, D.C., 1992.

Laane, R. W., Sources of dissolved organic carbon in the Ems-Dollart estuary: The rivers and phytoplankton, *Neth. J. Sea Res.*, *15*, 331-339, 1982.

Lampitt, R. S., R. Raine, D. S. Billet, and A. L. Rice, Material supply to the European continental slope: a budget based on benthic oxygen demand, DOC efflux and organic supply, *Deep-Sea Res.*, in press, 1994.

Loder, T. C., Distribution of dissolved and particulate organic carbon in sub-polar, Alaska polar, and estuarine waters, Ph.D.thesis, pp. 1-234, University of Alaska, Fairbanks, Alaska, 1971.

Malone, T. C., T. S. Hopkins, P. G. Falkowski, and T. E. Whitledge, Production and transport of phytoplankton biomass over the continental shelf of the New York Bight, *Cont. Shelf Res.*, *1*, 305-337, 1983.

Mantoura, R. F., and E. M. Woodward, Conservative behavior of riverine dissolved organic carbon in the Severn estuary: Chemical and geochemical implications, *Geochim. Cosmochim. Acta, 47*, 1293-1309, 1983.

Martinez, R., Biomass and respiratory ETS activity of microplankton in the Barents Sea, *Polar Res.*, *10*, 193-200, 1991.

McRoy, C. P., D. W. Hood, L. K. Coachman, J. J. Walsh, and J. J. Goering, Processes and resources of the Bering Sea shelf (PROBES):the development and accomplishments of the project, *Cont. Shelf Res.*, *5*, 5-21, 1986.

Melnikov, I. A., and G. L. Pavlov, Characteristics of organic carbon distribution in water and ice of the Arctic Basin, *Oceanology, 18*, 163-167, 1978.

Morel, A., Optical modeling of the upper ocean in relation to its biogenous matter content (Case I waters), *J. Geophys. Res.*, *93*, 10,749-10,768, 1988.

Morel, A., and A. Bricaud, Theoretical results concerning light absorption in a discrete medium, and application to specific absorption of phytoplankton, *Deep-Sea Res.*, *28*, 1375-1393, 1981.

Nagle, P. A., *Results of the Third Joint US-USSR Bering & Chukchi Seas Expedition (BERPAC), Summer 1988*, pp. 1-415, US Fish and Wildlife Service, Washington, D.C., 1992.

Park, P. K., L. I. Gordon, S. W. Hager, and M. C. Cissell, Carbon dioxide partial pressure in the Columbia River, *Science, 166*, 867-868, 1969.

Penta, B., and J. J. Walsh, A one-dimensional ecological model of summer oxygen distribution within the Chukchi Sea, *Cont. Shelf Res.*, in press, 1994.

Pomeroy, L. R., and D. Deibel, Temperature regulation of bacterial activity during the spring bloom in Newfoundland coastal waters, *Science, 233*, 359-361, 1986.

Pomeroy, L. R., S. A. Macko, P. H. Ostrom, and J. Dunphy, The microbial food web in Arctic seawater: concentration of dissolved free amino acids and bacterial abundance and activity in the Arctic Ocean and in Resolute Passage, *Mar. Ecol. Progr. Ser.*, *61*, 31-40, 1990.

Redfield, A. C., B. H. Ketchum, and F. A. Richards, The influence of organisms on the composition of seawater, in *The Sea*, Vol. 2, edited by M. N. Hill, pp. 26-77, Wiley, New York, 1963.

Reid, P. C., C. Lancelot, W. W. Gieskes, E. Hagmeier, and G. Weichart, Phytoplankton of the North Sea and its dynamics: a review, *Neth. J. Sea Res.*, *26*, 295-331, 1990.

Rowe, G. T., and W. C. Phoel, Nutrient regeneration and oxygen demand in Bering Sea continental shelf sediment, *Cont. Shelf Res.*, *12*, 439-450, 1992.

Sambrotto, R. N., J. J. Goering, and C. P. McRoy, Large yearly production of phytoplankton in western Bering Strait, *Science, 225*, 1147-1150, 1984.

Sambrotto, R. N., H. J. Niebauer, J. J. Goering, and R. L. Iversen, Relationships among vertical mixing, nitrate uptake, and phytoplankton growth during the spring bloom in the southeast Bering Sea middle shelf, *Cont. Shelf Res., 5*, 161-198, 1986.

Sambrotto, R. N., G. Savidge, C. Robinson, P. Boyd, T. Takahashi, D. M. Karl, C. Langdon, D. Chipman, J. Marra, and L. Codispoti, Elevated consumption of carbon relative to nitrogen in the surface ocean, *Nature, 363*, 248-250, 1993.

Sarmiento, J. L., and E. T. Sundquist, Revised budget for the oceanic uptake of anthropogenic carbon dioxide, *Nature, 356*, 589-593, 1992.

Slagstad, D., and K. Stole-Hansen, Dynamics of plankton growth in the Barents Sea: model studies, *Polar Res., 10*, 173-186, 1991.

Smith, S. L., Growth, development, and distribution of the euphausiids *Thysanoessa raschi* (M. Sars) and *Thysanoessa inermis* (Kroyer) in the southeastern Bering Sea, *Polar Res., 10*, 461-478, 1991.

Smith, S. V., and F. T. Mackenzie, The ocean as a net heterotrophic system: implications from the carbon biogeochemical cycle, *Glob. Biogeochem. Cycles, 1*, 187-198, 1987.

Smith, W. O., L. A. Codispoti, D. M. Nelson, T. Manley, E. J. Buskey, H. J. Niebauer, and G. F. Cota, Importance of *Phaeocystis* blooms in the high-latitude ocean carbon cycle, *Nature, 352*, 514-516, 1991.

Stein, R., S.-I. Nam, C. Schubert, C. Vogt, D. Futterer, and J. Heinemeier, The last deglaciation event in the eastern Central Arctic Ocean, *Science, 264*, 692-696, 1994.

Subba Rao, D. V., and T. Platt, Primary production of Arctic waters, *Polar Biol., 3*, 191-201, 1984.

Suzuki, Y., On the measurement of DOC and DON in seawater, *Mar. Chem., 41*, 287-288, 1993.

Suzuki, Y., E. Tanoue, and H. Ito, A high-temperature catalytic oxidation method for the determination of dissolved organic carbon in seawater: analysis and improvement, *Deep-Sea Res., 39*, 185-198, 1992.

Sverdrup, H. U., The waters on the north Siberian shelf, *Sci. Res. Norwegian North Polar Exped., 4*, 1-131, 1929.

Takahashi, T., J. Olafsson, J.G. Goddard, D. W. Chipman, and S. C. Sutherland, Seasonal variation of CO_2 and nutrients in the high-latitude surface oceans: a comparative study, *Glob. Biogeochem. Cycles, 7*, 843-878, 1993.

Tans, P. P., I. Y. Fung, and T. Takahashi, Observational constraints on the global atmospheric CO_2 budget, *Science, 247*, 1431-1438, 1990.

Taylor, A. H., J. R. Harris, and J. Aiken, The interaction of physical and biological processes in a model of the vertical distribution of phytoplankton under stratification, in *Ecohydrodynamics*, edited by J. J. Nihoul, pp. 313-330, Elsevier, Amsterdam, 1986.

Taylor, A. H., A. J. Watson, and J. E. Robertson, The influence of the spring phytoplankton bloom on carbon dioxide and oxygen concentrations in the surface waters of the northeast Atlantic during 1989, *Deep-Sea Res., 39*, 137-152, 1992.

Telang, S. A., R. Pocklington, A. S. Naidu, E. A. Romankevich, I. I. Gitelson, and M. I. Gladishev, Biogeochemistry of arctic rivers, in *Biogeochemistry of Major World Rivers*, edited by E. T. Degens, S. Kempe, and J. Richey, pp. 150-171, Wiley, New York, 1990.

Thingstad, T. F., and I. Martinussen, Are bacteria active in the cold pelagic ecosystem of the Barents Sea? *Polar Res., 10*, 255-266, 1991.

Verity, P. G., T. A. Villareal, and T. J. Smayda, Ecological investigations of blooms of colonial *Phaeocystis pouchetii*. I. Abundance, biochemical composition, and metabolic rates, *J. Plankt. Res., 10*, 219-248, 1988.

Verity, P. G., T. S. Smayda, and E. Sakshaug, Photosynthesis, excretion, and growth rates of *Phaeocystis* colonies and solitary cells, *Polar Res., 10*, 117-129, 1991.

Vidal, J., and S. L. Smith, Biomass, growth, and development of populations of herbivorous zooplankton in the southeastern Bering Sea during spring, *Deep-Sea Res., 33*, 523-556, 1986.

Wallace, D. W., R. M. Moore, and E. P. Jones, Ventilation of the Arctic Ocean cold halocline: rates of diapycnal and isopycnal transport, oxygen utilization, and primary production inferred using chlorofluoromethane distribution, *Deep-Sea Res., 34*, 1957-1980, 1987.

Walsh, J. J., *On the Nature of Continental Shelves*, 520 pp., Academic Press, San Diego, Calif., 1988.

Walsh, J. J., Arctic carbon sinks: Present and future, *Glob. Biogeochem. Cycles, 3*, 393-411, 1989.

Walsh, J. J., Importance of continental margins in the marine biogeochemical cycling of carbon and nitrogen, *Nature, 350*, 53-55, 1991.

Walsh, J. J., and D. A. Dieterle, CO_2 cycling in the coastal ocean. I. A numerical analysis of the southeastern Bering Sea, with applications to the Chukchi Sea and the northern Gulf of Mexico, *Prog. Oceanog.*, in press, 1994.

Walsh, J. J., and C. P. McRoy, Ecosystem analysis in the southeastern Bering Sea, *Cont. Shelf Res., 5*, 259-288, 1986.

Walsh, J. J., T. E. Whitledge, F. W. Barvenik, C. D. Wirick, S. O. Howe, W. E. Esaias, and J. T. Scott, Wind events and food chain dynamics within the New York Bight, *Limnol. Oceanogr., 23*, 659-683, 1978.

Walsh, J. J., D. A. Dieterle, and W. E. Esaias, Satellite detection of phytoplankton export from the Mid-Atlantic Bight during the 1979 spring bloom, *Deep-Sea Res., 34*, 675-703, 1987.

Walsh, J. J., D. A. Dieterle, and M. B. Meyers, A simulation analysis of the fate of phytoplankton within the Mid-Atlantic Bight, *Cont. Shelf Res., 8*, 757-787, 1988.

Walsh, J. J., C. P. McRoy, L. K. Coachman, J. J. Goering, J. J. Nihoul, T. E. Whitledge, T. H. Blackburn, P. L. Parker, C. D. Wirick, P. G. Shuert, J. M. Grebmeier, A. M. Springer, R. D. Tripp, D. A. Hansell, S. Djenidi, E. Deleersnijder, K. Henriksen, B. A. Lund, P. Andersen, F. E. Muller-Karger, and K. Dean, Carbon and nitrogen cycling within the Bering/Chukchi Seas: source regions of organic matter effecting AOU demands of the Arctic Ocean, *Prog. Oceanog., 22*, 279-361, 1989a.

Walsh, J. J., D. A. Dieterle, M. B. Meyers, and F. E. Muller-Karger, Nitrogen exchange at the continental margin: a numerical study of the Gulf of Mexico, *Prog. Oceanog., 23*, 245-301, 1989b.

Walsh, J. J., K. L. Carder, and F. E. Muller-Karger, Meridional fluxes of dissolved organic matter in the North Atlantic Ocean, *J. Geophys. Res., 97*, 15,625-15,637, 1992.

Ward, B. B., R. J. Olsen, and M. J. Perry, Microbial nitrification rates in the primary nitrite maximum off southern California, *Deep-Sea Res., 29*, 247-255, 1982.

Wassman, P., M. Vernet, B. G. Mitchell, and F. Rey, Mass sedimentation of *Phaeocystis pouchetii* in the Barents Sea, *Mar. Ecol. Progr. Ser., 66*, 183-195, 1990.

Whitledge, T. E., W. S. Reeburgh, and J. J. Walsh, Seasonal inorganic nitrogen distributions and dynamics in the southeastern Bering Sea, *Cont. Shelf Res., 5*, 109-132, 1986.

Wollast, R., and F. T. Mackenzie, Global biogeochemical cycles and climate, in *Climate and Geosciences*, edited by A. Berger, S. Schneider, and J. C. Duplessey, pp. 410-453, Kluwer, Dordrecht, 1989.

7

Biological Processes on Arctic Continental Shelves: Ice-Ocean-Biotic Interactions

Jacqueline M. Grebmeier, Walker O. Smith, Jr., and Robert J. Conover

Abstract

The biological processes of the Arctic are among the most temporally pulsed of the world's oceans, and such dependency on time-related processes greatly structures the physical-biological interactions that occur. A conceptual model is proposed, based on past and current data for ice-covered regions of the Arctic continental shelf and associated marginal sea regions, that links seasonal changes in irradiance and ice cover with phytoplankton biomass and primary productivity, bacterial activity, herbivorous grazing, vertical flux from the euphotic zone, benthic activity, and benthic-pelagic coupling. Data are presented from three geographically separate regions to support/modify the conceptual model: the northern Bering/Chukchi Seas, Lancaster Sound in the Canadian Archipelago, and the Northeast Water polynya region on the continental shelf off Northeast Greenland. Certain factors have strong influence on temporal patterns in all three regions, including ice retreat/break-up in the spring/summer, the onset of biological activity in the water column, and the degree of pelagic-benthic coupling, but additional factors such as depth, latitude, and advective processes are also important in regulating the biological interactions on Arctic continental shelves.

Introduction

The Arctic Ocean and its adjacent ice-covered seas have had limited ecosystem studies compared with other regions of the world's oceans. This is in part due to the seasonal ice cover, which precludes intensive field-based investigations for much of the year, particularly in the Arctic Basin itself. As a result, much of our knowledge of the Arctic is based on ice-free and/or reduced ice conditions, and it is assumed that biological activity during much of the ice-covered period is minimal.

A number of multi-year, process-oriented studies have been conducted in the Arctic and associated ice-covered marginal seas. PROBES (Processes and Rates on the Bering Sea

Arctic Oceanography: Marginal Ice Zones and Continental Shelves
Coastal and Estuarine Studies, Volume 49, Pages 231–261
Copyright 1995 by the American Geophysical Union

Shelf) and ISHTAR (Inner Shelf Transport and Recycling) included ecological investigations of the Bering and Chukchi Sea shelves during summer [May-September; Walsh and McRoy, 1986; Walsh et al., 1989; McRoy, 1993]. This work provided significant insights into the functioning of the associated biological communities. Similarly, multi-year studies within the Canadian Archipelago, where fast ice (defined as first year ice formed along land) predominates, have examined primary production from April to September, including that on the undersurface of the ice, its use by zooplankton and that available for benthic production [Legendre et al., 1992; Welch et al., 1992]. In the eastern Arctic, studies of air-sea-ice interactions, such as MIZEX (Marginal Ice Zone Experiment) and CEAREX (Coordinated Eastern Arctic Experiment) were only able to examine short-term patterns. As a result of limited time and space scales within these investigations, many studies could not link short-term responses, such as primary production, with longer-term effects on zooplankton population dynamics and benthic production over annual cycles.

A recent model for biological-physical processes on Bering/Chukchi shelves provides an example of the major processes occurring in polar systems [Shuert and Walsh, 1993]. These authors indicate through both data and modeling efforts that zooplankton and microbial consumption in highly productive regions of the Bering/Chukchi Sea continental shelf cannot effectively deplete the large carbon source in the water column, resulting either in the direct coupling of primary production to the underlying benthos or advection of this material downstream to deposition centers in the Arctic. In contrast, on continental shelves with low water column production, zooplankton and microbial processes may be able to effectively deplete these resources, structuring food webs in these waters. Wassman et al. [1991] described a conceptual model for the Arctic where the spring bloom starts in/after April and terminates by September, due to light limitation, resulting in one main bloom in the summer. Zooplankton biomass is low at the end of winter, and subsequently limits the ability of zooplankton to rapidly respond to rising food supply in the water column, resulting in extensive sedimentation of ungrazed phytoplankton and phytodetritus as senescence occurs at the end of the growing season.

We describe a simple conceptual model that links the seasonal trends of the major physical and biological processes operating in polar ecosystems (seasonal ice coverage, irradiance, nutrients, primary and secondary biomass and production, and vertical carbon flux; Figure 1). We acknowledge that these interactions are dependent on a number of factors not fully incorporated in the model, including location, depth, and advective regime. However, the factors emphasized are probably the most significant for describing biological interactions in the Arctic. We will then discuss the environmental parameters discussed in the model in relation to three cases studies in Arctic shelf areas: the northern Bering/Chukchi Seas, the Canadian Archipelago, and the Northeast Water polynya (an open-water area in ice-covered seas).

Conceptual Model

Seasonal ice cover

One temporal sequence that is well known for the entire Arctic and its adjacent seas is the seasonal change in ice cover. The physical environment, including globally-driven

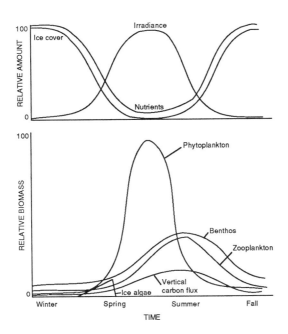

Figure 1. Conceptual model for the Arctic continental shelves.

circulation patterns and distribution of fresh water delivered to the sea as run-off influences the distribution and type of ice, which affects the availability of light seasonally and regionally. The Bering Sea is a well-studied example. Seasonal ice in the Bering Sea generally extends to the shelf break (about 200 m; Figure 2), and ice melt begins around the beginning of April [Parkinson et al., 1987]. By the end of June the ice edge generally has receded through Bering Strait into the Arctic Ocean basin, and the continental shelves of Russia, Alaska and Canada become exposed. Ice melt continues through September (the seasonal minimum), when ice formation begins and the ice edge advances rapidly, so that by late February it is again at the Bering Sea continental shelf break.

The seasonal ice pattern in the central and eastern Arctic is quite different. Around Greenland ice ablation begins in May-June, but the eastern continental shelf during some years remains substantially ice-covered throughout the summer, although in one region we will discuss in detail, the Northeast Water polynya, there is always substantial melt along portions of the northeastern coast. Within the Canadian Archipelago, fast ice reaches its maximum thickness near the end of May, and, with the exception of a few polynyas, breakup does not usually occur before the second half of July. Much of the area remains at least partially ice-covered throughout the summer. Interannual variability for all ice-covered regions is substantial [Parkinson et al., 1987].

Irradiance

Irradiance varies substantially on a seasonal basis, and it is the driving force for the large variations in the heat budget of much of the Arctic. Irradiance is also the major

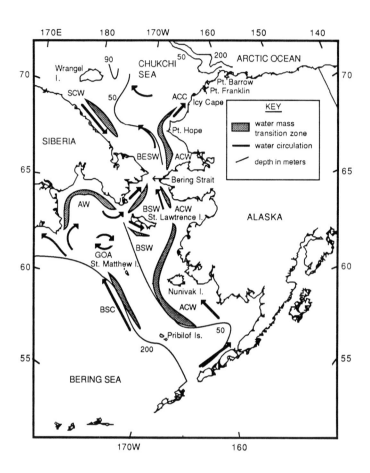

Figure 2. Study map of the Bering and Chukchi Seas continental shelves showing general bathymetry, water masses, and current directions [modified from Coachman et al., 1975; Walsh et al., 1989; Grebmeier, 1993]. The following abbreviations refer to water masses and currents in the region: ACC=Alaska Coastal Current; ACW=Alaska Coastal Water; AW=Anadyr Water; BESW=Bering Sea Water; BSC= Bering Slope Current; BSW=Bering Shelf Water; GOA=Gulf of Anadyr water; SCW: Siberian Coastal Water.

environmental factor controlling phytoplankton productivity in the Arctic on a seasonal basis [Smith and Sakshaug, 1990]. The amount of utilizable irradiance in turn depends on cloud cover, snow and ice cover, depth of vertical mixing, and season. Irradiance is a function of the angle at which the sun's energy hits the earth, thus the total energy is also a function of latitude and the surface albedo. Photoperiod (the total number of hours of sunlight per day) also is a function of latitude, and regions above the Arctic Circle (66° 45 N) receive 24 hours of continuous sunlight for varying lengths of time. Hence the growing season for Arctic phytoplankton is strongly structured on both a seasonal basis as well as on shorter time scales. How phytoplankton respond to these various changes in irradiance is only partially understood, and very little is known about how the rest of the ecosystem responds to the temporal changes in phytoplankton growth and biomass.

Phytoplankton and ice algae

Most incident irradiance in the Arctic that is not reflected is absorbed by sea ice and its associated snow cover [Smith and Sakshaug, 1990]. Where the sea is seasonally affected by pack ice, phytoplankton production may be stimulated simply by reduction of ice cover during a period of increasing irradiance. But ice is both a barrier to light penetration, and also a substrate on which ice algae may grow [Legendre et al., 1992]. So for phytoplankton in ice-covered seas the irradiance actually penetrating to the depth of a suspended cell first depends on three extinction coefficients: that for snow, that for ice, and that for the chlorophyll-containing ice algal layer, in addition to that for the sub-ice water itself [Smith et al., 1988]. Of the four, snow depth is the most variable and it alone can account for 70 to 85% of the variance in sub-ice chlorophyll concentration in an area of fast ice [Bergmann et al., 1991]. Assuming a fast ice environment rather than pack ice, the most important factor regulating phytoplankton production could be snow melt, which may increase light penetration by more than 10 times [Bergmann et al., 1991]. This first release of melt water and the release of ice algae from the underice surface more or less coincide, which has the combined effect of increasing water column stability, while eliminating the irradiance-absorbing ice-algal layer and providing an initial seeding of pennate diatoms to a potential phytoplankton bloom [Michel et al., 1993]. Ice algae grow to substantial concentrations in various polar environments [Harrison and Cota, 1991; Smith et al., 1988], and support early season growth for pelagic grazers [Tremblay et al., 1989; Conover and Siferd, 1993] and a diverse ice fauna [Horner et al., 1992; Grainger, 1991]. On the other hand, at the beginning of the melt season, much of the remaining sympagic (i.e., ice-related) production may not be efficiently transferred to the pelagic food web, but could pass directly to the benthic food web [Legendre et al., 1992]. Ice algae, when released from the melting ice in a relatively shallow environment, tend to sink quickly thus providing temporally limited input to the benthos [Carey, 1987].

As irradiance increases, phytoplankton growth is initiated shortly after the onset of melting, perhaps seeded by ice algae but with rapid changeover to predominantly centric diatoms characteristic of high latitude phytoplankton [Conover et al., 1991; Michel et al., 1993]. Nutrients are abundant at the start of the bloom and food for pelagic grazers is usually in considerable excess, but eventually nutrients can become limiting, at which time primary production is considerably reduced. Loss processes, especially rapid sinking of intact cells, then regulate the standing stock of phytoplankton. Macrozooplankton grazing may have greater impact at this time, but generally is not a quantitatively important loss term [Cooney and Coyle, 1982; Smith et al., 1988; Smith, 1994]. Microbial food webs operate at high latitudes, but do not seem as quantitatively important as in temperate neritic environments [Pomeroy and Deibel, 1986; Andersen, 1988].

Zooplankton

The amount and kind of ice, water mass characteristics, and irradiance influence the onset and magnitude of primary production of northern continental shelves, but they seem to have a relatively small influence on the structure of the pelagic communities that successfully exploit such ecosystems. Calanoid copepods are everywhere dominant, usually of several sizes and belonging to several trophic levels: there are probably three species of lipid-storing herbivores, one or two omnivores and at least one robust carnivore. Euphausiids are not found in the high Arctic, although they occur in the surrounding marginal seas. Amphipods and mysids, probably omnivorous, are frequently present and are usually associated with the under-ice surface or the littoral zone. Gelatinous herbivores are often rare, perhaps because they have limited energy storage capacity and no alternative food source in the dark season. Gelatinous carnivores are relatively common, however, including two species of chaetognaths and three species of ctenophores, each with a different feeding mechanism and belonging to a different order [Conover and Huntley, 1991].

Vertical carbon flux

Vertical particle flux on arctic continental shelves and marginal seas has rarely been measured over a full year primarily because of problems of access to and recovery of collected materials under ice-covered seas. Where these measurements have been successfully carried out with time-series controlled traps, rates have varied from <0.1 mg C $m^{-2} d^{-1}$ under an ice island in the central Arctic Ocean off Axel Heiberg Island, the lowest rates recorded for any ocean [Hargrave et al., 1989], to approximately 500 mg C $m^{-2} d^{-1}$ on a productive, shallow shelf in the northern Bering Sea [Fukuchi et al., 1993]. Although probably influenced by resuspension to some degree, the higher rates were in general agreement with independently determined sediment carbon requirements for the underlying benthos in the same area [Grebmeier and McRoy, 1989]. An estimate of vertical carbon flux in the Northeast Water polynya for a month in summer using ^{234}Th as a tracer yielded extremely low values [Cochran et al., 1995], similar to those in the central Arctic Ocean. Presumably most estimates of carbon flux in the high arctic will fall between these extremes.

Benthos

The benthic communities of the Arctic vary tremendously due to food supply, depth, disturbance (via ice), and predation. Most benthic studies in the Arctic have been carried out in the summer, and secondary production is thought to increase from spring through summer due to the increased food input from the surface. Høpner Petersen and Curtis [1980] hypothesized that in polar regions, benthic production is more tightly linked to pelagic production than in temperate and tropical areas, and a number of field studies have confirmed this hypothesis [e.g., Stoker, 1981; Grebmeier and McRoy, 1989; Grebmeier, 1993]. Regions of high overlying water column production over the shelves have a direct influence on underlying benthic biomass [Pfannkuche and Thiel, 1987; Grebmeier et al., 1988; Rowe and Phoel, 1991; Grebmeier, 1993]. However, it is not known whether temporal patterns of vertical flux (continuous vs. pulsed) give rise to different benthic communities in polar systems.

Even within individual regions, such as the Bering and Chukchi Seas, great variability can occur due to heterogeneous sediment regimes. For example, bivalves and polychaetes dominate shelf regions with fine sand and mud regimes [Haflinger, 1981; Stoker, 1981; Grebmeier and Cooper, 1995], whereas amphipods dominate in more coarse-grained sandy regimes [Grebmeier et al., 1989; Highsmith and Coyle, 1992]. In comparison, the continental slope areas in both the Bering and Chukchi Seas are dominated by a variety of polychaetes [Stoker, 1981, Grebmeier, 1993; Feder et. al., 1994]. High benthic biomass in the northern Bering and Chukchi Seas is a result of high carbon supply to the shallow benthos over a 5-7 month ice free period (Grebmeier and Barry, 1991]. In less fertile areas of the Arctic, ice coverage and other physical factors may play a greater role. In Frobisher Bay in the eastern Canadian Arctic 30-50% of the water column production settles to the benthos to support an active benthic food web [Atkinson and Wacasey, 1987]. Highest benthic biomass in Lancaster Sound is found in seasonally ice-free waters, with lowest values in central Baffin Bay, where ice cover and a deep mixed layer are thought to limit primary production [Thomson, 1982]. Off the northeastern coast of Greenland within the Northeast Water polynya the benthos contains large populations of epibenthic echinoderms (primarily brittle stars) and bivalves on the shallow bank regions, with low abundances of brittle stars and infaunal polychaetes under the pack ice and on the slope [Piepenburg, 1988].

Case Studies

Continental shelves of the Northern Bering and Chukchi Seas

Seasonal ice cover and irradiance

The seasonal ice coverage of the Bering and Chukchi Seas plays an important role in heat exchange and water mass development (Figure 2). Maximum ice cover occurs in late winter-early spring (March-April), extending as far as the shelf break in the SE Bering Sea in cold years [Niebauer and Alexander, 1985]. Ice-free conditions occur in summer (July-September) over the whole Bering Sea as well as most of the Chukchi Sea shelf [Niebauer, 1983]. Ice coverage and the resulting control on light penetration influence levels of annual primary production.

Ice-edge production

Although studies on ice-edge primary production have been conducted in the Bering Sea, little information is available on ice edge production in the northern Bering and Chukchi Seas. The Bering Sea ice-edge bloom precedes the open water bloom (normally occurring in early May for the SE Bering Sea and progressing northward with ice retreat) and can deplete nutrients in the surface layer that have been stratified by meltwater [Alexander and Niebauer, 1981; Niebauer and Alexander, 1985]. Upwelling at the ice edge as well as in situ recycling may introduce nutrients into the mixed layer, thus maintaining ice edge production [Niebauer, 1982; Müller Karger and Alexander, 1987], but little evidence exists to establish its quantitative importance [Niebauer and Smith, 1989]. The area in the Bering Sea between the 50-100 m isobath (known as the middle domain) is normally a region of an intense spring phytoplankton bloom [Walsh and McRoy, 1986].

It is not certain whether ice-edge production provides an important component of the annual water column production. If nutrient limitation persists into the open-water season, the ice edge bloom (annual production estimates range from 1-22 g C m^{-2} y^{-1}) [Subba Rao and Platt, 1984] could be a substantial fraction of the total annual primary production. Annual primary production in the SE Bering Sea ranges from 50-80 g C m^{-2} in the inner domain (0-50 m), which becomes depleted in nutrients after the spring bloom, to 166 gC m^{-2} in the middle domain (50-100 m) and offshore domain [100-160 m; Walsh and McRoy; Table 1]. What seems likely is that the importance of ice edge blooms is proportionally highest in those regions with limited annual primary production and diminishes moving northward in those regions of the northern Bering and Chukchi Seas influenced by Anadyr Water, since they are not nutrient-limited and annual primary production is high.

Open water primary production and water mass structure

The northern Bering and southern Chukchi Seas are characterized by three major water masses: the most saline and productive (average annual production of 300-500 g C m^{-2}) is Anadyr Water (AW; Figure 2; Table 1). AW is Bering Sea and slope water which is upwelled onto the Gulf of Anadyr shelf and is found on the western side of the continental shelf region from Cape Navarin to the Chukchi Sea [Coachman et al., 1975]. Nutrient-poor Alaska Coastal Water (ACW) is found in the eastern region near the Alaska coast, with an annual production of 50-80 g C m^{-2}. Between these major water types is Bering Shelf water (BSW) formed by entrainment of southeastern Bering Shelf water flowing northward and Bering Sea water that has overwintered in the Gulf of Anadyr [Coachman et al., 1975), and which has variable primary production (Table 1).

The combined BSW and AW, defined as Bering Sea water (BESW) [Coachman et al., 1975], moves north through Bering Strait and offshore to the northwest near Pt. Hope, Alaska (Figure 2). The ACW transits as a distinct water type through Bering Strait and northward along the Alaska coast. Nutrients are depleted in the surface waters moving northward from Bering Strait, with annual primary production decreasing from near 500-700 g C m^{-2} in the southern Chukchi Sea [Springer and McRoy, 1993] to 50-100 g C m^{-2} in the northern Chukchi Sea [Hameedi, 1978] and 10-25 g C m^{-2} y^{-1} in the Beaufort Sea [Parrish, 1987]. From the northern Chukchi Sea shelf to the slope region of the Arctic Ocean, surface water is composed of a complex mixture of water types, freshened by ice melt and river runoff in the spring/summer [Johnson, 1989].

Variations in nutrient input to different shelf regions of the Arctic have an important influence on levels of primary and secondary production. The highest nitrate and silica input to the Arctic Ocean is carried in Pacific waters entering through Bering Strait [Walsh et al., 1989; Jones and Anderson, 1986]. Deep, nutrient-rich Bering Sea water is upwelled off Cape Navarin and moves northward (Figure 2). This water remains near the western edge of the Gulf of Anadyr and flows northward through Anadyr Strait, with a small amount transiting to the east just south of St. Lawrence Island [Coachman et al., 1975; Shuert and Walsh, 1993]. Despite extremely high primary production, this nutrient signature extends into the Chukchi Sea and Arctic Ocean proper. Anderson et al. [1986] suggested that water moving northward over the Bering, Siberian and Chukchi shelves is modified, with additional nutrients being entrained from regeneration of organic matter at the sediment-water interface. Jones and Anderson [1986] proposed that the halocline of the Arctic Ocean was formed when brine-injected waters move down slope and then move northward at a depth determined by their individual densities.

TABLE 1. Comparison of phytoplankton and zooplankton production, benthic biomass and sediment oxygen uptake rates for different regions on the continental shelves of the Bering and Chukchi Seas (ACW=Alaska Coastal Water; BSW/AW= Bering Shelf Water/Anadyr Water; BESW=Bering Sea Water).

Area	Phytoplankton primary production ($g\ C\ m^{-2}\ y^{-1}$)	Zooplankton secondary production ($g\ C\ m^{-2}\ y^{-1}$)	Benthic biomass ($g\ C\ m^{-2}$)	Sediment oxygen uptake rates ($mmol\ O_2\ m^{-2}\ d^{-1}$)
Southeastern Bering Sea				
- inner domain (0-50 m)	50-80 (1)	-	<10 (2)	3-8 (3)
- middle domain (50-100 m)	166 (1)	8-30 (4, 5)	<20 (2, 6)	3-8 (3)
- outer domain (100-160 m)	162 (1)	30-50 (4, 5)	<11 (2, 6)	5-10 (3)
- oceanic domain (>160 m)	50 (7)	20 (4)	<5 (2)	-
Northern Bering Sea				
- ACW	80 (8)	5 (9)	<10 (10, 11)	<10 (11, 12)
- BSW/AW	80-480 (8, 13)	9 (9)	10-30 "	10-30 "
Southern Chukchi Sea				
- ACW	80 (8)	-	<10 (11, 12)	<10 (11, 12)
- BESW	470-720 (8, 14)	-	10-60 (10, 11)	10-40 (11, 12)
Northern Chukchi Sea				
"southern group"	50-100 (15, 16)	-	1-11 (17)	5-10 (11, 18)
"northern group"	50-100 (15, 16)	-	2-20 (17)	5-10 (18)

References
(1) Walsh and McRoy, 1986
(2) Haflinger, 1981
(3) Rowe and Phoel, 1991
(4) Cooney, 1981
(5) Vidal and Smith, 1986
(6) Stoker, 1981
(7) Frost, 1987
(8) Springer and McRoy, 1993
(9) Springer et al., 1989
(10) Grebmeier et al., 1988
(11) Grebmeier and McRoy, 1989
(12) Grebmeier, 1993
(13) Sambrotto et al., 1984
(14) Hansell et al., 1993
(15) Hameedi, 1978
(16) Parrish, 1987
(17) Feder et al., 1994
(18) J. Grebmeier, unpublished data

Zooplankton

Waters of the northwestern Bering Sea, containing high nutrients and large populations of offshore zooplankton (primarily *Neocalanus* spp.) become entrained in a western boundary current following the 60-70 m isobaths around the Gulf of Anadyr and through Anadyr Strait on the western side of St. Lawrence Island (Figure 2). The large *Neocalanus* copepods of the northern Bering Sea grow rapidly during spring and early summer, increasing their biomass by more than ten times during April and May alone. In summer more than 60% of the northerly transport through Bering Strait, about 1.1 +/-0.2 Sv, is AW [Walsh et al., 1989]. About 1.8×10^{12} g C of zooplankton are transported annually into the Arctic Ocean by this mechanism, and 70 to 90% of these animals are from the outer shelf community [Springer et al., 1989]. None of these species are normally associated with any continental shelf and all become expatriate, shortly after they enter the Chukchi Sea.

The eastern shelf community along the Alaska coastline in both the northern Bering and southern Chukchi seas is dominated by *Calanus marshallae* and small *Pseudocalanus* spp. [Springer et al., 1989]. Although zooplankton and bacterial studies have been limited in the northern regions of the Bering and Chukchi seas, these studies indicate that bacteria and zooplankton have a proportionally greater influence in utilizing the limited water column production in ACW than in the highly productive AW and BSW [Walsh et al., 1989; Shuert and Walsh, 1993]. The copepods of the outer shelf domain of the SE Bering Sea routinely ingest 20 to 30% of the daily primary production; by contrast, the middle shelf domain copepods rarely ingest more than 5% [Cooney and Coyle, 1982]. During an ice-edge bloom over the middle shelf, zooplankton ingestion rates yielded carbon flux measurements less than 2% of daily production [Coyle and Cooney, 1988]. It therefore seems accurate to say that zooplankton grazing does not deplete ice-edge production or open-water production in the middle domain in the SE Bering Sea. Most of the organic carbon produced in the surface waters sinks to the benthos intact.

There is still uncertainty regarding the role and fate of zooplankton populations in the northern Bering/Chukchi area. Copepods produced in the deep waters of the Bering Sea are carried within the Anadyr Stream northward onto the continental shelf [Springer et al., 1989]. Once on the Bering shelf they are trapped by the lack of deep water in which to overwinter, but they nonetheless serve as food for the benthos, birds and fish. Grazing by the copepods is seasonally important in the sub-Arctic Pacific [Parsons and Lalli, 1988], but a recent model, based upon ISHTAR data, suggests that their role in the north Bering-Chukchi system is small [Shuert and Walsh, 1993]. Nonetheless, Fukuchi et al. [1993], using time-series sediment traps, found that 25% of the primary production (2000 mg C $m^{-2} d^{-1}$) was lost from the euphotic zone, with 25% sinking as large fecal pellets and 10% as marine snow derived from appendicularian houses. Despite this, it is unclear where the particle flux in the Fukuchi et al. [1993] report originated from, given the variable advective transport into the area.

Vertical carbon flux

In the northern Bering and southern Chukchi Seas benthic standing stock is well correlated with overlying water column production [Grebmeier and McRoy, 1989; Grebmeier, 1993]. Highest standing stocks occur under the AW and BSW in regions influenced by high water column production, low zooplankton grazing, and presumably high carbon flux of phytoplankton to the benthos. Sediment respiration experiments on shallow continental shelves provide an indication of organic carbon deposition to the benthos, which in the "hot spots" of the northern Bering and southern Chukchi Sea ranged from 20-30 mmol O_2 $m^{-2} d^{-1}$, or an average organic carbon mineralization requirement of 0.5-1.0 g C $m^{-2} d^{-1}$. These rates are of the same order of magnitude as those obtained in a sediment trap deployment in the the Chirikov Basin [Fukuchi et al., 1993], where carbon flux averaged 0.5 g C $m^{-2} d^{-1}$ in a location where sediment respiration rates indicated an average benthic requirement of 0.48 g C $m^{-2} d^{-1}$ [Grebmeier and McRoy, 1989; Fukuchi et al., 1993]. Sediment oxygen uptake under the waters near the Alaska coastline in both the Bering and Chukchi Seas indicate a much lower carbon flux to the benthos, with rates less than 10 mmol O_2 $m^{-2} d^{-1}$ [Grebmeier, 1993].

Benthos

The composition and biomass of the benthos of the Bering and Chukchi Seas varies on- and off-shore and with latitude. Benthic biomass in the SE Bering Sea averages 1-10 g C m^{-2} in the inner domain, 1-20 g C m^{-2} in the middle domain, and 1-11 g C m^{-2} in the outer domain [Table 1]. The lowest benthic biomass in the northern Bering and Chukchi Seas occurs in the east along the Alaska coastline in ACW, averaging 1-10 g C m^{-2}, and is due to ice gouging, river runoff, high current scouring and low water column production [Stoker, 1981; Grebmeier et al., 1988, 1989]. Benthic biomass increases to 20-40 g C m^{-2} southwest of St. Lawrence Island and in various shelf regions of the Gulf of Anadyr [Sirenko and Koltun, 1992; Grebmeier, 1993; Table 1]. The benthic biomass north of St. Lawrence Island ranges from 10-30 g C m^{-2} [Table 1]. Benthic biomass reaches a maximum of 50-60 g C m^{-2} in the southern Chukchi Sea, where a current-driven deposition of organic materials occurs [Grebmeier and McRoy, 1989; Grebmeier, 1993].

In general, benthic biomass decreases as one moves to the north in the Chukchi Sea [1-11 g C m^{-2}; Stoker, 1981; Feder et al., 1994], although there are exceptions, such as the region north of Icy Cape where despite low algal production, benthic standing stock can reach up to 20 g C m^{-2} [Feder et al., 1994]. They suggested that particulate organic carbon (POC), advected in Bering Sea water, is directed northeastward north of Icy Cape, and provides POC to suspension-feeding organisms (primarily amphipods) that dominate these nearshore regions. In addition, polynyas occur regularly along this section of the Alaska coastline, possibly enhancing ice-edge production in the spring and summer that would provide another organic carbon source to the underlying benthos. Finally, near-bottom currents moving northward may be an important mechanism transporting organic carbon produced further south on the Bering/Chukchi shelves across the northern shelves and into the Arctic Ocean via canyons along the continental slope. Anoxic sediments have been observed periodically under areas of high water column production on the shelf [J. Grebmeier, personal observation] as well as in deeper sediments in Barrow Canyon at the Chukchi Sea-Arctic Ocean interface [J. Grebmeier and L. Cooper, personal observation]. The quantitative importance of this mechanism for transporting organic carbon into the Arctic Ocean remains uncertain.

Higher trophic levels

A characteristic of the shallow shelf ecosystem of the northern Bering and Chukchi seas is the presence of both water column feeding mammals (bowhead whales, seals) and birds (kittiwakes, fulmars, auklets) as well as benthic-feeding mammals (walruses, gray whales) and birds (murres). Although nekton and demersal fish, along with macroinvertebrate epifaunal predators, are important consumers in the SE Bering Sea, they become less important in the north as bottom water temperatures decrease [Zenkevitch, 1963; Nieman, 1963; Jewett and Feder, 1980]. This undoubtedly influences the production of the extremely high biomass of benthic fauna in the northern regions [Neiman, 1963].

Recent Study: St. Lawrence Island Polynya

The influence of a Bering Sea polynya (Figure 3a) on water column production and subsequent transport to the shallow continental shelf benthos of the Bering Sea was evaluated in June 1990 by studying the spatial patterns of organic material deposition, benthic biomass, sediment community metabolism, benthic population structure, and other potential indicators of enhanced carbon transport to the benthos [Grebmeier and Cooper, 1995]. Although polynyas may be important localized centers for primary production in polar waters, this study did not conclusively find that the polynya (south of St. Lawrence Island) enhanced underlying benthic communities. Instead, a more complicated interaction appeared to occur between the highly productive Anadyr Stream to the west and the cyclonic, baroclinic currents set up by winter brine formation southwest of St. Lawrence Island that could entrain polynya production and transport it to the southwest.

Benthic standing stocks were relatively high (10-20 g C m^{-2}) south of St. Lawrence Island, under the location of the polynya during winter/spring, with the highest values to the southwest of the island under the Gulf of Anadyr cold pool (20-50 g C m^{-2}; Figure 3b). The lowest benthic biomass values were to the southeast of the island under ACW (1-10 g C m^{-2}). Total sediment respiration rates were lowest south of the island and in ACW, but rates increased steadily to the southwest in an offshore, downstream direction from the island, indicating a deposition zone under the Gulf of Anadyr cold pool, suggesting that currents, rather than the polynya, are most important in influencing benthic production (Figure 3c).

However, the presence of the St. Lawrence Island polynya may provide an early organic carbon input to the region, based on surface sediment ^{7}Be inventories (^{7}Be is a particle-reactive tracer with a 53-d half-life; Figure 3d), and this carbon source may well be an important stimulus to the benthic population [Grebmeier and Cooper, 1995]. Also, measurements of the oxygen isotope composition of tunicate cellulose [an experimental technique for monitoring longer-term, winter-spring, benthic fluxes; Grebmeier et al., 1990], are consistent with benthic production overwinter. Nevertheless, it appears the Anadyr Stream, flowing west to east and bringing high nutrients into the area south of St. Lawrence Island, is the major forcing function for high production in the region. The interaction of Anadyr Water with the winter/spring ephemeral polynya and associated baroclinic currents, combine to positively influence benthic communities.

Relevance to Conceptual Model

The ecosystem in the nearshore, eastern regions of the Bering and Chukchi seas, where ACW occurs, is the area most similar to the conceptual model. Nutrients become

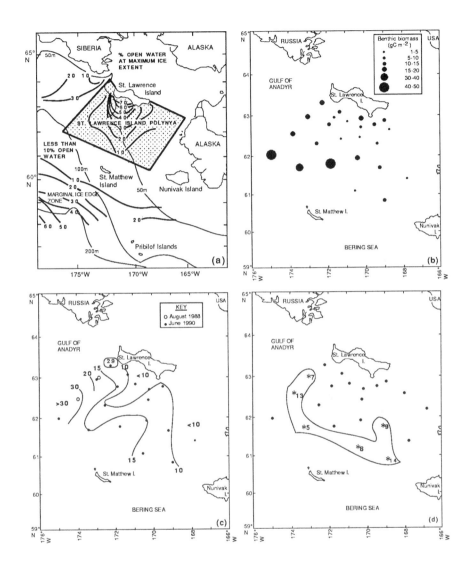

Figure 3. (a) Map showing study area for 1990 in relation to the average percent open water at maximum ice extent in the St. Lawrence Island Polynya (SLIP) region;
(b) Distribution of macrofaunal benthic biomass (g C m^{-2}) in the SLIP region during June 1990;
(c) Distribution of sediment oxygen uptake rates (mmol O$_2$ m^{-2} d^{-1}) in the SLIP region during June 1990. The open white circles indicate values obtained during summer 1988;
(d) Detection of ^{7}Be (∗) in surface sediments near St. Lawrence Island polynya during June 1990. Ordinal numbers indicate gamma activity associated with ^{7}Be in mBq cm^{-2} in surface sediments. Assays conducted where no ^{7}Be activity was detected are indicated by (•). [All figures from Grebmeier and Cooper, 1995].

limiting after the spring bloom, thus producing only a low annual primary production. Zooplankton and microbial processes become more important with the limited organic material available. Organic carbon supply to the benthos is limited and benthic biomass is generally low [Grebmeier, 1993]. A modification of the conceptual model is in order for the western regions of the northern Bering Sea and southern Chukchi Sea where primary production is high, being maintained through the open-water season by upwelling of high nutrient water in the Anadyr Stream and, most likely, summer recycling of ammonium at the sediment-water interface [Walsh et al., 1989; Shuert and Walsh, 1993]. Neither zooplankton grazing or the microbial loop have much influence on carbon utilization in these extremely phytoplankton rich waters [Andersen, 1988; Springer et al., 1989].

The Canadian Archipelago

The physical environment

On arctic shelves within the Canadian Archipelago, fast ice is seasonally a major factor in primary production dynamics. In the Lancaster Sound region fast ice forms in September, but may not become consolidated until March. The fast ice front can lie as far east as Bylot Island or even west of Resolute, but more typically it forms between Somerset and Devon Island across Barrow Strait (Figure 4). Breakup is initiated at the eastern extension of Lancaster Sound about June 1 and continues westward into Barrow Strait reaching Resolute after mid-July. However, the ice does not melt completely and wind-driven pack ice can cover from 0.4 to >0.9 of the surface of Barrow Strait into September [Dirschl, 1980; LeDrew et al., 1992].

The prevailing, non-tidal flow in Barrow Strait is easterly or southeasterly [Prinsenberg and Bennett, 1987]. Because the Strait is constrained by a sill at 125 m west of Cornwallis Island and shallower sills to the northwest, sea water entering the region has near-surface origins. Waters from the south and west have approximately the same salinity (31.5-32.5 psu) due to mixing at the sills, while water from the north is slightly more saline (>32.8 psu). In the Strait, with the normal easterly flow and fast ice present, there is upwelling of higher salinity water on the north side and an accumulation of lower salinity water on the south side. This estuarine-like circulation also results in westerly penetration of warmer, more saline Atlantic Water at depth from eastern Lancaster Sound, resulting in upward advection of heat and formation of a polynya and flaw leads westward along the southeast coast of Devon Island. As the season progresses these open water areas become continuous with the Northwater polynya in northern Baffin Bay [Stirling and Cleator, 1981]. When the ice front is west of Resolute, before the new ice has become consolidated, or following breakup, circulation patterns in Barrow Strait may be altered by atmospheric forcing resulting in eddy formation and current reversal, causing the introduction of different phytoplankton and zooplankton populations into the area [Bedo et al., 1990].

Primary and secondary production: ice

Irradiance available for photosynthesis is a function of incident radiation, and attenuation

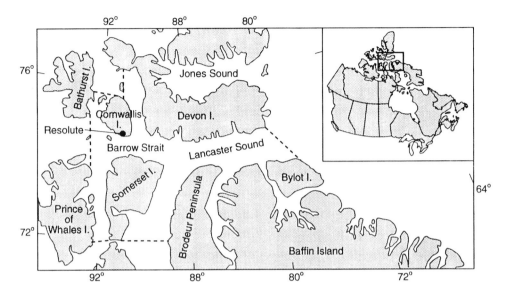

Figure 4. Diagram of Lancaster Sound region (LSR), arctic Canada, giving place names. LSR, as used here, is defined by the dashed lines [after Welch et al., 1992].

by snow and ice, both of which vary with thickness. Using an empirical model, Bergmann et al. [1991] estimated primary production of ice algae from incident radiation and snow depth and calculated an average annual productivity for the Resolute area of 5 g C m^{-2} y^{-1}, or about 10% of the total of phytoplankton, kelp and ice algal production [Welch et al., 1992]. This estimate is similar to previous direct measurements, although on the low side, for the same area at low snow cover [Smith et al., 1988].

Associated with the ice is a community of macrozooplankton consisting primarily of amphipods. At the ice edge and at break up these prey play an important trophic role as food for a number of bird species, for the arctic cod *Boreogadus saida* and for young ring seals *Phoca hispida* [Bradsteet and Cross, 1982]. Even so, their biomass and energetic contribution to the total ecosystem of the Lancaster Sound region was not large [Welch et al., 1992]. Although these organisms are accessible to avian predators at the ice edge, no spatial correlation between standing stock of amphipods and the ice edge was observed by Cross [1982] or Pike [1987]. Cross [1982] did show that amphipods were numerically more abundant and had higher biomass under rough ice than under smooth. In the Resolute area these amphipods were more abundant in water less than 50 m deep and, when the ice broke up, most sought the underlying sediment [Pike, 1987].

There is also a large and varied meiofaunal community associated with the ice algae. In the Frobisher Bay area the mean standing crop of ice algae contained 257 mg C m^{-2} and the estimated meiofaunal carbon biomass was about 73 mg C m^{-2}. Assuming that relatively small organisms have a daily ration about equal to their carbon weight, they might consume carbon equivalent to the daily primary production by the ice algae [Grainger and Hsiao, 1990].

The ice algae are also an important nutritional source for calanoid copepods in the early spring before water-column primary production increases. In May 1984 the small arctic calanoid *Pseudocalanus acuspes* was observed swarming under the fast ice and feeding on ice algae [Conover et al., 1986]. Sometimes they also showed diel migratory behavior and fluctuations in gut contents [Conover et al., 1988]. This species passes through several developmental stages and reaches sexual maturity using ice algae as a nutritional source and would not be able to complete its life cycle in a single season otherwise [Conover et al., 1991].

The importance of ice algal production under fast ice to the developing populations of pelagic copepods was further demonstrated by Harris [1992]. Between May 30 and June 21, 1988 primary production in the water column of Resolute Passage averaged only 2.13 mg C m^{-2} d^{-1} while the combined grazing of the major copepod species demanded 11.9 mg C m^{-2} d^{-1}. This need was satisfied by sedimenting ice algae equivalent to about 55 mg C m^{-2} d^{-1}. Heavy usage of ice algae by diel-migrating, night feeding populations of *Calanus* and *Pseudocalanus* was also observed in Hudson Bay, accompanied by egg production. Once the ice melt began in mid-May, the same species continued feeding on the ice algae in the water column but ceased to migrate to the ice [Runge and Ingram, 1991]. Carbon flux also reflected the importance of ice algae to copepod nutrition in Resolute Bay in early spring in that more carbon sedimented as fecal pellets than as diatoms, but this relationship changed by the end of May when the ice algae began to melt off the ice [Anning, 1989; Figure 5].

Primary and secondary production: water column

In Barrow Strait, the spring bloom of phytoplankton begins three to four weeks before the ice breaks up, and by mid-August about 1 g C m^{-2} d^{-1} is produced [Harris, 1992]. The physical process of making salt water ice also creates an environment favorable for the growth of algae. As ice crystals grow, algal cells, mostly pennate diatoms, are incorporated and grow, and nutrients and brine concentrated from sea water accumulate between the developing ice crystals. However, as the ice-algal bloom develops, the nutrients in the ice are quickly depleted and the ice algae, being limited by light from the top and by nutrients from the bottom, form a relatively thin layer on the under-ice surface. Even in an environment such as in Barrow Strait with strong tidal currents (>50 cm s^{-1}), a nutrient gradient between the ice colonized by algae and the bottom of the mixed layer can be demonstrated [Cota et al., 1987]. The spring-neap cycle of the lunar tides also affects the growth of ice algae, apparently modifying the nutrient gradient and flux into their micro-environment [Gossellin et al., 1985; Conover et al., 1990].

Welch et al. [1992] estimated that mean annual primary production was probably not very different from that in the Resolute area and estimated 55 g C m^{-2} y^{-1} as typical. Harrison and Cota [1991] summarized Canadian observations from Baffin Bay and the eastern archipelago and estimated a mean daily primary production of about 300 mg C m^{-2} d^{-1}. Moreover, this estimate was apparently independent of nitrate concentration, suggesting that nutrient limitation did not affect primary production seasonally here, in marked contrast to conditions in the Bering Sea [Walsh et al., 1989; Henriksen et al., 1993].

Once the phytoplankton bloom gets under way, chlorophyll concentrations near Resolute usually remain high through July and August, and the copepod community can only use

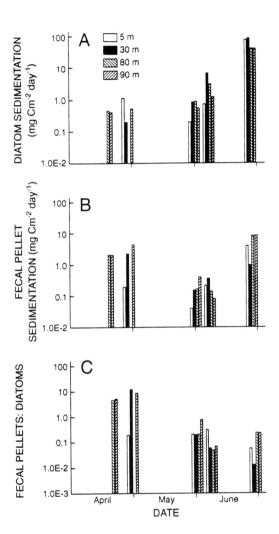

Figure 5. A. Daily sedimentation rates of carbon, calculated from estimates of diatom volume in sediment trap collections at Station 5, LSR, 1984.
B. Daily sedimentation rates of carbon, calculated from fecal pellet volume in sediment trap collections at Station 5, LSR, 1984;
C. Ratio of fecal pellet carbon to diatom carbon in sediment trap collections at Station 5, LSR, 1984 (after Anning, 1989).

1 to 3% of net primary production [Harris,1992]. During this period the dry biomass of total zooplankton can increase seven-fold (0.2 to 1.4 g m^{-2}) [Conover and Huntley, 1991]. The numbers of some species may increase by an order of magnitude as the population structure shifts from adults to the young of the next generation, but individuals of a given stage may also more than double in dry weight [Conover and Siferd, 1993]. Most of this new weight was lipid, primarily wax ester, which will serve as an energy store during the

dark season, and probably as a source of raw material for reproduction as well. *Calanus hyperboreus* can survive without food for at least three months by using its stored lipid [Conover, 1964; Lee, 1974] and it reproduces in winter or early spring without feeding [Conover and Siferd, 1993]. However, at prevailing winter temperatures (-1.8 °C), even early copepodids with little energy reserves can overwinter [Kosobokova, 1990; Conover and Siferd, 1993].

The role of the zooplankton in the Lancaster Sound region can best be visualized as part of an energy flow model [Welch et al., 1992]. The maximum zooplankton dry weight (>2 g m^{-2}), occurred in 1984 but was biased by unusually high concentrations of small *Pseudocalanus* in spring (71% of the total herbivore biomass). With better seasonal coverage in 1986, mean dry weight biomass was 0.69 g m^{-2} and *Pseudocalanus* and *Calanus hyperboreus* were co-dominant, each contributing about 33% to total dry weight. Respiration was calculated for mixed arctic copepods [Conover and Cota, 1985] and used to calculate growth, assuming 60% assimilation and 60% net growth efficiency [Conover, 1978: see Table 5-27 within]. The herbivorous copepods were estimated to consume 888 kJ m^{-2} annually or 31% of the total net production (2885 kJ m^{-2} y^{-1}). Of the net growth of copepods, an estimated 319 kJ m^{-2} y^{-1}, 269 were consumed by chaetognaths and other gelatinous zooplankton and only 1.12 by arctic cod. These estimates of primary and secondary production appear reasonable, but the present estimates of arctic cod production and biomass are low by more than an order of magnitude. Apparently current hydro-acoustic methods are too limited in areal coverage to detect the rare but very dense schools of cod which are occasionally observed visually. One such well-documented school seen near Resolute was estimated to contain 75,000 tonnes of arctic cod or 12.5 times the acoustic extrapolation for all of LSR (Welch et al., 1992).

Some regions in the Canadian archipelago have large carbon fluxes to the sediments, resulting in high benthic biomass. Northwestern Baffin Bay and Lancaster Sound have standing stocks similar to the SE Bering Sea middle shelf (300-500 g wet wt m^{-2}), but the central Baffin Bay biomass is only 40-80 g wet wt m^{-2} [Thomson, 1982]. In the Resolute area, bivalves, dominated by *Mya truncata*, contribute about 30 g m^{-2} of dry biomass and utilize about 25.6 g O$_2$ y^{-1}(364 kJ y^{-1}). However, the presence of smaller bivalves, and other invertebrate species not quantitatively surveyed, suggest that the benthos may utilize more of the annual primary production than the herbivorous zooplankton [Welch et al., 1992]. In regions of low pelagic primary productivity, the benthos is dependent on advected POC or rapid settling of ice algae. Dunbar [1981] proposed that the high biological productivity of Lancaster Sound could be partially due to nutrients advected from the North Water polynya via estuarine circulation and Coriolis-driven meanders and eddies that result in upwelling. Benthic standing stock and production is variable in the Canadian Archipelago, but generally lower than in the Bering and Chukchi seas.

Relevance to Conceptual Model

The dominance and persistance of fast ice in the Canadian Archipelago is surely the major deviation from the conceptual model (Figure 1). Here one year fast ice virtually completely controls the production cycle for 8 or 9 months, but even after breakup the drifting pack ice, mostly derived from the previous winter's fast ice, persists in variable concentrations (10 to 95% coverage) that never completely melts [Conover and Siferd, 1993]. As fast ice is thinner and generally accumulates less snow than broken pack or multi-year ice, the light-limited sympagic production is initiated earlier (generally by April 1) and may persist later than that for the ice algal populations elsewhere. Also light-limitation continues to regulate phytoplankton production in the Archipelago because the period between breakup and freeze up is so short. Hence a greater proportion of the total primary production is sympagic [Smith et al., 1988], and nutrient-limitation of primary production would seem to be less important than in high-latitude environments with a longer open water period [Harrison and Cota, 1991]. Zooplankton, of necessity, have adjusted to a shorter period of intense primary production by lengthening their life cycles, by making greater use of sympagic production for early season development [Conover et al., 1986; Runge and Ingram, 1991], and by storing energy during the brief but intense phytoplankton bloom to facilitate survival during the polar winter [Conover and Huntley, 1991]. Despite its short growth period much of the phytoplankton production is still lost to the benthos, where it supports large populations of bivalves [Welch et al., 1992], while the zooplankton biomass shows less seasonal fluctuation than is shown in Figure 1 [Conover et al., 1991].

Continental Shelf off Northeast Greenland

The polynya on the Greenland continental shelf at ca. 80 °N, 14 °W was studied intensively in July - August, 1992 (Figure 6). Work previous to this intense study suggested that the region's bathymetry was dominated by a trough system that surrounded a central bank [Bourke et al., 1987], with circulation characterized by a clockwise gyre which flowed through the depression and around the bank. However, it was also possible that the circulation was estuarine in character [similar to that of other Greenland trough systems; de Vernal et al., 1993] and that the circulation regimes in the northern and southern troughs were not connected. In 1992 it was found that the circulation was indeed dominated by a gyre, although the mean flow during summer was sluggish and dominated by tides [Johnson and Niebauer, 1995]. The polynya's zooplankton community was dominated by calanoid copepods [primarily *Calanus glacialis*, *C. hyperboreus*, and *C. finmarchicus* [Ashjian et al., 1995]. The benthic faunal abundance was large (up to 10,000 individuals m^{-2}) and appeared to be weakly correlated with autotrophic pigments in the surface sediments [Ambrose and Renaud, 1995]. Benthic respiration was active [Rowe et al., 1993], although up to 6-fold less than rates observed in the Bering Sea [Grebmeier, 1993; Grebmeier and Cooper, 1995].

Although the same geographic location was re-occupied a number of times, no clear temporal trends were observed, due to the three-dimensional spatial variability of the region. An AVHRR satellite image revealed that a cold water filament flowed northward from under a fast ice sheet anchored to shore and entered the polynya [Figure 6; Wallace et al., 1995]. By following this water parcel, one could begin to delineate the biological consequences and patterns through time. However, the thermal image showed the temperature distribution on July 23, 1992, but the stations which were sampled within the area of the plume were not sampled on that date. Also, because there were only a few stations taken within the observable limits of the plume, the total data to construct a time-series are limited. Therefore, to construct a time-series, the assumption was made that the flow of the cold water from under the fast ice was in steady state over the time

Figure 6. Station locations used in the analysis in the Northeast Water polynya. The approximate location of the cold water mass emanating from under the fast ice shelf is indicated by the dashed line.

periods that the stations were sampled. Based on the temperature and density data, this assumption appears reasonable [Smith et al., 1995]. Finally, the data analysis is further complicated by the fact that at some stations within the plume, only the basic hydrographic parameters were measured, whereas at others a full suite of variables (rate process information, particulate concentrations, etc.) were assessed.

The waters initially flowing into open water from under the fast ice initially had low surface temperatures (-0.8 to -1.6 °C; Figure 7a), non-limiting surface nutrient concentrations (nitrate >3 µM; Figure 7b) and substantial integrated nitrate concentrations (>220 mmol m^{-2}). As the water parcel was exposed to higher irradiance and mixed laterally with surrounding waters, the phytoplankton assemblage grew and reduced the nitrate concentrations to undetectable levels throughout the upper 20 m (Figure 7b). Surprisingly, no evidence of phosphate uptake was observed, and silica uptake was minor (Figure 7b). Ammonium concentrations in the euphotic zone were in all cases low (less than 0.2 µM). Because the surface temperature signal was obliterated by either surface heating or lateral mixing, the parcel could not be traced any farther, and additional biological changes are unknown.

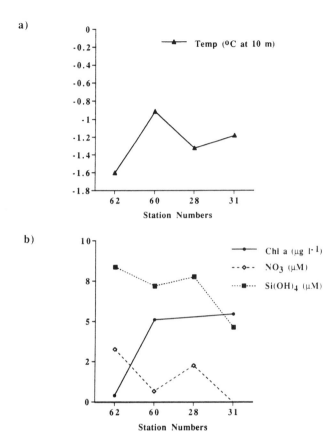

Figure 7. The a) temperature at 10 m and b) nitrate, silicic acid and fluorometric chlorophyll between the stations within the cold water plume of the Northeast Water polynya. Based on the location of those stations, a temporal sequence is presumed.

The phytoplankton assemblage which developed during this temporal sequence was dominated by diatoms (as shown by the dominance of fucoxanthin in the pigment profile; Figure 8 a,b), although chrysophytes also contributed substantially to phytoplankton biomass, based on the 19'-butanoyloxyfucoxanthin (19-BUT) concentrations. The relative ratios of fucoxanthin to chlorophyll increased more than the ratio of 19-BUT to chlorophyll, which suggests that the diatom fraction became more dominant during community development, although the cellular pigment levels also can change in response to variations in irradiance. Maximum fluorometric chlorophyll levels increased from ca. 1.5 to 5.6 μg l^{-1} (Table 2).

Primary productivity increased from 0.22 to 1.14 mg C m^{-2} d^{-1} as the plume "aged" (Table 2), reflecting the increased biomass and acclimation to near optimal growth conditions. Chlorophyll-specific productivity at the surface was also elevated at Station 31, which probably was the result of photoadaption to the higher irradiances experienced at the surface. Nitrate uptake increased dramatically between the two stations, and f-ratios

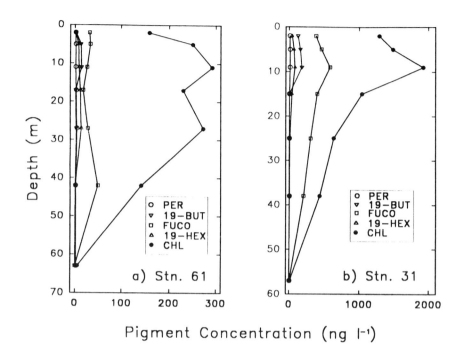

Figure 8. The vertical distribution of pigments in Northeast Water polynya at a) Station 61 and b) Station 31. PER = peridinin; 19-BUT = 19'- butanoyloxyfucoxanthin; FUCO = fucoxanthin; 19-HEX = 19'-hexanoyloxyfucoxanthin; CHL = chlorophyll a.

TABLE 2. Phytoplankton biomass and productivity within the cold water plume observed in the Northeast Water polynya on July 23, 1992. All integrations are from the surface to the bottom of the euphotic zone. POC = particulate organic carbon.

Variable	Station 61	Station 31
Euphotic zone (m)	63	57
Surface chlorophyll ($\mu g\ l^{-1}$)	1.53	5.60
Integrated chlorophyll (mg m^{-2})	32.5	122.9
Surface POC (mg m^{-3})	15.0	39.2
Integrated POC (mg m^{-3})	135.2	297.4
Surface primary production (mg C m^{-3} h^{-1})	1.48	8.30
Integrated productivity (mg C m^{-2} d^{-1})	0.22	1.14
Surface nitrate uptake (mmol N l^{-1} h^{-1})	0.009	0.19
Integrated nitrate uptake (mmol N^{-1} h^{-1})	0.08	2.41
Surface f-ratio	0.99	0.85
Integrated f-ratio	0.94	0.56

were very high at both (0.99 and 0.85, respectively). Hence rates of new production were significant. Integrated f-ratios for the entire water column were 0.94 for Station 61 and 0.56 for Station 31, which indicates that much more of the uptake was attributable to ammonium in the "aged" water parcel than initially. Ammonium concentrations remained low (less than 0.2 μM throughout the water column), and hence the rate of ammonium turnover was greater at Station 31. Apparently the activity and quantitative importance of the microbial loop had increased, and water column nitrogen cycling was much more active than initially. However, the f-ratio remained very high, and, as a result, a majority of the production was still new production and available for export. Absolute quantities of nitrate-based production were greater at Station 31 than at Station 61 (Table 2).

By assuming that horizontal exchanges with surrounding waters were minimal, it was possible to construct crude nutrient budgets for the water parcel. Similar approaches for the Greenland Sea [Smith et al., 1991; Smith, 1994] have demonstrated that substantial losses occurred from the surface layer. The technique compares the "initial" nutrient concentrations (those at Station 62, which are assumed to be negligibly influenced by biological uptake) and those at stations removed from the source (e.g., Stations 60, 28 and 31) and presumably "later" in the biological sequence. It also inventories the particulate material which was formed from that nutrient (i.e., particulate nitrogen from nitrate plus ammonium; biogenic silica from silicic acid). Unfortunately, only two stations within the plume had particulate matter concentrations assessed (Stations 61 and 31). By summing all the measured forms of nutrients and the associated particulate material formed from each nutrient and comparing them, the difference between integrated dissolved nutrients and the particulate inventories is the amount of material which exited the layer. Based on the budget of nitrogen and silica for the Northeast Water polynya, it appears that little vertical flux of biogenic material occurred during movement of the cold water tongue through the region (Table 3); in fact, any vertical flux that occurred is obscured by the variations in initial concentrations of silicic acid and nitrate, as well as the assumptions of equilibrium and negligible horizontal mixing. This implies that fluxes during short periods were only a few percent of productivity, which is not surprising given that the plume probably had been exposed to surface irradiances for less than one week. It also suggests that the coupling between the pelagic system and the benthos is restricted to a shorter time period, one which probably occurred later in the season (although the ice-free period lasted only one more month). Vertical flux may in fact be mediated by the life cycles of the phytoplankton present [e.g., by diatom resting spore formation: Karl et al., 1991] or by a rapid change in the environment, such as a reduction in irradiance by the rapid growth of ice in September, or by a change in physiological state, such as those induced by nutrient limitation.

Relevance to Conceptual Model

In the conceptual model it was suggested that primary productivity is largely dependent on nitrate, with the microbial loop being quantitatively less important in Arctic waters than in temperate and tropical regions [Pomeroy and Diebel, 1986]. Given the low concentrations of ammonium which were found throughout the region, the large concentrations of diatom pigments (relative to other forms) and the dependence of production on nitrate (f-ratios greater than 0.80), it does appear that the water column in

TABLE 3. Nutrient and particulate matter concentrations within the cold water plume. All integrations done between the surface and the 0.1% isolume (63 and 57 m for Stations 61 and 31, respectively).

Station Number	Integrated Nitrate (mmol m^{-2})	Particulate Nitrogen (mmol m^{-2})	Integrated Silicic Acid (mmol m^{-2})	Biogenic Silica (mmol m^{-2})	Total Nitrogen (mmol m^{-2})	Total Silica (mmol m^{-2})
61	187	22.4	513	19.5	209	532
31	171	71.6	478	71.6	243	549

the Northeast Water polynya is functioning in a manner similar to that proposed (Figure 1). Vertical fluxes [based on the nutrient budget calculations from ^{234}Th data; Cochran et al., 1995] apparently are low, and may reflect the temporally restricted nature of vertical flux in polar regions. Studies in the Antarctic [DeMaster et al., 1992] have also found that the flux of biogenic material is restricted to short time periods. Such patterns would provide the benthos with a high quality food source in a short period of time, and may be responsible for the induction of benthic reproduction. Ambrose and Renaud [1995] found that benthic biomass was most closely correlated with sediment pigment concentrations, suggesting a strong benthic-pelagic coupling in the polynya. Zooplankton removed only a small portion of daily primary productivity [Ashjian et al., 1995], and hence the vertical flux apparently was supported by sinking of phytoplankton cells or aggregates. Bacterial activity was also reduced, and nutrient (nitrate) limitation of phytoplankton growth and productivity was observed (Smith et al., 1995). Therefore, the conceptual model proposed (Figure 1) seems to describe the biological interactions found in the open waters of the Northeast Water polynya during summer, 1992.

Discussion and Conclusions

The three studies presented (the northern Bering/Chukchi Sea continental shelves, Lancaster Sound in the Canadian Archipelago, and the Northeast Water polynya off Greenland) indicate both similarities and differences with the proposed conceptual model (Figure 1). The eastern regions of the Bering and Chukchi Sea continental shelves and the NEW polynya off Greenland share basic affinities with the conceptual model, including: 1) high water column production in the late spring/summer, which becomes limited when nutrients are depleted, 2) minor amounts of ice-edge production relative to annual open-water primary production, 3) delayed development of zooplankton communities until after the peak of primary production, and 4) variable degrees of pelagic-benthic coupling, resulting in variability in the underlying benthic standing stocks.

The northwestern Bering Sea and Lancaster Sound appear to be at variance with the model, although the variance is due to widely differing forcing functions. In the northwestern Bering and southern Chukchi Seas, the complete absence of ice during the late spring and summer, the high open-water season primary production (due to substantial nutrients and light), the restricted abilities of zooplankton or bacteria to deplete this carbon source, and shallow shelf depth enable a strong coupling of pelagic production and secondary benthic production. This system functions throughout the open-water season, allowing the development of an extremely productive food web. In contrast, Lancaster Sound is characterized by high variability in the concentration of ice cover and associated ice-algal production. Ice recedes from the shelf later in the growing season than in the Bering Sea, resulting in reduced water column production. This variability in ice coverage results in a mosaic of water column and benthic production, depending more on current regimes, water column depth, and physical forcing functions.

The Northeast Water polynya also offers contradictions to the model. Because nutrients of the polynya's source waters are so low initially relative to the Bering Sea (5 vs. 30 μM nitrate), the overall productivity is limited on a seasonal basis. Benthic biomass is lower, and benthic regeneration rates are significantly lower [Grebmeier, 1993; Rowe et al., 1993]. One possible explanation for this is that passive vertical flux (that mediated by the sinking of intact phytoplankton rather than in fecal pellets) is relatively greater in the high biomass (and shallower) Bering and Chukchi Seas. This might be explained not only by the relatively shallow water column of the Bering Sea (ca. 50 m), but also by the formation of aggregates which significantly increase settling velocities. Aggregate formation has been hypothesized to be the dominant mode of vertical flux in regions with high phytoplankton standing stocks [Hill, 1992]. Furthermore, polar regions such as the Bering and Chukchi Seas, which are known for extremely high annual primary productivity, are some of the few areas which attain sufficient biomass so that aggregate formation could be quantitatively important. The Northeast Water region never attains a biomass as high as that of the northern Bering and southern Chukchi Seas and has an average depth of ca. 300 m, and hence the relative contribution to vertical flux by zooplankton fecal pellet production may be greater.

However, if in the NEW polynya the microbial loop is of minor importance, zooplankton grazing is low, and aggregate formation is limited, what is the fate of the excess organic matter? One explanation may be that losses were minimal during the period of observation and occur later in the season. Perhaps algal senescence occurs at the end of the season, resulting in aggregate formation and rapid flux from the surface layer. Finally this organic matter might be entrained in offshore-flowing currents and be entrained as part of a shelf-slope exchange via canyons as was hypothesized for the Chukchi shelf-Arctic Ocean interface. Presently available data do not allow us to distinguish among these potential pathways and fates.

In conclusion, the controls on biological productivity of the shelves of the Arctic Ocean and its marginal seas are complex, and depend on ice coverage, seasonal irradiance and nutrient inputs for primary production. Ice-algal production is most important in regions of higher ice concentrations and longer durations of coverage. Zooplankton populations may be important as conduits of organic matter to higher trophic levels, either in the water column or benthos, although they have a minor role in regulating the overall standing stocks of phytoplankton. Pelagic-benthic coupling is more important on the shallow, productive continental shelves, resulting in high benthic standing stocks, with water column processes and associated recycling pathways dominating in the deeper Arctic shelf regions. Further studies into the different ecosystems of the Arctic will assist in our overall understanding of the complexities of biological interactions in polar regions.

Acknowledgments. We would like to thank S. Henrichs, P. Matrai, L. Cooper and T.H. Peng for constructive comments on an earlier version of the manuscript. S. Polk and M.L. Garrett provided technical assistance. Financial support during preparation of this paper was provided by the National Science Foundation to J.M. Grebmeier (OPP-9300694) and W.O. Smith, Jr. (DPP-9113754). R.J. Conover would like to thank his staff of the Polar Continental Shelf Project for logistical and technical support during years of Arctic research. Publication No. 4341, Environmental Sciences Division, Oak Ridge National Laboratory.

References

Alexander, V., and H.J. Niebauer, Oceanography of the eastern Bering Sea ice-edge zone in spring, *Limnol. Oceanogr.*, *26*, 1111-1125, 1981.

Ambrose, W.G., and P.E. Renaud, Collaborative Research in the Northeast Water Polynya: Soft-bottom community structure and function, *J. Geophys. Res.*, *100*, 4411-4423, 1995.

Andersen, P., The quantitative importance of the "microbial loop" in the marine pelagic: a case study for the North Bering/Chukchi Sea, *Arch. Hydrobiol. Beih.*, *31*, 243-251, 1988.

Anderson, L.G., P.O.J. Hall, A. Iverfeldt, M.M. Rutgers van der Loeff, B. Sundby, and S.F.G. Westerlund, Benthic respiration measured by total carbonate production, *Limnol. Oceanogr.*, *31*, 319-329, 1986.

Anning, J.L., The development and decline of the epontic algal bloom community in Barrow Strait, N.W.T., M.S. thesis, University of Guelph, Ontario, 86 pp., 1989.

Ashjian, C.J., S.L. Smith, and P.V.Z. Lane, The Northeast Water polynya during summer 1992: secondary production and the distribution of crustacean zooplankton (Copepods), *J. Geophys. Res.*, *100*, 4371-4388, 1995.

Atkinson, E.G., and J.W. Wacasey, Sedimentation in Arctic Canada: particulate organic carbon flux to a shallow marine benthic community in Frobisher Bay, *Polar Biol.*, *8*, 3-7, 1987.

Bedo, A.W., E.J.H. Head, R.J. Conover, E.P.W. Horne, and L.R. Harris, Physiological adaptations of an under-ice population of *Pseudocalanus* in Barrow Strait (N.W.T.) to increasing food supply in spring, *Polar Biol.*, *10*, 561-570, 1990.

Bergmann, M.A., H.E. Welch, J. Walker-Butler, and T.D. Siferd, Ice algal photosynthesis at Resolute and Saqvaqjuac in the Canadian arctic, *J. Mar. Syst.*, *2*, 43-52, 1991.

Bourke, R.H., J.L. Newton, R.G. Paquette, and M.D. Tunnicliffe, Circulation and water masses of the East Greenland Shelf, *J. Geophys. Res.*, *92*, 6729-6740, 1987.

Bradstreet, M.S.W., and W.E. Cross, Trophic relationships at high arctic ice edges, *Arctic, 35*, 1-12, 1982.

Carey, A.G., Jr., Particle flux beneath fast ice in the shallow Beaufort Sea, Arctic Ocean, *Mar. Ecol. Prog. Ser.*, *40*, 247–257, 1987.

Coachman, L.K., K. Aagaard, and R.B. Tripp, *Bering Strait: The Regional Physical Oceanography*, Univ. of Wash. Press, Seattle, 172 pp., 1975.

Cochran, K.C., C. Barnes, D. Achman, and D. Hirschberg, ^{234}Th/^{238}U disequalibrium: an indicator of scavenging rates and carbon fluxes in the Northeast Water Polynya, Greenland, *J. Geophys. Res.*, *100*, 4399-4410, 1995.

Conover, R.J., Food relations and nutrition of zooplankton, Proceedings of Symposium on Experimental Marine Ecology, Occasional Publication No. 2, p. 81-91, Univ. Rhode Island, Kingston, R.I., 1964.

Conover, R. J., Transformation of organic matter, in *Marine Ecology, Vol. IV.* Dynamics, edited by O. Kinne, pp. 221-499, John Wiley & Sons, Chichester, U.K., 1978.

Conover, R.J., and G.F. Cota, Balance experiments with arctic zooplankton, in *Marine Biology of Polar Regions and Effects of Stress on Marine Organisms*, edited by J.S. Gray, M.E. Christiansen, pp. 217-236, John Wiley & Sons, Chichester, U.K., 1985.

Conover, R.J., and M. Huntley, Copepods in ice-covered seas distribution-adaptations to seasonally limited food, metabolism, growth patterns and life cycle strategies in polar seas, *J. Mar. Syst.*, *2*, 1-42, 1991.

Conover, R.J., and T.D. Siferd, Dark-season survival strategies of coastal-zone zooplankton in the Canadian arctic, *Arctic*, *46*, 303-311, 1993.

Conover, R.J., A.W. Herman, S.J. Prinsenberg, and L.R. Harris, Distribution of and feeding by the copepod *Pseudocalanus* under fast ice during the arctic spring, *Science*, *232*, 1245-1247, 1986.

Conover, R.J., A.W. Bedo, A.W. Herman, E.J.H. Head, L.R. Harris, and E.P.W. Horne, Never trust a copepod-some observations on their behavior in the Canadian arctic, *Bull. Mar. Sci.*, *43*, 650-662, 1988.

Conover, R.J., G.F. Cota, W.G. Harrison, E.P.W. Horne, and R.E.H. Smith, Ice/water interactions and their effect on biological oceanography in the Arctic Archipelago, in *Canada's Missing Dimension-Science and history in the Canadian Arctic Islands*, Vol. I, edited by C.H. Harrington, pp. 204-228, Canadian Musuem of Nature, Ottawa, Canada, 1990.

Conover, R.J., L.R. Harris, and A.W. Bedo, Copepods in cold oligotrophic waters: how do they cope? Proceedings of the Fourth International Conference on Copepoda, *Bull. Plankton Soc. Japan, Spec. Vol.*, pp. 177-199, 1991.

Cooney, R.T., Bering Sea zooplankton and micronekton communities with emphasis on annual production, in *The Eastern Bering Sea Shelf: Oceanography and Resources, Vol. 2*, edited by D.W. Hood and J.A. Calder, pp. 947-974, Univ. Wash. Press, Seattle, 1981.

Cooney, R.T., and K.O. Coyle, Trophic implications of cross shelf copepod distributions in the southeastern Bering Sea, *Mar. Biol.*, *70*, 187-196, 1982.

Cota, G.F., S.J. Prinsenberg, E.B. Bennett, J.W. Loder, M.R. Lewis, J.L. Anning, N.H.F. Watson, and L.R. Harris, Nutrient fluxes during extended blooms of arctic ice algae, *J. Geophys. Res.*, *92*, 1951-1962, 1987.

Coyle, K.O., and R.T. Cooney, Estimating carbon fluxes to pelagic grazers in the ice-edge zone of the eastern Bering Sea, *Mar. Biol.*, *95*, 299-30, 1988.

Cross, W.E., Under-ice biota at the Pond Inlet ice edge and in adjacent fast ice areas during spring, *Arctic*, *35*, 13-27, 1982.

DeMaster, D.J., R.B. Dunbar, L.I. Gordon, A.R. Leventer, J.M. Morrison, D.M. Nelson, C.A. Nittrouer, and W.O. Smith, Jr., Cycling and accumulation of biogenic silica and organic matter in high-latitude environments: the Ross Sea, *Oceanography*, *5*, 146-153, 1992.

de Vernal, A., J. Guiot, and J.-L. Turon, Late and postglacial terrestrial palynological evidence, in *Geographie physique et Quaternaire*, *47*, 167-180, 1993.

Dirschl, H.J., Preliminary Data Atlas. Lancaster Sound Regional Study, Indian and Northern Affairs Canada, 83 maps, 1980.

Dunbar, M.J., Physical causes and biological significance of polynyas and other open water in sea ice, in *Polynyas in the Canadian Arctic*, edited by I.A. Stirling and H. Cleator, Can. Wildl. Serv., Ottawa, Occassional Paper 45, pp. 29-43, 1981.

Feder, H.M., A.S. Naidu, J.M. Hameedi, S.C. Jewett, and W.R. Johnson, The Chukchi Sea continental shelf: benthos environmental interactions, *Mar. Ecol. Prog. Ser.*, *111*, 171-190, 1994.

Frost, B., Grazing control of phytoplankton stocks in the open subarctic Pacific Ocean: a model assessing the role of mesozooplankton, particularly the large calanoid copepods, *Neocalanus* spp., *Mar. Ecol. Prog. Ser.* , *39*, 49-68, 1987.

Fukuchi, M., H. Sasaki, H. Hattori, O. Matsuda, A. Tonimura, N. Handa, and C.P. McRoy, Temporal variability of particulate flux in the northern Bering Sea, *Cont. Shelf Res.*, *13*, 693-704, 1993.

Gosselin, M., L. Legendre, S. Demers, and R.G. Ingram, Responses of sea-ice microalgae to climate and fortnightly tidal energy inputs (Manitounuk Sound, Hudson Bay), *Can. J. Fish. Aquat. Sci.*, *42*, 999-1006, 1985.

Grainger, E.H., Exploitation of arctic sea ice by epibenthic copepods, *Mar. Ecol. Prog. Ser.*, *77*, 119-124, 1991.

Grainger, E.H., and S.I.C. Hsiao, Trophic relationships of the sea ice meiofauna in Frobisher Bay, arctic Canada, *Polar Biol.*, *10*, 283-292, 1990.

Grebmeier, J.M., Studies of pelagic-benthic coupling extended onto the Soviet continental shelf in the northern Bering and Chukchi Seas, *Cont. Shelf Res.*, *13*, 653-668, 1993.

Grebmeier, J.M., and C.P. McRoy, Pelagic-benthic coupling on the shelf of the northern Bering and Chukchi Seas. III. Benthic food supply and carbon cycling, *Mar. Ecol. Prog. Ser.*, *53*, 79-91, 1989.

Grebmeier, J.M., and J. P. Barry, The influence of oceanographic processes on pelagic-benthic coupling in polar regions: A benthic perspective, *J. Mar. Syst.*, 2, 495-518, 1991.

Grebmeier, J.M., and L.W. Cooper, Influence of the St. Lawrence Island Polynya upon the Bering Sea benthos, *J. Geophys. Res.*, *100*, 4439-4460, 1995.

Grebmeier, J.M, C.P. McRoy, and H.M. Feder, Pelagic-benthic coupling on the shelf of the northern Bering and Chukchi Seas. I. Food supply source and benthic biomass, *Mar. Ecol. Prog. Ser.*, *48*, 57-67, 1988.

Grebmeier, J.M., H.M. Feder, and C.P. McRoy, Pelagic-benthic coupling on the shelf of the northern Bering and Chukchi Seas. II. Benthic community structure, *Mar. Ecol. Prog. Ser.*, *51*, 253-268, 1989.

Grebmeier, J.M., L.W. Cooper, and M.J. DeNiro, Oxygen isotope composition of bottom seawater and tunicate cellulose used as indicators of water masses in the northern Bering and Chukchi Seas, *Limnol. Oceanogr.*, *35*, 1182-1195, 1990.

Haflinger, K., A survey of benthic infaunal communities of the southeastern Bering Sea, in *The Eastern Bering Sea Shelf: Oceanography and Resources, Vol. 2*, edited by D.W. Hood and J.A. Calder, Univ. Wash. Press, Seattle, pp. 1091-1104, 1981.

Hameedi, J.M., Aspects of water column primary productivity in the Chukchi Sea during summer, *Mar. Biol.*, *48*, 37-46, 1978.

Hansell, D.A., T.E. Whitledge, and J.J. Goering, Patterns of nitrate utilization and new production over the Bering-Chukchi shelf, *Cont. Shelf Res.*, *13*, 601-627, 1993.

Hargrave, B.T., B. von Bodungen, R.J. Conover, A.J. Fraser, G. Phillips, and W.P. Vass, Seasonal changes in sedimentation of particulate matter and lipid content of zooplankton collected by sediment trap in the Arctic Ocean off Axel Heiberg Island, *Polar Biol.*, *9*, 467-475, 1989.

Harris, L.R., The importance of ice algae as an early season food source for arctic pelagic copepods. M.S. thesis, Department of Biology, Dalhousie University, Halifax, 98 pp., 1992.

Harrison, W.G., and G.F. Cota, Primary production in polar waters: relation to nutrient availability, *Polar Res.*, *10*, 87-104, 1991.

Henriksen, K., T.H. Blackburn, B.A. Lomstein, and C.P. McRoy, Rates of nitrification, distribution of nitrifying bacteria and inorganic N fluxes in northern Bering-Chukchi shelf sediments, *Cont. Shelf. Res.*, *13*, 629-651, 1993.

Highsmith, R.C., and K.O. Coyle, Productivity of arctic amphipods relative to gray whale energy requirements, *Mar. Ecol. Prog. Ser.*, *83*, 141-150, 1992.

Hill, P.S., Reconciling aggregation theory with observed vertical fluxes following phytoplankton blooms, *J. Geophys. Res.*, *97*, 2295-2308, 1992.

Høpner Petersen, G., and M.A. Curtis, Differences in energy flow through major components of subarctic, temperate and tropical marine shelf ecosystems, *Dana*, *1*, 53-64, 1980.

Horner, R., S.F. Ackley, G.S. Dieckmann, B. Gulliksen, T. Hoshiai, L. Legendre, I.A. Melnikov, W.S. Reeburgh, M. Spindler, and C.W. Sullivan, Ecology of sea ice biota, 1. Habitat, terminology, and methodology, *Polar Biol.*, *12*, 417-427, 1992.

Jewett, S.C., and H.M. Feder, Autumn food of adult starry flounder *Platichthys stellatus* from the NE Bering Sea and SE Chukchi Sea, *J. Cons. Int. Explor. Mer.*, 39, 7-14, 1980.

Johnson, W.R., Current response to wind in the Chukchi Sea: A regional coastal upwelling event, *J. Geophys. Res.*, *94*, 2057-2064, 1989.

Johnson, M., and H.J. Niebauer, The 1992 summer circulation in the Northeast Water Polynya from acoustic doppler current profiler measurements, *J. Geophys. Res.*, *100*, 4301-4307, 1995.

Jones, E.P., and L.G. Anderson, On the origin of the chemical properties of the Arctic Ocean halocline, *J. Geophys. Res.*, *91*, 10759-10767, 1986.

Karl, D.M., B.D. Tilbrook, and G. Tien, Seasonal coupling of organic matter production and particle flux in the western Bransfield Strait, Antarctica, *Deep-Sea Res.*, *38*, 1097-1126, 1991.

Kosobokova, K.N., Age-related and seasonal changes in the biochemical makeup of the copepod *Calanus glacialis* as related to the characteristics of its life cycle in the White Sea, *Oceanology*, *30*, 103-109, 1990.

LeDrew, E., D. Barber, T. Agnew, and D. Dunlop, Canadian sea ice atlas from microwave remotely sensed imagery: July 1987 to June 1990, *Climatological Studies*, No. 44, 80 pp. Canada Communications Group-Publishing, Ottawa, 1992.

Lee, R.F., Lipid composition of the copepod *Calanus hyperboreas* from the Arctic Ocean. Changes with depth and season, *Mar. Biol.*, *26*, 313-318, 1974.

Legendre, L., S.F. Ackley, G.S. Dieckmann, Bjorn Gulliksen, R. Horner, T. Hoshiai, I.A. Melnikov, W.S. Reeburgh, M. Spindler, and C.W. Sullivan, Ecology of sea ice biota, 2. Global significance, *Polar Biol.*, *12*, 429-444, 1992.

McRoy, C.P., ISHTAR, the project: an overview of Inner Shelf Transfer and Recycling in the Bering and Chukchi seas, *Cont. Shelf. Res.*, *13*, 473-479, 1993.

Michel, C., L. Legendre, J.-C. Therriault, S. Demers, and T. Vandevelde, Springtime coupling between ice algal and phytoplankton assemblages in southeastern Hudson Bay, Canadian arctic, *Polar Biol.*, *13*, 441-449, 1993.

Müller-Karger, F., and V. Alexander, Nitrogen dynamics in a marginal sea ice zone, *Cont. Shelf. Res.*, *7*, 805-823, 1987.

Niebauer, H.J., Wind and melt driven circulation in a marginal sea ice edge frontal system: a numerical model, *Cont. Shelf. Res.*, *1*, 49-98, 1982.

Niebauer, H.J., Multiyear sea ice variability in the eastern Bering Sea: an update, *J. Geophys. Res.*, *88*, 2733-2742, 1983.

Niebauer, H.J., Wind and melt driven circulation in a marginal sea ice edge frontal system: a numerical model, *Cont. Shelf Res.*, *1*, 49-98, 1991.

Niebauer, H.J., and V. Alexander, Oceanographic frontal structure and biological production at an ice edge, *Cont. Shelf Res.*, *4*, 367-388, 1985.

Niebauer, H.J., and W.O. Smith, Jr., A numerical model of mesoscale physical-biological interactions in the Fram Strait marginal ice zone, *J. Geophys. Res.*, *94*, 16151-16175, 1989.

Nieman, A.A., Quantitative distribution of benthos on the shelf and upper continental slope in the eastern part of the Bering Sea, in *Soviet Fisheries Investigations in the Northeast Pacific, Part 1*, edited by P.A. Moiseev, pp. 143-217, 1963 (Israel Program for Scientific Translations, 1968).

Parkinson, C.L., J.C. Comiso, H.J. Zwally, D.J. Cavalieri, P. Gloersen, and W.J. Campbell, Arctic Sea ice, 1973-1976: satellite passive-microwave observations, *NASA Sp. Publ.*, *489*, Washington, DC, 296 pp., 1987.

Parsons, T.R., and C.M. Lalli, Comparative oceanic ecology of the plankton communities of the subarctic Atlantic and Pacific Oceans, *Oceanogr. Mar. Biol. Ann. Rev.*, *26*, 317-359, 1988.

Parrish, D.M., An estimate of annual primary production in the Alaska Arctic Ocean, M.S. thesis, Univ. Alaska, Fairbanks, Alaska, 166 pp., 1987.

Pfannkuche, O., and H. Thiel, Meiobenthic stocks and benthic activity on the NE-Svalbard shelf and in the Nansen Basin, *Polar Biol.*, *7*, 253-266, 1987.

Piepenburg, D., On the composition of the benthic fauna of the western Fram Strait, *Ber. Polarforsch.*, *52*, 1-118, 1988.

Pike, D.O., The distribution and abundance of sub-ice macrofauna in the Barrow Strait area, N.W.T., M.S. thesis, 96 pp., Univ. Manitoba, Winnipeg, Manitoba, 1987.

Pomeroy, L.R., and D. Deibel, Temperature regulation of bacterial activity during the spring bloom in Newfoundland coastal waters, *Science, 233*, 359-361, 1986.

Prinsenberg, S.J., and E.B. Bennett, Mixing and transports in Barrow Strait, the central part of the Northwest Passage, *Cont. Shelf Res., 7*, 913-935, 1987.

Rowe, G.T., and W.C. Phoel, Nutrient regeneration and oxygen demand in Bering Sea continental shelf sediments, *Cont. Shelf. Res., 12*, 439-449, 1991.

Rowe, G., G. Boland, L. Cruz-K., A. Newton, E. Escobar, and W. Ambrose, Sediment oxygen demand in the NE Greenland Polynya (abstract), Third Scientific Meeting, Oceanography Society, April, p. 122, 1993.

Runge, J.A., and R.G. Ingram, Under-ice feeding and diel migration by the planktonic copepods *Calanus glacialis* and *Pseudocalanus minutus* in relation to the ice algal production cycle in southeastern Hudson Bay, Canada, *Mar. Biol., 108*, 217-225, 1991.

Sambrotto, R.N., J.J. Goering, and C.P. McRoy, Large yearly production of phytoplankton in the western Bering Strait, *Science, 225*, 1147-1150, 1984.

Shuert, P.G., and J.J. Walsh, A coupled physical-biological model of the Bering/Chukchi Seas, *Cont. Shelf Res., 13*, 1993.

Sirenko, B.I., and V.M. Koltun, Characteristics of benthic biocenoses of the Chukchi and Bering Seas, in *Results of the Third Joint U.S.-U.S.S.R. Bering and Chukchi Sea Expedition (BERPAC)*, edited by P.A. Nagel, pp. 251-261, U.S. Fish and Wildlife Service, Washington, DC, 1992.

Smith, R.E.H., J. Anning, P. Clement, and G. Cota, Abundance and production of ice algae in Resolute Passage, Canadian arctic, *Mar. Ecol. Prog. Ser., 48*, 251-263, 1988.

Smith, W.O., Jr., Primary productivity of a *Phaeocystis* bloom in the Greenland Sea during spring, *Proceedings of the Nansen Symposium*, edited by J. Overland and R. Muench, American Geophysical Union, Washington, DC, in press, 1994.

Smith, W.O., Jr., and E. Sakshaug, Polar phytoplankton, in *Polar Oceanography*, edited by W.O. Smith, Jr., pp. 477-525, Academic Press, San Diego, 1990.

Smith, W.O., Jr., L.A. Codispoti, D.M. Nelson, T. Manley, E.J. Buskey, H.J. Niebauer, and G.F. Cota, Importance of *Phaeocystis* blooms in the high-latitude ocean carbon cycle, *Nature, 352*, 514-516, 1991.

Smith, W.O., Jr., I.D. Walsh, B.C. Booth and J.W. Deming, Particulate matter and phytoplankton biomass distributions in the Northeast Water Polynya during summer, 1992, *J. Geophys. Res., 100*, , 4341-4356, 1995.

Springer, A.M., C.P. McRoy, and K.R. Turco, The paradox of pelagic food webs in the northern Bering Sea-II. Zooplankton communities, *Cont. Shelf Res., 9*, 359-386, 1989.

Springer, A.M., and C.P. McRoy, The paradox of pelagic food webs in the northern Bering Sea-III. Patterns of primary production, *Cont. Shelf Res., 13*, 575-599, 1993.

Stirling, I., and H. Cleator, Polynyas in the Canadian arctic, *Occasional Paper No. 45*, 72 pp., Canadian Wildlife Service, 1981.

Stoker, S.W., Benthic invertebrate macrofauna of the eastern Bering/Chukchi continental shelf, in *The Eastern Bering Sea Shelf: Oceanography and Resources, Vol. 2*, edited by D.W. Hood and J.A. Calder, pp. 1069-1090, Univ. Wash. Press, Seattle, 1981.

Subba Rao, D.V., and T. Platt, Primary production of Arctic waters, *Polar. Biol., 3*, 191-201, 1984.

Thomson, D.H., Marine benthos in the Eastern Canadian High Arctic: multivariate analysis of standing crop and community structure, *Arctic, 35*, 61-74, 1982.

Tremblay, C., J.A. Runge, and L. Legendre, Grazing and sedimentation of ice algae during and immediately after a bloom at the ice water interface, *Mar. Ecol. Prog. Ser., 56*, 291-300, 1989.

Vidal, J., and S.L. Smith, Biomass, growth, and development of populations of herbivorous zooplankton in the southeastern Bering Sea during spring, *Deep-Sea Res., 33*, 523-556, 1986.

Wallace, D.R., P.J. Minnett, and T.S. Hopkins, Oxygen generation and new production in the Northeast Water Polynya in summer, *J. Geophys. Res.*, 100, 4423-4438, 1995.

Walsh, J.J., and C.P. McRoy, Ecosystem analysis in the southeastern Bering Sea, *Cont. Shelf Res.*, 5, 259-288, 1986.

Walsh, J.J., C.P. McRoy, L.K. Coachman, J.J. Goering, J.J. Nihoul, T.E. Whitledge, T.H. Blackburn, P.L. Parker, C.D. Wirick, P.G. Shuert, J.M. Grebmeier, A.M. Springer, R.D. Tripp, D.A. Hansell, S. Djenidi, R. Deleersnijder, K. Henriksen, B.A. Lund, P. Andersen, F.E. Müller-Karger, and K. Dean, Carbon and nitrogen cycling within the Bering/Chukchi Seas: source regions for organic matter affecting AOU demands of the Arctic Ocean, *Prog. Oceanog.*, 22, 277-359, 1989.

Wassman, P., R. Peinert, and V. Smetacek, Patterns of production and sedimentation in the boreal and polar Northeast Atlantic, *Polar Res.*, 10, 209-228, 1991.

Welch, H.E., M.A. Bergmann, T.D. Siferd, K.A. Martin, M.A. Curtis, R.E. Crawford, R.J. Conover, and H. Hop, Energy flow through the marine ecosystem of the Lancaster sound region, arctic Canada, *Arctic*, 45, 343-357, 1992.

Zenkevitch, L., *Biology of the Seas of the U.S.S.R.*, 955 pp., Allan and Unwin, London, 1963.

8

Resolved: The Arctic Controls Global Climate Change

Richard B. Alley

Abstract

Arguments are presented here for the affirmative side of the proposition that the Arctic controls global climate change over lower frequencies than the ENSO signal. Paleoclimatic records of the most recent million years show strong variability in the Arctic, and nearly in-phase variability of similar or smaller magnitude elsewhere. The timing of climate variability relative to changes in the seasonality and strength of insolation reaching the Earth (Milankovitch forcing) shows that much of the global response is controlled by conditions at high northern latitudes. Physical modeling of this system requires some important climatic element with a slow time constant; Arctic or subarctic continental ice sheets are the most likely candidates. Higher-frequency (Heinrich/Bond and Dansgaard/Oeschger) climate oscillations are strongest in the North Atlantic region but appear elsewhere. Internal oscillations of ice sheets and of the North Atlantic ocean are the leading hypotheses for controlling these higher-frequency oscillations. The global climate system is probably linked to Arctic forcing and oscillations through deep-water formation in the North Atlantic and its effects on global atmospheric circulation, sea ice, carbon dioxide, methane and dust.

Introduction

At first glance, the affirmative side of the proposition in the title might seem absurd. After all, the tropics cover about five times the area of the polar regions. Even if we stretch the Arctic to include the subarctic North Atlantic and the former Northern Hemisphere ice sheets (which I intend to do), the tropics dominate. Also, the tropics are energy-rich— compared to the poles, a temperature change in the tropics has a much larger effect on the energy of the troposphere through latent heat terms and through the greater thickness of the

Arctic Oceanography: Marginal Ice Zones and Continental Shelves
Coastal and Estuarine Studies, Volume 49, Pages 263–283
Copyright 1995 by the American Geophysical Union

troposphere in the tropics (e.g. Peixoto and Oort, 1992, ch. 13). Furthermore, the tropics play a critical role in the few-year El Niño/Southern Oscillation (ENSO) variability.

Such "obvious" physical arguments are at the root of much of the global-change research strategy for the U.S. and the world, as embodied in such major climatic programs as Tropical Oceans Global Atmosphere (TOGA) or the World Ocean Circulation Experiment (WOCE), which focus primarily on the warmer parts of the globe. Research in the polar regions then relies on the argument that the Earth is an interlocked system, so we need to study everything, even if we "know" that the real action is somewhere else. The probability of polar amplification of climate changes owing to ice-albedo feedback or other processes also is of interest (Hansen et al., 1984; Mitchell et al., 1990).

However, many of those who study the polar regions are evaluating a different idea: that the polar regions, and especially the North Atlantic basin, actually control the major climate changes. This idea is especially attractive among those paleoclimatologists who have examined the records of climate change over the last glacial cycles.

Here, I summarize evidence of Arctic influence on global climate, and I argue the affirmative for the proposition in the title. I focus on changes over the most recent million years and especially over the most recent glacial cycle, although I suspect that similar arguments could be made for the onset of the late Cenozoic ice ages. I then briefly discuss possible mechanisms by which the Arctic could control the larger, more energy-rich mid-latitudes and tropics.

Milankovitch Cycles

Long, high-resolution records of climate change typically show strong variability with periods of about 100,000 years, 41,000 years, and 19,000/23,000 years (variability in a band with power concentrated at 19,000 and 23,000 years; the 23,000-year signal dominates so I will refer only to it hereafter). This is true of many records from many regions: deep and shallow oceans, tropical and polar oceans, and various polar and non-polar terrestrial sites (see the major synthesis papers by Broecker and Denton, 1989; and Imbrie et al., 1992; 1993).

These periodicities correspond, within dating accuracy, to periodicities for variation in incoming solar radiation (insolation) reaching different regions of the Earth at different seasons. The 23,000-year periodicity is the variation in the time of occurrence of seasons relative to the distance of the Earth from the sun as the Earth traverses its elliptical orbit. Today, the Earth is closest to the sun (perihelion) during northern hemisphere winter and farthest from the sun during northern hemisphere summer, giving cool summers and warm winters in the north but warm summers and cool winters in the south; however, about 10,000 years ago, perihelion occurred during northern-hemisphere summer, giving stronger northern seasonality but weaker southern seasonality. This orbital variation is called precession.

The 41,000 year periodicity is the variation in the tilt of the Earth, or obliquity—the angle between the rotational axis of the Earth and the plane containing the Earth's orbit. When the obliquity is high, both poles receive much sunshine and seasonality is stronger than when the obliquity is low. The 100,000-year periodicity may be related to the variation in the ellipticity or eccentricity of the Earth's orbit, as it changes from being more-nearly circular to more elongated and back. The main climatic effect of changes in the eccentricity is to modulate the precessional cycle; the timing of perihelion has little effect on seasonal insolation if the orbit is nearly circular but has much effect if the orbit is strongly elongated.

A difficulty is that the climate records typically show small signals at 23,000 years and 41,000 years, and a large signal at 100,000 years, whereas the insolation varies significantly at 23,000 years and 41,000 years but trivially at 100,000 years. The main questions then are how the insolation variations affect the Earth (directly, or by affecting some regions that then affect others), and why the 100,000 year climate signal is so large. Interestingly, the answers for both seem to involve the Arctic.

Milankovitch into the Climate System

When precession causes warm summers in the northern hemisphere, it also causes cool summers in the southern hemisphere (Broecker and Denton, 1989). However, the global system responds almost synchronously and equally to Milankovitch forcing (Broecker and Denton, 1989); the southern and northern hemispheres warm and cool at almost the same times, with leads and lags much shorter than the periodicities (Imbrie et al., 1992; 1993).

Many mechanisms have been suggested that may contribute to the in-phase behavior of the hemispheres. Colder times tend to have higher global albedo through the effects of increased snow and ice (e.g. Hansen et al., 1984), greater atmospheric loading of sunlight-blocking dust because of increased aridity or wind speed (e.g. Harvey, 1988), and lower concentrations of atmospheric greenhouse gases (e.g. Barnola et al. 1987; Raynaud et al., 1988). Planetary waves may be altered owing to changes in ice-sheet topography (e.g. Manabe and Broccoli, 1985).

However, these may not be sufficient by themselves to synchronize the hemispheres fully. GCM simulations (e.g. Manabe and Broccoli, 1985) show that changes in the atmosphere of one hemisphere, such as might be caused by enhanced snowcover, tropospheric dustiness, or topographically forced changes in planetary waves in one hemisphere, do not propagate efficiently across the equator to the other hemisphere through the troposphere. Methane is controlled largely by terrestrial sources in the northern hemisphere; however, it is not a sufficiently potent greenhouse gas in preanthropogenic concentrations to have a controlling effect on global temperatures (e.g. Shine et al., 1990). CO_2 almost certainly plays an important part (e.g. Lorius et al., 1990), but because it is controlled primarily by the oceans, I consider it with the oceans, below.

Regardless of the mechanism by which the temperature variations are caused to be in-phase between the hemispheres, the data show clearly that global temperature is linked to insolation at high northern latitude, with warming occurring when summers are warm and winters cold there, and cooling occurring with cool summers/warm winters there (Imbrie et al., 1992; 1993). The relative response of the climate to the obliquity and precession cycles shows the importance of the northern hemisphere. The effect of obliquity on insolation in a season (such as summertime in the northern hemisphere and summertime in the southern hemisphere) is in-phase between the hemispheres, whereas the effect of precession is out-of-phase between the hemispheres. Global warming is observed to occur when insolation is large in northern-hemisphere summers, whether insolation is large or small in southern hemisphere summers (Imbrie et al., 1992).

The northern high latitudes are believed to be more important than the tropics based on a better match between climate records and calculated histories of high-latitude insolation compared to low-latitude insolation. In addition, if the tropics played a dominant role in climate change, one would expect a significant cycle at about 10,000 years. During a precession cycle, perihelion moves from northern hemisphere summer across the equator to southern hemisphere summer and back across the equator to northern hemisphere summer, so that strong seasonality occurs in one hemisphere or the other twice per precession cycle, or roughly once per 10,000 years, in a split-precession cycle. This split-precession cycle is not strong in most climate records. Certainly, many records lack the resolution to characterize it, and some workers truncate data for periods as short as 10,000 years when showing power spectra. However, a recent search for this split-precession cycle in high-resolution climate records showed that it is present, and thus that the records can preserve it, but that it is quite weak (Hagelberg et al., 1994).

Thus, the first argument for the importance of the Arctic in global climate change is that the comparison of paleoclimatic records and orbital insolation forcing indicates that the climatic variability with 23,000-year and 41,000-year periods is driven by insolation at high northern latitudes, with reduced insolation during summers there linked to global cooling, and increased insolation during summers there linked to global warming.

The 100,000-year Cycle

The climate responses to the two shorter Milankovitch periodicities (23,000 and 41,000 years) can be treated as linear functions—larger amplitude forcing produces larger amplitude response, proportional to the forcing (Imbrie et al., 1992). The 100,000-year cycle cannot be modeled in this way, however, as this largest climate variability in the Milankovitch band is produced by virtually no forcing.

Imbrie et al. (1993) identify seven classes of models that attempt to explain the 100,000-year cycle. Approaches taken have ranged from statistical through various levels of increasingly realistic physical representations of the system, and have emphasized different aspects of the system.

As described above, the 100,000-year eccentricity cycle serves mainly to modulate the precessional cycle—the timing of perihelion affects seasonal insolation significantly when the Earth's orbit is highly eccentric but insignificantly when the orbit is nearly circular. The amplitude of the precession cycle thus varies with a 100,000-year periodicity, causing the power spectrum of the envelope of insolation to show a clear peak at 100,000 years. Attempts to explain the 100,000-year cycle of climate records in this way are confounded, however, by the observation that this envelope exhibits other peaks, including an equally large one at about 400,000 years, that are absent or only weakly present in most climate records (Imbrie et al., 1993).

It proves relatively easy to model the 100,000-year cycle in climate records from the calculated Milankovitch forcing if the climate system includes an important and moderately "slow" element—something that responds to climatic forcing over 10,000-20,000 years. The 100,000-year cycle in climate records appears strongly asymmetric, with slow onset of ice-age conditions followed by more-rapid termination of ice-age conditions (e.g. Broecker and Denton, 1989), and can be modeled accurately if the slow-response element in the climate system includes an asymmetry such that ice-age termination can be rapid if and only if some glaciation threshold is first exceeded. Any such slow element might suffice to explain the climate response, but the only physically likely candidates are linked to ice sheets, in particular the mid-latitude ice sheets such as the Laurentide in North America (e.g. Weertman, 1976; Pollard, 1982; 1983; Birchfield and Grumbine, 1985; DeBlonde and Peltier, 1991). Several possible mechanisms exist to give ice sheets an asymmetric response to climate forcing. Many of these are based on the fact that following some threshold, ice sheets may become physically much more dynamic than before crossing the threshold.

Ice-sheet build-up occurs at the slow rate of snowfall (order of 0.1 m/year in ice-sheet thickness); ice-sheet drawdown occurs at the faster rate of surface melting (order of 1 m/year) or the even faster rate of dynamic collapse (order of 10 m/year; MacAyeal, 1993a; 1993b). Ice-sheet growth occurs on isostatically adjusted bedrock that is high in the atmosphere; ice-sheet collapse occurs after isostatic depression from the weight of the ice has moved the ice sheet into a warmer atmosphere (faster melting) and given it marine or lacustrine margins (faster flow) (Weertman, 1974; Pollard, 1982; 1983; Peltier; 1987). Ice-sheet growth occurs when the ice is cold, frozen to its bed and deforming slowly; ice-sheet collapse occurs after geothermal heat has warmed that ice and thawed it from its bed, allowing very rapid flow (MacAyeal, 1993a; 1993b). Numerous positive feedbacks exist that tend to reinforce ongoing trends (Hughes, 1992). Thus, a large ice sheet may collapse dynamically and disappear more rapidly than a smaller one under similar geological and meteorological conditions, once the larger one passes certain thresholds leading to a thawed bed and perhaps to a marine margin (MacAyeal, 1993a; 1993b).

The time scale for a large ice sheet to adjust to changes in its environment or to changes forced by its own evolution is on the order of ten thousand years (Whillans, 1981). Oceanic, atmospheric and biospheric responses are faster than the ice-sheet time scale, and lithospheric changes such as mountain-building are slower. Ice sheets thus are the leading candidates as the mechanism by which the dominant 23,000- and 41,000-year periodicities in insolation are translated into the dominant 100,000-year climate response (Imbrie et al., 1993). And, based on their dramatic variation in size through ice-age cycles, the Arctic or Arctic-derived Laurentide and Fennoscandian ice sheets appear to occupy center stage. It remains unclear, however, whether direct response to the precession cycle contributes significantly to the observed climate records, or whether the 100,000-year cycle is produced entirely from the slow response of the ice sheets to the shorter-period forcings (Imbrie et al., 1993).

Shorter-Lived Events

Heinrich Events

In their simplest definition, Heinrich events are occasional features of the North Atlantic sedimentary record, consisting of layers of ice-rafted debris (Heinrich, 1988; Broecker et al., 1992). The layers are widespread, reaching across the North Atlantic in the broad belt of iceberg drift (Bond et al., 1992; Grousset et al., 1993). They thicken toward the northwest into the Labrador Sea. Cores from near the mouth of Hudson Strait have penetrated through Heinrich events 1, 2, and possibly 3 (and the Younger Dryas, which also appears to be a Heinrich-type event or Heinrich event 0; Andrews et al., 1993; 1994; Bond et al., 1993a) showing that the sediment from these events is thickest near the mouth of Hudson Strait (Andrews et al., 1993; 1994) and thins monotonically away to the east (Bond et al., 1992; Grousset et al., 1993). Sediments of the other Heinrich events also exhibit a pattern of being thicker to the west, although data are not available close to possible sources such as Hudson Bay or the St. Lawrence River (Grousset et al., 1993).

The sediments in the Heinrich layers have characteristics consistent with a Hudson Bay source, although other sources are almost certainly involved, and detailed fingerprinting is not yet complete (e.g. Broecker et al., 1992; Bond et al., 1992; 1993a; Grousset et al., 1993). A volume of sediments of roughly 4×10^{11} m^3 per Heinrich event is estimated, equivalent to a 10-cm-thick layer across the entire North Atlantic belt of iceberg drift (Alley and MacAyeal, 1994).

The layers contain very few contemporaneous marine fossils. Those pelagic microfossils that are present are of cold-water varieties, and have shells with isotopic compositions indicative of significant dilution by light ice-sheet melt-water (to perhaps 3% melt-water in the surface mixed layer; Bond et al., 1992).

Heinrich events represent greatly increased rates of ice-rafted-debris sedimentation, but the times of rapid sedimentation of ice-rafted debris may occur within longer-lasting periods of low oceanic productivity shown by reduced fluxes of foraminiferal deposition (Bond et al., 1992). The Heinrich layers are quite short-lived, lasting perhaps centuries to a millennium (Bond et al., 1992; Andrews et al., 1994).

The thickening of the most recent Heinrich layers towards Hudson Strait indicates that the Heinrich events are true increases in sedimentation in the ocean, rather than shifts in sedimentation from near-coastal to open-ocean sites through increased survival of icebergs. Based solely on the North Atlantic record, one cannot absolutely rule out shifts from sedimentation in Hudson Bay to sedimentation beyond Hudson Strait, however. Similar proximal data are not in hand for other possible sources, so the presence of ice-rafted debris from these sources in Heinrich layers or other Heinrich-type layers could indicate either increased discharge or increased survival of icebergs.

If we assume that Heinrich events indeed are increases in the rate of delivery of ice-rafted debris and of glacial melt-water to the North Atlantic, we are left asking what they might be. Several possibilities exist. Perhaps they are surges of the ice sheet in Hudson Bay through Hudson Strait driven by ice-sheet processes (MacAyeal, 1993a; 1993b; Alley and MacAyeal, 1994). In this view, non-Hudson Strait ice-rafted debris in Heinrich layers represents either increased survival of icebergs because the Hudson Strait surge cooled the North Atlantic (in which case one should see a change in locus of sedimentation between near-shore and off-shore for these other sources), or else rapid response of this other ice to forcing from the Hudson Strait surge (perhaps because of sea-level rise disturbing marine ice, or colder temperatures shifting the balance between surface melting and iceberg calving for small, fast-reacting ice masses in Iceland or other warmer regions). Rapid sea-level rise, on the order of meters in centuries, may have occurred during Heinrich events (MacAyeal, 1993b; Alley and MacAyeal, 1993).

Alternatively, Heinrich events might be externally forced, with the Hudson Strait and other ice responding to some external perturbation. The split-precession cycle of Milankovitch variation is one possible non-polar cause (Bond et al. 1993b; Hagelberg et al., 1994). The non-periodicity of the Heinrich layers rules out simple models of periodic external forcing, although interactions of several forcings with different periodicities or of periodic and non-periodic forcings remain possible. The difficulty of causing a large ice mass such as the Laurentide ice sheet in Hudson Bay to respond rapidly to external forcing also complicates external-forcing models of Heinrich events (Oerlemans, 1993). Most ice-sheet models produce slow response to external forcing, with response time scaling with ice-sheet size, so that an external forcing with a periodicity of a few thousand years would produce complicated responses at very different times from the different-sized ice masses around the North Atlantic. However, the possibility remains of a marine instability giving very rapid response to a small sea-level forcing from elsewhere (Weertman, 1974; Mercer, 1978).

The Heinrich events might be the break-off of floating ice shelves around the North Atlantic, shifting melt-water and ice-rafted-debris sedimentation from basin-marginal positions beneath ice shelves to basin-central positions beneath freely floating icebergs (I. Whillans, pers. comm., 1994). Certainly, ice shelves in the modern world can calve large pieces

(Jacobs et al., 1992) or even disintegrate entirely in short times (Doake and Vaughan, 1991; Vaughan, 1993). However, the observation that the Heinrich events involve enhanced sedimentation near the mouth of Hudson Strait requires that any large ice shelf there would have been inside Hudson Bay at times when grounded ice is believed to have occupied the Bay, and that the icebergs discharged rapidly through the narrow mouth of the Bay, possibly a mechanically difficult task.

Also, modern large ice shelves seem to be characterized by basal melting near their grounding lines (Jacobs et al., 1992), producing ice containing little debris. Were such an ice shelf to break free, it would produce melt-water with little ice-rafted debris. However, break-off of such an ice shelf followed by surging of the grounded ice behind would produce melt-water and a cold ocean followed by ice-rafted debris, in agreement with some observations. The MacAyeal model for surges can produce melt-water without debris following an ice-rafted-debris event if certain dynamical criteria are satisfied (Alley and MacAyeal, 1994), so this model allows for an ice-rafted-debris peak to occur embedded in a longer-lasting interval of a cold, melt-water-laden mixed layer.

The most likely story, in my opinion, is that the Heinrich events represent surges of the ice sheet in Hudson Bay, caused by internal dynamics of that ice rather than by external forcing. If so, then these features require Arctic participation (the Laurentide ice sheet) and are triggered by Arctic processes (also the ice sheet). Much work remains to confirm this, however.

Significance of Heinrich Events

It then remains to ask whether the Heinrich events are coupled to globally significant climate changes, or whether the recent flurry of papers about them arises from a North-Atlantic-centered point of view. Here, the jury is still out, but the evidence is pointing toward a worldwide significance for these events (Broecker, 1994). The first piece of evidence is that the Younger Dryas event appears to have the sedimentary signature of a Heinrich event (Bond et al., 1993a; Andrews et al., 1994). The Younger Dryas is the best-known and best-dated rapid-climate-change event. Questions still remain about its occurrence and magnitude in some regions, but it is now clear that the Younger Dryas affected regions scattered widely across the globe. Younger-Dryas age climate changes are known from western Canada (Mathewes, 1993; Reasoner et al., 1994), the Santa Barbara Basin off the western U.S. (Kennett and Ingram, 1994), the Sulu Sea (Linsley and Thunell, 1990), East Africa (Roberts et al., 1993), the South Island of New Zealand (Denton and Hendy, 1994), and elsewhere (see Kennett, 1990 and Peteet, 1993). If the Younger Dryas is a product of the Laurentide ice sheet in some way, then that signal is being propagated far from its source.

The search is on for the signature of older Heinrich events outside of the North Atlantic. At least a regional signal is suggested by correlation of Heinrich events with continental-dust peaks in Greenland ice cores, as the dust was derived from beyond Greenland (Mayewski et al., 1994). Clark (1994) found a match between timing of advances of the southern margin of the Laurentide ice sheet and Heinrich events 1 and 2. Grimm et al. (1993) report Heinrich events 1-5 in pollen records from Florida, which probably are responding to water availability. A true believer can find the Younger Dryas in their data as well. Clark and Bartlein (1994; in review) correlated fluctuations of glaciers in the western U.S. to Heinrich events. Various other possible correlations exist, although dating uncertainties remain (Bond et al., 1993a; 1993b; Broecker, 1994).

The Heinrich events seem to be related to a higher-frequency oscillation: the Dansgaard-Oeschger events. As seen in Greenland ice cores and in high-resolution North Atlantic ocean-sediment cores, the Dansgaard-Oeschger oscillations and the similar Younger Dryas oscillation are alternating cold and warm periods, lasting roughly a millennium each but with considerable variability, with a nearly square-wave appearance owing to changes of a significant fraction of the glacial/interglacial amplitude in as little as 1-3 years (Johnsen et al., 1992; Taylor et al., 1992; Alley et al., 1993). Synchronous changes are observed in methane (Chappellaz et al., 1993; Bender et al., 1993) and local and regional climatic factors such as stable-isotopic composition of precipitation and atmospheric loading of dust and sea salt (Mayewski et al., 1994).

Curve-matching, supported by absolute dating of records by ice-core layer-counting and sediment-core radiocarbon translated to a linear time scale using U-Th calibration, allows the Dansgaard-Oeschger oscillations to be identified in ocean and ice cores as synchronous or near-synchronous events (Bond et al., 1993a). The Dansgaard-Oeschger events seem to be bundled into longer cycles of gradual cooling over millennia. Each of these "Bond" cycles ends with a Heinrich event and an abrupt warming (Bond et al., 1993a).

Gas analyses, and especially the $\delta^{18}O$ of O_2, allow ice-core records to be correlated between Antarctica and Greenland, and allow correlation into marine records (Bender et al., 1993; 1994) The fascinating result is that Dansgaard-Oeschger events lasting about one-two millennia or more in the North Atlantic are evident in the isotopic composition of precipitation on East Antarctica, which primarily reflects Antarctic temperature, but shorter-lived North Atlantic events are absent or only equivocally identifiable in the Antarctic record (Bender et al., 1993; 1994). The general appearance is that the North Atlantic events have large amplitudes and some element of square-wave behavior, but the Antarctic equivalents show smaller amplitudes and more gradual onset and decay. The available data cannot reveal leads and lags of southern *versus* northern events.

This bestiary of North Atlantic climate features—Heinrich, Bond, Dansgaard/Oeschger—is complex. The simplest picture appears to be that the North Atlantic records show large Dansgaard-Oeschger oscillations with some square-wave character bundled into Bond cycles, at sub-Milankovitch frequencies. Periodicities may exist, but the "obvious" one of a split-precession cycle is not clear in the records despite vague similarity between its periodicity (10,000 years) and the average Heinrich-layer spacing of a few thousand years (but ranging from roughly 4,000 to 15,000 radiocarbon years). At least some of the Dansgaard-Oeschger events and the Heinrich events/Bond cycles of the North Atlantic appear in climate records elsewhere, but they typically show smaller amplitude and less square-wave character in regions beyond the North Atlantic. This is the pattern one would expect if large, abrupt North Atlantic climate events were being transmitted to the rest of the globe through some mechanism(s) with a time constant longer than that for atmospheric mixing.

Mechanisms

We observe that much global climatic variability is in-phase, despite some forcings that are out-of-phase between northern and southern hemispheres. At orbital time scales, global warming and cooling track increases and decreases in insolation at extra-tropical northern latitudes. The response is greatly amplified at the 100,000-year time scale. At sub-orbital time scales, longer-lived North Atlantic climate events (millennial for the Younger Dryas, to perhaps greater than two millennia for others) seem to have a widespread or global effect, whereas shorter North Atlantic events may not. The North Atlantic events have some square-wave character and large amplitude, but their global signature appears more muted and less abrupt.

Tropospheric dust, atmospheric waves, and regional atmospheric temperature anomalies propagate across the equator with difficulty if at all. Leading candidates for causing the northern and southern hemispheres to warm and cool together are tropical signals, globally mixed greenhouse gases, and oceanic circulation. The power spectra of climate variability argue against a tropical mechanism controlling the translation of the Milankovitch variations into the global climate, and the pattern of observed variability (strongest and squarest-wave behavior in the North Atlantic) argues against a tropical source for sub-Milankovitch variations. Here, the lack of numerous long, highly-resolved climate records from regions beyond the North Atlantic is of some concern, and better records may change our views. For now, we focus on extra-tropical mechanisms.

Greenhouse gases certainly are related to global climate changes. The Vostok ice-core record from East Antarctica clearly shows the covariance of southern hemisphere temperature, carbon dioxide and methane (Raynaud et al., 1993). The southern hemisphere temperature record can be explained largely as a combination of greenhouse gases, plus high-latitude northern-hemisphere insolation or else ice volume responding to that insolation (Genthon et al., 1987; Lorius et al., 1990). However, at the cooling from the previous interglacial (marine oxygen substage 5e), CO_2 lags temperature at Vostok, so CO_2 cannot be the primary cause. In addition, much of the control of CO_2 lies in the ocean, so I lump it into the oceanic signal discussed below. Methane has less atmospheric effect than CO_2, and methane variations do not seem to match Vostok temperatures quite as well as do CO_2 variations, so I omit further consideration here. Methane probably is controlled by changes in wetland areas linked to the other processes discussed below (Chappellaz et al., 1993).

Oceanic Circulation

The leading candidate for an Arctic control on global climate is related to oceanic circulation. Today, deep-water formation in the North Atlantic, largely in the Greenland, Iceland and Norwegian seas, drives a vigorous oceanic circulation pattern—the "conveyor belt"—that spans the world oceans (e.g. Broecker and Denton, 1989). Cold, salty water sinks in restricted regions in the North Atlantic, travels south in the deep Atlantic, upwells more diffusely in regions including the Southern Ocean and the North Pacific, and returns to the North Atlantic near the surface. The North Atlantic deep-water formation is quite sensitive to the local fresh-water balance. Addition of fresh-water can reduce the surface-water density and prevent deep-water formation. The North Atlantic fresh-water budget is sensitive to sea-ice export from the Arctic Ocean to the North Atlantic, runoff or storage of water in lakes or ice sheets around the North Atlantic, and similar factors (Aagaard and Carmack, 1989).

Large changes in deep-water formation are modeled to occur in response to entirely plausible changes in the North Atlantic fresh-water budget (e.g. Broecker et al., 1990; Birchfield and Broecker, 1990). Such changes might be related to tropical processes (export of water vapor from the Atlantic basin across the Isthmus of Panama), or more likely, are related to processes closer to the sites of deep-water formation (water transport through the atmosphere from the low-latitude to the high-latitude North Atlantic, ice export from the Arctic Ocean, runoff from the continents around the North Atlantic, etc.) (Aagaard and Carmack, 1989; Teller, 1990). Whether large changes in deep-water formation require the presence of large ice sheets around the North Atlantic remains an open question.

Model results (e.g. Covey and Barron, 1988; Rind and Chandler, 1991; Barron et al., 1993) indicate an imperfect trade-off between equator-to-pole heat transport by the oceans and the atmosphere. If the oceanic heat transport is reduced, the poles cool, and the resulting steepened equator-to-pole temperature gradient causes stronger atmospheric circulation and enhanced atmospheric heat transport, partially offsetting the reduced oceanic heat transport.

Qualitatively, if deep-water formation were slowed or stopped in the North Atlantic owing to a local forcing, there is no tightly coupled mechanism to guarantee initiation of a similarly strong source of deep-water elsewhere. Model results do allow deep-water formation elsewhere, under more extreme perturbations (Stocker et al., 1992), but the main result of reduced deep-water formation in the North Atlantic is reduced global thermohaline circulation (also see Paillard and Labeyrie, 1994; Rahmstorf, 1994). This slowed oceanic circulation would reduce oceanic heat transport, causing global changes in the atmosphere toward compensation.

Global average cooling is a probable result of reduced oceanic heat transport; colder polar oceans would allow increased sea-ice growth, increasing the global albedo and thus cooling the globe. However, tropical warming or little tropical change might result because the ocean is not exporting tropical heat as efficiently as before.

CO_2 Changes

Changes in ocean processes would also affect atmospheric CO_2. A not-unreasonable climate sensitivity to CO_2, including such positive feedbacks as ice-albedo, atmospheric dust loading, some contribution from CH_4 and possibly changes in cloud albedo, then may be sufficient to explain much or all of the global temperature signal (Broecker and Denton, 1989).

It is virtually certain that the ocean exerts the major control on CO_2 at these time scales. Many models exist to explain the glacial-interglacial changes in CO_2. A fully satisfactory story does not yet exist (see Broecker and Peng, in press), but changes in oceanic circulation and in delivery of nutrients to the ocean and nutrient cycling in the ocean probably were important.

Increased glacial-age productivity in low-latitude oceans owing to the increased wind speed of glacial times (e.g. Petit et al., 1981) may have played a role by removing CO_2 from the surface ocean and transferring it to the deep ocean as particulate matter (the biological pump). Increased wind speed may have increased nutrient supply and thus productivity in two ways: through increased wind-blown dust, or through increased wind-forced upwelling of nutrient-rich deeper waters (Martin, 1990; Berger and Wefer, 1991).

An intriguing new hypothesis (Archer and Maier-Reimer, 1994; also see Broecker and Peng, 1989) suggests that glacial-age cooling of polar and subpolar surface waters led to a switch from calcareous to siliceous planktonic organisms there. This in turn increased the ratio of organic carbon to inorganic carbonate deposited on the sea floor, increasing dissolution of sea-floor carbonate. The resulting dissolved carbonate ions reacted with aqueous CO_2 to form bicarbonate, lowering the CO_2 partial pressure of the ocean and thus of the atmosphere.

Ice-sheet growth reduces sea level, which may lower CO_2 in at least two ways. Lower sea level might increase fertilization of the ocean by reducing sequestration of riverine nutrient fluxes in deltas (Broecker, 1982). Alternatively, it might reduce coral-reef growth on continental shelves and the associated release of CO_2 to the atmosphere (Berger, 1982; Opdyke and Walker, 1992).

Questions remain about whether glacial-age temperature depression was nearly uniform globally (Broecker and Denton, 1989; Stute et al., 1992; 1993; Beck et al., 1992; Guilderson et al., 1994), or whether there was little change in low-latitude, low-altitude regions but strong polar amplification (CLIMAP, 1976; Ohkouchi et al., 1994). Polar amplification might argue that the direct effects of changes in oceanic heat transport are most important, whereas more nearly latitudinally uniform changes might argue for the importance of changes in greenhouse gases.

Synthesis of Mechanisms

To explain the variety of behavior observed in the global climate record through changes in ocean circulation driven from the North Atlantic, the deep-water formation there must have considerable variability and the potential for rapid switches—gradual as well as abrupt changes are observed. The system must contain an oscillator with a short period, or experience short-period forcing (the Dansgaard-Oeschger events). It must have a slower forcing or variability (the Heinrich events/Bond cycles), some linear-response elements (shorter Milankovitch periodicities), and a very slow element (to give the 100,000-year cycle).

Models exist for a self-oscillation of the North Atlantic at 1000-year time scales (e.g. Broecker et al., 1990; Birchfield and Broecker, 1990; Birchfield et al., 1990). Reduction of deep-water formation leads to salt accumulation in the North Atlantic, density increase, and eventually increased deep-water formation; vigorous deep-water formation removes salt more rapidly than it is supplied and slows or stops deep-water formation. Both gradual and abrupt changes are modeled. Such oscillations might explain Dansgaard-Oeschger oscillations. The abrupt changes would be transmitted rapidly through the atmosphere to the North Atlantic region, explaining the dramatic North Atlantic climate records (Fawcett et al., 1994), and more slowly through the ocean to the southern hemisphere (Bender et al., 1994).

The Heinrich events may be related to processes internal to the Laurentide ice sheet (MacAyeal, 1993a; 1993b), with the Bond cycles being the downwind effect of the ice sheet on the regional climate (MacAyeal, 1993a; 1993b). The linear-response elements then are either the direct effect of Milankovitch variations on deep-water formation, or the effect of Milankovitch variations on ice-sheet growth and of ice-sheet growth on fresh-water fluxes and deep-water formation. The long-time-period behavior is the inertia of large ice sheets to climatic forcing.

Such an explanation asks much of the North Atlantic system and the adjacent Laurentide ice sheet. The ice sheet must exhibit two time scales—MacAyeal-type surges to generate Heinrich/Bond oscillations, and slower responses, perhaps in other areas of the ice sheet, to pace the 100,000 cycle. Deep-water formation must exhibit at least a fast oscillation (Dansgaard/Oeschger) and slower responses to ice-sheet and Milankovitch forcing. However, there is little doubt that ice sheets can exhibit more than one time scale of variability (MacAyeal, 1993a; 1993b), perhaps with marine-based regions and regions with soft, deformable beds oscillating more rapidly than regions on continental shields (Alley, 1991; MacAyeal, 1993b; Clark, 1994; Clark, 1995). Deep-water formation is modeled to run at various rates, can exhibit switches, and probably can switch in some regions while varying more slowly in other regions (Broecker and Denton, 1989; Imbrie et al., 1992; 1993).

Summary

Our understanding of climate change remains poor. Some of the truly exciting signals (Heinrich events, Bond cycles) have been recognized for less than a decade. High-resolution records are lacking for much of the world, and dating has only recently approached a level to allow comparison of sub-Milankovitch features worldwide.

The picture emerging thus far is one of climate variability in various bands. I have concentrated on variability between about 1,000 and 1,000,000 years. In this range, we see large signals at three Milankovitch periodicities (100,000, 41,000 and 23,000 years), and at two shorter lengths (Heinrich/Bond cycles at a few thousand years, and Dansgaard/Oeschger at 1000-2000 years), with the probability that the Heinrich/Bond and possibly the Dansgaard/ Oeschger variations represent quasi-periodic events rather than true periodicities; small variability also occurs at the 10,000-year split-precession cycle, and in other bands (see Table).

If the data available to us accurately reflect the true state of the paleoclimatic record, then variability occurs worldwide in all of these bands. However, the signal is stronger in the North Atlantic than elsewhere for the shorter periodicities. The worldwide response is broadly in-phase—warming in Antarctica and in the tropics occurs at about the same time as warming in the North Atlantic.

The analysis of the Milankovitch periodicities by Imbrie et al. (1992; 1993) shows that global temperature is responding to extratropical, northern-hemisphere insolation—when summertime insolation increases at high northern latitudes, the world warms. The 10,000-year periodicity of the split-precession cycle, a possible fingerprint of tropical control, is very small in most climate records. When summertime insolation increases at high southern latitudes, the world may warm or cool depending on the northern-hemisphere behavior, so the southern hemisphere is not controlling global climate. (This includes an assumption

Table. Summary of observed climate variability, millenial to million-year time scales.

Oscillation	Timing (ka)	Probable Source	Size
Dansgaard/ Oeschger	≈1-3	N. Atl. deep-water self-oscillation	large N. Atl., small-large global?
Heinrich/Bond	≈4-15	Ice-sheet self-oscillation	large N. Atl., small-large global?
Split Precession	9-11	Milankovitch	absent-small global?
Precession	19-23	Milankovitch	small global
Obliquity	41	Milankovitch	small global
Eccentricity?	100	Milankovitch through ice sheet	large global

that the lags in the system are much less than the periods of the forcing, which physically seems to be a good assumption.) The dominance of 100,000-year climate change in response to a trivial variation in insolation at that periodicity requires some nonlinear element with a long time constant in the climate system; the most plausible explanation is the behavior of northern-hemisphere ice sheets.

Variability in the Heinrich/Bond band (few millennia) can plausibly be explained as an instability in the Laurentide ice sheet in Hudson Bay. Variability in the Dansgaard/Oeschger band (millennial) may represent a free oscillation of the ocean, may be paced by other short-period events in the North Atlantic system, or may represent ice-sheet oscillations (P.U. Clark, pers. comm., 1995; Bond and Lotti, in press). Again, the aperiodicity of the Heinrich/Bond cycles argues against a split-precession origin in the tropics.

It is disturbing that we lack well-developed models for possible tropical causes of many of these features, and that we lack many long, highly-resolved records of tropical climate change. Nagging worries thus remain about any conclusions we might reach. For now, the only hypothesis, and one consistent with many of the observations, is that the main climate changes represent reorganizations of the ocean/atmosphere system originating in the North Atlantic (Broecker and Denton, 1989).

Most models agree that the voluminous deep-water formation in the North Atlantic is a sensitive process, that plausible variations in the fresh-water fluxes of the North Atlantic could greatly change the volumes of deep-water formed, that there is no direct compensating process elsewhere in the world ocean to balance North Atlantic changes, and that the oceanic thermohaline circulation has significant implications worldwide. If the deep-water formation were slowed, the circulation of the global ocean would slow, and the atmosphere would assume a greater share of the global heat transport.

Reduced oceanic heat flux would cool the poles. This would allow expansion of sea ice, land ice and seasonal snow cover, increasing the global albedo and cooling the planet further. CO_2 would decrease in the atmosphere owing to oceanic processes, and CH_4 would decrease in response to freezing of high-latitude wetlands and possible desiccation of low-latitude wetlands. Enhanced windiness would increase the atmospheric loading of aerosols capable of blocking insolation. These various processes would serve to cause climate change between the hemispheres to be nearly in-phase (see the Figure).

Variations in the North Atlantic fresh-water budget could be forced from beyond the Arctic (say, by changes in water-vapor exchange across the Isthmus of Panama), or by a host of processes in the Arctic (changes in ice export from the Arctic Ocean, water storage and release in continental ice sheets, water storage and release in ice-marginal lakes, or periodic oscillations of the North Atlantic itself, perhaps involving the atmospheric moisture transport from mid-latitudes to high-latitudes). The timing of variability involved implicates self-oscillations in the North Atlantic and ice-sheet processes rather than water-vapor transport across Panama, but uncertainties remain.

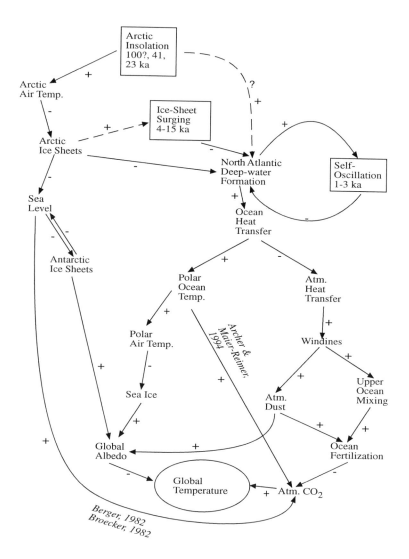

Figure. Flow chart showing some of the linkages by which the major forcing processes in the Arctic (Milankovitch variations, ice-sheet surges, and oscillations in deep-water formation; boxed) produce the main output of global temperature changes (circled). Dashed lines indicate indirect or uncertain linkages. A "+" beside an arrow indicates that an increase in the variable at the tail of the arrow causes an increase in the variable at the head; a "-" indicates that an increase at the tail causes a decrease at the head. Two "-" in a sequence have the same net effect as one "+". The figure shows that any increase in deep-water formation in the North Atlantic leads to global warming through various paths, and that the major forcings also cause parallel warmings through other paths such as sea level and ice-albedo feedbacks. The system is much more complex than this, and further paths could be added. Paths that are too complex to be diagrammed in full are labeled with a reference.

In a community raised on the beautiful ENSO signal and dazzled by the size and energy-richness of the tropics, it approaches heresy to suggest that the tropics are a slave to smaller regions in the Arctic and sub-Arctic. The paleo-record, however, argues that this indeed is the case—the "accident" of highly variable deep-water formation in the North Atlantic causes this region to control global climate on the timescales of its variation. The rapidity (as little as one to three years) and "flickering" or oscillating character of many of these changes (Taylor et al., 1992; Alley et al., 1993) certainly place them within the "timescale of human interest". External (Milankovitch) and internal (ice, ocean) oscillations probably are important, and ice-sheet processes are needed to explain the longer Milankovitch variations. (Milankovitch variations longer than 100,000 years are weak or absent because nothing in the system responds at the appropriate time scales.) The control is probably transmitted from the North Atlantic to the Southern Ocean, the next stop for North Atlantic deep-water, and from the poles to the tropics through the atmosphere or ocean.

This model is far from complete. No coupled climate model can yet produce all of the observed signals from the observed forcings. And the incompleteness of paleoclimatic records is troubling. The burgeoning interest in coral records (e.g. Beck et al., 1992; Guilderson et al., 1994), low-latitude ice cores (Mosley-Thompson et al., 1993) and other low-latitude records presumably will reduce the tropical uncertainties; better high-resolution southern records certainly are needed. Pending collection of such records, research focus on the North Atlantic system and on its mixing into the global oceans (especially around the Antarctic, in the Northern Pacific, and in the Arctic Ocean) seems well-justified by the paleoclimatic records showing an Arctic control of climate.

Acknowledgments. I thank the NSF Office of Polar Programs, NASA EOS, and the D. and L. Packard Foundation for financial support, P.U. Clark, P.F. Fawcett and J.W.C. White for helpful suggestions, and C.R. Alley for preparation of camera-ready copy.

References

Aagaard, K. and E.C. Carmack, The role of sea ice and other fresh water in the Arctic circulation, *J. Geophys. Res., 94C*, 14,485-14,498, 1989.

Alley, R.B., Deforming-bed origin for southern Laurentide till sheets?, *J. Glaciol., 37*, 67-76, 1991.

Alley, R.B., D.A. Meese, C.A. Shuman, A.J. Gow, K.C. Taylor, P.M. Grootes, J.W.C. White, M. Ram, E.D. Waddington, P.A. Mayewski and G.A. Zielinski, Abrupt increase in snow accumulation at the end of the Younger Dryas event, *Nature, 362*, 527-529, 1993.

Alley, R.B. and D.R. MacAyeal, 1993, West Antarctic ice sheet collapse: Chimera or clear danger? *Antarctic Journal of the U.S., 28*, 59-60, 1993.

Alley, R.B. and D.R. MacAyeal, 1994, Ice-rafted debris associated with binge/purge oscillations of the Laurentide Ice Sheet, *Paleoceanography, 9*, 503-511, 1994.

Andrews, J.T., H. Erlenkeuser, K. Tedesco, A.E. Aksu and A.J.T. Jull, Late Quaternary (Stage 2 and 3) meltwater and Heinrich events, northwest Labrador Sea, *Quat. Res., 41*, 26-34, 1994.

Andrews, J.T., K. Tedesco and A.E. Jennings, Heinrich events: Chronology and processes, East-central Laurentide Ice Sheet and NW Labrador Sea, in Ice in the Climate System, *NATO ASI Ser., Ser.*

I, 12, 167-186, 1993.

Archer, D. and E. Maier-Reimer, Effect of deep-sea sedimentary calcite preservation on atmospheric CO_2 concentration, *Nature, 367*, 260-263, 1994.

Barnola, J.M., D. Raynaud, Y.S. Korotkevich and C. Lorius, Vostok ice core provides 160,000-year record of atmospheric CO_2, *Nature 329*, 408-414, 1987.

Barron, E.J. W.H. Peterson, D. Pollard and S. Thompson, Past climate and the role of ocean heat transport: model simulations for the Cretaceous, *Paleoceanography, 8*, 785-798, 1993.

Beck, J.W., R.L. Edwards, E. Ito, F.W. Taylor, J. Recy, F. Rougerie, P. Joannot and C. Henin, Sea-surface temperature from coral skeletal strontium/calcium ratios, *Science, 257*, 644-647, 1992.

Bender, M., T. Sowers, M.L. Dickson, J. Orchardo, P. Grootes, P.A. Mayewski and D. Meese, Climate connections between Greenland and Antarctica throughout the last 100,000 years, *Nature, 372*, 663-666, 1994.

Bender, M., T. Sowers and J. Orchardo, Rapid CH_4 variations during the Younger Dryas and "Dansgaard-Oeschger"' events in the GISP2 ice core, *Eos Trans. AGU, 74*(43), Fall Meeting suppl., 78, 1993.

Berger, W.H. Increase of carbon dioxide in the atmosphere during deglaciation: The coral reef hypothesis, *Naturwissenschaften, 69*, 87-88, 1982.

Berger, W.H. and G. Wefer, Productivity of the glacial ocean: Discussion of the iron hypothesis, *Limnol. Oceanogr., 36*, 1899-1918, 1991.

Birchfield, G.E. and W.S. Broecker, A salt oscillator in the glacial Atlantic? 2. A "scale analysis" model, *Paleoceanography, 5*, 835-843, 1990.

Birchfield, G.E. and R.W. Grumbine, "Slow" physics of large continental ice sheets and underlying bedrock, and its relation to the Pleistocene ice ages, *J. Geophys. Res., 98*, 11,294-11,302, 1985.

Birchfield, G.E., H. Wang and M. Wyant, A bimodal climate response controlled by water vapor transport in a coupled ocean-atmosphere box model, *Paleoceanography, 5*, 383-395, 1990.

Bond, G.C. and R. Lotti, Iceberg discharges into the North Atlantic on millennial time-scales during the last glaciation, *Science*, in press.

Bond, G., H. Heinrich, W. Broecker, L. Labeyrie, J. McManus, J. Andrews, S. Huon, R. Jantschik, S. Clasen, C. Simet, K. Tedesco, M. Klas, G. Bonani and S. Ivy, Evidence for massive discharges of icebergs into the North Atlantic ocean during the last glacial period, *Nature, 360*, 245-249, 1992.

Bond, G., W. Broecker, S. Johnsen, J. McManus, L. Labeyrie, J. Jouzel, and G. Bonani, Correlations between climate records from North Atlantic sediments and Greenland ice, *Nature 365*, 143-147, 1993a.

Bond, G., W. Broecker, M. Prentice, J. McManus, and S. Higgins, Global-scale imprints of abrupt climate shifts in Greenland ice cores, *Eos Trans. AGU, 74*(43), Fall Meeting suppl., 84, 1993b.

Broecker, W. Ocean chemistry during glacial time, *Geochim. Cosmochim. Acta, 46*, 1689-1705, 1982.

Broecker, W. Massive iceberg discharges as triggers for global climate change. *Nature, 372*, 421-424, 1994.

Broecker, W.S., G. Bond and M. Klas, A salt oscillator in the glacial Atlantic? 1. The concept, *Paleoceanography, 5*, 469-477, 1990.

Broecker, W., G. Bond, M. Klas, E. Clark and J. McManus, Origin of the northern Atlantic's Heinrich events, *Clim. Dyn., 6*, 91-109, 1992.

Broecker, W.S. and G.H. Denton, The role of ocean-atmosphere reorganizations in glacial cycles, *Geochim. Cosmochim. Acta, 53*, 2465-2501, 1989.

Broecker, W.S. and T.-H. Peng, The cause of the glacial to interglacial atmospheric CO_2 change: A polar alkalinity hypothesis, *Global Biogeochemical Cycles, 3*, 215-239, 1989.

Broecker, W.S. and T.-H. Peng. What caused the glacial to interglacial CO_2 change? *NATO volume on the Global Carbon Cycle*, in press.

Chappellaz, J., T. Blunier, D. Raynaud, J.M. Barnola, J. Schwander and B. Stauffer, Synchronous changes in atmospheric CH_4 and Greenland climate between 40 and 8 kyr b.p., *Nature, 366*, 443-445, 1993.

Clark, P.U. Unstable behavior of the Laurentide ice sheet over deforming sediment and its implications for climate change, *Quat. Res. 41*, 19-25, 1994.

Clark, P.U. Fast glacier flow over soft beds, *Science, 267,* 43-44, 1995.

Clark, P.U. and P.J. Bartlein, Correlation of late-Pleistocene mountain glaciation in western North America with North Atlantic Heinrich Events, *Geological Society of America Abstracts with Programs, 26*(7), A255, 1994.

Clark, P.U. and P.J. Bartlein, Correlation of late-Pleistocene glaciation in the western U.S. with North Atlantic Heinrich Events, *Geology,* in review.

CLIMAP, The surface of the ice-age earth, *Science, 191,* 1131-1136, 1976.

Covey, C. and E. Barron, The role of ocean heat transport in climate change, *Earth-Science Reviews, 24,* 429-445, 1988.

DeBlonde, G. and W.R. Peltier, A one-dimensional model of continental ice volume fluctuations through the Pleistocene: Implications for the origin of the mid-Pleistocene climate transition, *J. Clim., 4,* 318-344, 1991.

Denton, G.H. and C.H. Hendy, 1994, Younger Dryas age advance of Franz Josef Glacier in the Southern Alps of New Zealand, *Science, 264,* 1434-1437.

Doake, C.S.M. and D.G. Vaughan, Rapid disintegration of Wordie Ice Shelf in response to atmospheric warming, *Nature, 350,* 328-330, 1991.

Fawcett, P.F., A.M. Agustsdottir, and R.B. Alley, Change in North Atlantic ocean heat transport as a cause of the rapid termination of the Younger Dryas climate event: results from GENESIS climate model studies, *Eos Trans. AGU, 75*(43), Fall Meeting suppl., 1994.

Genthon, C., J.M. Barnola, D. Raynaud, C. Lorius, J. Jouzel, N.I. Barkov, Y.S. Korotkevich and V.M. Kotlyakov, Vostok ice core: Climatic response to CO_2 and orbital forcing changes over the last climatic cycle, *Nature, 329,* 414-418, 1987.

Grimm, E.C., G.L. Jacobson, Jr., W.A. Watts, B.C.S. Hansen and K.A. Maasch, A 50,000-year record of climate oscillations from Florida and its temporal correlation with the Heinrich events, *Science, 261,* 198-200, 1993.

Grousset, F.E., L. Labeyrie, J.A. Sinko, M. Cremer, G. Bond, J. Duprat, E. Cortijo and S. Huon, Patterns of ice-rafted detritus in the glacial North Atlantic (40-55°N), *Paleoceanography, 8,* 175-192, 1993.

Guilderson, T.P., R.G. Fairbanks and J.L. Rubenstone, Tropical temperature variations since 20,000 years ago: modulating interhemispheric climate change, *Science, 263,* 663-665, 1994.

Hagelberg, T.K., G. Bond and P. deMenocal, Milankovitch band forcing of sub-Milankovitch climate variability during the Pleistocene, *Paleoceanography, 9,* 545-558, 1994.

Hansen, J., A. Lacis, D. Rind, G. Russell, P. Stone, I. Fung, R. Ruedy, and J. Lerner, Climate sensitivity: analysis of feedback mechanisms, in *Climate Processes and Climate Sensitivity,* edited by J.E. Hansen and T. Takahashi, pp. 130-163, AGU, Washington, D.C., 1984.

Harvey, L.D.D., Climatic impact of ice-age aerosols, *Nature 334,* 333-335, 1988.

Heinrich, H., Origin and consequences of cyclic ice rafting in the northeast Atlantic Ocean during the past 130,000 years, *Quat. Res., 29,* 143-152, 1988.

Hughes, T.J. Abrupt climatic change related to unstable ice-sheet dynamics: toward a new paradigm, *Palaeo. (Global and Planet. Change Section), 97,* 203-234, 1992.

Imbrie, J., E.A. Boyle, S.C. Clemens, A. Duffy, W.R. Howard, G. Kukla, J. Kutzbach, D.G. Martinson, A. McIntyre, A.C. Mix, B. Molfino, J.J. Morley, L.C. Peterson, N.G. Pisias, W.L. Prell, M.E. Raymo, N.J. Shackleton and J.R. Toggweiler, On the structure and origin of major glaciation cycles: 1. Linear responses to Milankovitch forcing, *Paleoceanography, 7,* 701-738, 1992.

Imbrie, J., A. Berger, E.A. Boyle, S.C. Clemens, A. Duffy, W.R. Howard, G. Kukla, J. Kutzbach, D.G. Martinson, A. McIntyre, A.C. Mix, B. Molfino, J.J. Morley, L.C. Peterson, N.G. Pisias, W.L. Prell, M.E. Raymo, N.J. Shackleton and J.R. Toggweiler, On the structure and origin of major glaciation cycles: 2. The 100,000-year cycle, *Paleoceanography, 8,* 699-735, 1993.

Jacobs, S.S., H.H. Helmer, C.S.M. Doake, A. Jenkins and R.M. Frolich, Melting of ice shelves and the mass balance of Antarctica, *J. Glaciol., 38*, 375-387, 1992.

Johnsen, S.J., H.B. Clausen, W. Dansgaard, K. Fuhrer, N. Gundestrup, C.U. Hammer, P. Iversen, J. Jouzel, B. Stauffer and J.P. Steffensen, Irregular glacial interstadials recorded in a new Greenland ice core, *Nature, 359*, 311-313, 1992.

Kennett, J.P, The Younger Dryas cooling event: An introduction, *Paleoceanography, 5*, 891-895, 1990.

Kennett, J.P. and B.L. Ingram. Paleoclimatic evolution of Santa Barbara Basin during the last 30 kyrs: Marine evidence from Hole 893A. *Eos Trans. AGU, 75*(44), Fall Meeting suppl., 347, 1994.

Linsley, B.K. and R.C. Thunell, The record of deglaciation in the Sulu Sea: evidence for the Younger Dryas event in the tropical western Pacific, *Paleoceanography, 5*, 1025-1039, 1990.

Lorius, C., J. Jouzel, D. Raynaud, J. Hansen and H. Le Treut, The ice-core record: climate sensitivity and future greenhouse warming, *Nature, 347*, 139-145, 1990.

MacAyeal, D.R., A low-order model of growth/purge oscillations of the Laurentide Ice Sheet, *Paleoceanography, 8*, 767-773, 1993a.

MacAyeal, D.R., Binge/purge oscillations of the Laurentide Ice Sheet as a cause of the North Atlantic's Heinrich events, *Paleoceanography, 8*, 775-784, 1993b.

Manabe, S. and A.J. Broccoli. The influence of continental ice sheets on the climate of an ice age. J. *Geophys. Res., 90*, 2167-2190, 1985.

Martin, J.H., Glacial-interglacial CO_2 change: The iron hypothesis, *Paleoceanography, 5*, 1-13, 1990.

Mathewes, R.W., Evidence for Younger Dryas-age cooling on the North Pacific coast of America, *Quat. Sci. Rev., 12*, 321-331, 1993.

Mayewski, P.A., L.D. Meeker, S. Whitlow, M.S. Twickler, M.C. Morrison, P. Bloomfield, G.C. Bond, R.B. Alley, A.J. Gow, P.M. Grootes, D.A. Meese, M. Ram, K.C. Taylor and W. Wumkes, Changes in atmospheric circulation and ocean ice cover over the North Atlantic during the last 41,000 years, *Science, 263*, 1747-1751, 1994.

Mercer, J.J., West Antarctic ice sheet and CO_2 greenhouse effect: a threat of disaster, *Nature, 271*, 321-325.

Mitchell, J.F.B., S. Manabe, V. Meleshko and T. Tokioka, Equilibrium climate change—and its implications for the future, in *Climate Change: The IPCC Scientific Assessment*, edited by J.T. Houghton, G.J. Jenkins and J.J. Ephraums, pp. 131-172, Cambridge University Press, Cambridge, U.K., 1990.

Mosley-Thompson, E., L.G. Thompson, J. Dai, M. Davis and P.N. Lin, Climate of the last 500 years: High resolution ice core records, *Quat. Sci. Rev., 12*, 419-430, 1993.

Oerlemans, J., Evaluating the role of climate cooling in iceberg production and the Heinrich events, *Nature, 364*, 783-786, 1993.

Ohkouchi, N., K. Kawamura, T. Nakamura and A. Taira, Small changes in the sea surface temperature during the last 20,000 years: molecular evidence from the western tropical Pacific, *Geophys. Res. Letters, 21*, 2207-2210, 1994.

Opdyke, B.N. and J.C.G. Walker, Return of the coral reef hypothesis: Basin to shelf partitioning of $CaCO_3$ and its effect on atmospheric CO_2, *Geology, 20*, 733-736, 1992.

Paillard, D. and L. Labeyrie, Role of the thermohaline circulation in the abrupt warming after Heinrich events, *Nature, 372*, 162-164, 1994.

Peixoto, J.P. and A.H. Oort. *Physics of Climate*, American Institute of Physics, New York, 1992.

Peteet, D.M., ed., Global Younger Dryas, a special issue of *Quat. Sci. Rev., 12*, 1993.

Peltier, W.R., Glacial isostasy, mantle viscosity, and Pleistocene climatic change, in *North America and Adjacent Oceans During the Last Deglaciation*, edited by W.F. Ruddiman and J.E. Wright, Jr., The Geological Society of America, Boulder, CO, p. 155-182, 1987.

Petit, J.R., M. Briat and A. Royer, Ice age aerosol content from East Antarctic ice core samples and past wind strength, *Nature, 293*, 391-394, 1981.

Pollard, D., A simple ice sheet model yields realistic 100 kyr glacial cycles, *Nature, 296*, 334-338, 1982.

Pollard, D., A coupled climate-ice sheet model applied to the Quaternary ice ages, *J. Geophys. Res., 88,* 7705-7718, 1983.

Rahmstorf, S., Rapid climate transitions in a coupled ocean-atmosphere model, *Nature, 372,* 82-85, 1994.

Raynaud, D., J. Chappellaz, J. Barnola, Y. Korotkevich and C. Lorius, Climatic and CH_4 implications of glacial-interglacial change in the Vostok ice core, *Nature 333,* 655-657, 1988.

Raynaud, D., J. Jouzel, J.M. Barnola, J. Chappellaz, R.J. Delmas and C. Lorius, The ice record of greenhouse gases, *Science, 259,* 926-934, 1993.

Reasoner, M.A., G. Osborn and N.W. Rutter, Age of the Crowfoot advance in the Canadian Rocky Mountains: A glacial event coeval with the Younger Dryas oscillation, *Geology, 22,* 439-442, 1994.

Rind, D. and M. Chandler, Increased ocean heat transports and warmer climate, *J. Geophys. Res., 96,* 7437-7461, 1991.

Roberts, N., M. Taieb, P. Barker, B. Damnati, M. Icole and D. Williamson, 1993, Timing of the Younger Dryas event in East Africa from lake-level changes, *Nature, 366,* 146-148, 1993.

Shine, K.P., R.G. Derwent, D.J. Wuebbles and J.-J. Morcrette, Radiative forcing of climate, in *Climate Change: The IPCC Scientific Assessment,* edited by J.T. Houghton, G.J. Jenkins and J.J. Ephraums, Cambridge University Press, Cambridge, U.K., 41-68, 1990.

Stocker, T.F., D.G. Wright and L.A. Mysak, A zonally averaged, coupled ocean-atmosphere model for paleoclimate studies, *Journal of Climate, 5,* 773-797, 1992.

Stute, M., J.F. Clark and W.S. Broecker, Paleotemperatures in the southwestern United States derived from noble gases in groundwater, *Science, 265,* 1000-1002, 1992.

Stute, M., J.F. Clark, P. Schlosser and W.S. Broecker, Reconstruction of the lapse rate in the western United States during the Last Glacial Maximum using the noble gas paleothermometer, *Eos Trans. AGU, 74*(43), Fall Meeting suppl., 365, 1993.

Taylor, K.C., G.W. Lamorey, G.A. Doyle, R.B. Alley, P.M. Grootes, P.A. Mayewski, J.W.C. White and L.K. Barlow, The 'flickering switch' of late Pleistocene climate change, *Nature, 361,* 432-436, 1992.

Teller, J.T., Meltwater and precipitation runoff to the North Atlantic, Arctic, and Gulf of Mexico from the Laurentide ice sheet and adjacent regions during the Younger Dryas, *Paleoceanography, 5,* 897-905, 1990.

Vaughan, D.G., Implications of the break-up of Wordie Ice Shelf, Antarctica for sea level, *Antarctic Science, 5,* 403-408, 1993.

Weertman, J., Stability of the junction of an ice sheet and ice shelf, *J. Glaciol., 5,* 145-158, 1974.

Weertman, J., Milankovitch solar radiation variations and ice-age ice-sheet sizes, *Nature, 261,* 17-20, 1976.

Whillans, I.M., Reaction of the accumulation zone portions of glaciers to climatic change, *J. Geophys. Res., 86,* 4274-4282, 1981.

EPILOGUE

It is now generally believed that the Arctic and its surrounding seas will be the first region to be influenced by a warming induced by anthropogenic changes in the atmosphere-ocean system. Whether such changes have already begun is, of course, a matter of controversy. Despite this widespread acceptance of the Arctic's sensitivity to global change, relatively little attention has been given to the role of the Arctic in regulating global processes. Papers in this volume, as well as other recently published results, suggest that instead of simply reacting to changes, the Arctic system responds to large-scale forcing in a highly complex manner, initiating changes which are in turn felt throughout the oceanic realm. We are only just beginning to understand these responses and their implications.

Future investigations in the Arctic will continue to provide answers to some of the questions posed in this volume. For example, the halocline in the Arctic Basin is a feature which controls many of the region's physical, chemical and biological features (for example, it caps the warm water in the deeper regions, thereby restricting heat flux and maintaining a permanent ice cover off the continental shelf). Although a plausible mechanism (i.e., brine rejection and cross-shelf flow) for its formation has been suggested, it is by no means clear if this process occurs frequently enough and generates a great enough volume of brine to maintain the halocline over the entire basin. Because such a process would be strongly influenced by global warming, understanding the formation and maintenance of the halocline is critical to our understanding of the Arctic in its present state. Similar problems pervade Arctic research, and it is hoped that the papers in this volume emphasize the paradoxes which remain in our understanding of the Arctic.

This book provides only a brief synopsis of some recent physical, biological, geological and atmospheric studies of the Arctic Ocean and its ice-covered marginal seas. No attempt was made to be complete or exhaustive, but rather to present provocative hypotheses and results of recent investigations. It is hoped that future studies of the continental shelves and marginal ice zones of the Arctic Ocean can test some of these hypotheses, extend the results, and provide further understanding of the role of the Arctic in global processes.

Walker O. Smith, Jr.
Jacqueline M. Grebmeier
University of Tennessee—Knoxville

LIST OF CONTRIBUTORS

Richard B. Alley
Earth System Science Center
Pennsylvania State Unversity
306 Deike Building
University Park, PA 16802

Leif G. Anderson
Univ. of Goteborg and Chalmers
Univ. of Tech.
S-412 96 Goteborg
Sweden

Josefino C. Comiso
Goddard Space Flight Center
NASA
Greenbelt, MD 20771

Robert Conover
Bedford Institute of Oceanography
Dartmouth, Nova Scotia
Canada

Kenneth L. Davidson
Meteorology Dept., Code MR/Ds
Naval Postgraduate School
589 Dyer Rd., Rm 252
Monterey, CA 93943

Paul Frederickson
Meteorology Dept., Code MR/Ds
Naval Postgraduate School
589 Dyer Rd., Rm 252
Monterey, CA 93943

Jean-Claude Gascard
LODYC-Universite de Paris 6
4 Place Jussieu
75252-Paris Cedex 05
France

Jacqueline M. Grebmeier
Graduate Program in Ecology
University of Tennessee
Knoxville, TN 37996

Peter S. Guest
Meteorology Dept., Code MR/Ds
Naval Postgraduate School
589 Dyer Rd., Rm 252
Monterey, CA 93943

James E. Overland
NOAA/PMEL C15700
7600 Sand Pt. Way NE
Seattle, WA 98115

Laurence Padman
Coll. of Oceanic and Atmospheric Sci.
Oregon State University
Corvallis, OR 97331

Claude Richez
LODYC-Universite de Paris 6
4 Place Jussieu
75252-Paris Cedex 05
FRANCE

Catherine Rouault
LODYC-Universite de Paris 6
4 Place Jussieu
75252-Paris Cedex 05
FRANCE

Walker O. Smith, Jr.
Graduate Program in Ecology
University of Tennessee
Knoxville, TN 37996

John J. Walsh
Department of Marine Science
University of South Florida
140 Seventh Ave. S.
St. Petersburg, FL 33701